H £9.50

THE
UNIVERSITY OF WINNIPEG
PORTAGE & BALMORAL
WINNIPEG, MAN. R3B 2E9
CANADA

THE OLD TELEGRAPHS

GEOFFREY WILSON

The Old Telegraphs

Phillimore

1976

Published by
PHILLIMORE & CO., LTD.,
London and Chichester

Head Office: Shopwyke Hall,
Chichester, Sussex, England.

© Geoffrey Wilson, 1976

ISBN 0 900592 79 6

PRINTED AND BOUND BY
W & J Mackay Limited, Chatham

Contents

	List of Illustrations	vii
	Acknowledgments	xiii
I	Introduction	1
II	The Admiralty Shutter Telegraph	11
III	The Admiralty Semaphore Telegraph	33
IV	The Admiralty Coast Signals	64
V	The Holyhead-Liverpool Telegraph	68
VI	Watson's Other Ventures	81
VII	Channel Islands	94
VIII	Ireland	102
IX	Other British Inventors	112
X	France	120
XI	Italy and Switzerland	152
XII	Netherlands	155
XIII	Germany	159
XIV	Sweden	166
XV	Denmark	173
XVI	Norway	175
XVII	Finland	177
XVIII	Russia	181
XIX	Spain and Portugal	183
XX	The British Army in the Peninsular War	186
XXI	Egypt	188
XXII	South Africa	190
XXIII	St. Helena	193
XXIV	India	195
XXV	Australia	202
XXVI	Canada	207
XXVII	United States of America	210
	Bibliography	219
	Index	228

List of Illustrations

between pages 32 and 33

PLATES
1. Lady's fan of the late 1790s, showing the operation of the shutter telegraph.
2. Lord George Murray's model of his shutter telegraph in the National Maritime Museum, Greenwich.
3. Cable and Wireless advertisement depicting the shutter telegraph and its operation.
4. The Admiralty and Horse Guards, showing shutter telegraph.
5. No. 36 West Square, Lambeth: the telegraph tower, 1810.
6. No. 36 West Square, Lambeth: recent view.
7. Shutter telegraph operating room, Nunhead.
8. Highwaymen being hanged at Blackheath, with representation of Shooter's Hill in the background.
9. Execution of Richard Parker at Sheerness, 30 June 1797, showing also shutter telegraph on Garrison Point.
10. Chelsea Hospital, showing shutter telegraph on roof.
11. Putney Heath in the early 1800s, showing shutter telegraph.
12. 'The Telegraph', Putney Heath.
13. Shutter telegraph at High Stoy, Dorset.
14. Shutter telegraph, Hampstead, 1808, from a water-colour by J. J. Park.
15. Clock tower, St. Albans, with shutter telegraph.
16. Shutter telegraph on Dunstable Downs, from a water-colour by George Shepherd.
17. Shutter telegraph on South Gate, Yarmouth.
18. Rear-Admiral Sir Home Riggs Popham, 1762-1820, from a portrait by M. Brown.
19. Sir Home Popham's Telegraph, London-Portsmouth line.
20. Guildford in the 1820s, showing Pewley Hill semaphore at top left.
21. Surviving semaphore telegraph house at Sherfield English.
22. Farley Chamberlayne: derelict semaphore telegraph station.
23. The Admiralty, showing semaphore telegraph on roof.
24. Worplesdon: semaphore telegraph tower (subsequently demolished) and church.
25. The Duke of York's School, Chelsea, showing semaphore telegraph on roof.
26. Semaphore telegraph, Putney Heath, in 1844.
27. Modern view of Cooper's Hill, Hinchley Wood, semaphore station.
28. Chatley Heath (Cobham) semaphore telegraph before dereliction.
29. Chatley Heath: operating mechanism in first floor room of tower.

VIII   THE OLD TELEGRAPHS

30. Semaphore telegraph station, Pewley, Guildford, from a water-colour by E. Hassell, 1826.
31. View of Pewley Hill semaphore station in 1939. The cupola is one of several later additions.
32. Older Hill semaphore telegraph station.
33. Compton, Sussex, semaphore telegraph station.
34. Portsmouth semaphore station, in the High Street, on Square Tower. The arms are at rest.

between pages 80 and 81

35. Liverpool waterfront, 1859, showing telegraph tower centre right.
36. Ruined telegraph station on Puffin Island off Anglesey.
37. Hilbre Island telegraph station house today.
38. Hilbre Island telegraph station observation room.
39. Llysfaer telegraph station today.
40. Point Lynas telegraph station today.
41. Great Orme, Llandudno, telegraph station.
42. Operating room of Watson's telegraph, Southwark.
43. The fire at Topping's Wharf, Southwark, 19 August 1843. Watson's telegraph is ablaze at centre left.
44. Alderney, surviving telegraph tower.
45. Woolwich church tower, with Gamble's radiated telegraph, 1798.
46. General Sir Charles Pasley, K.C.B.

between pages 128 and 129

47. Claude Chappe.
48. France: St. Pierre de Montmartre church in 1832, surmounted by telegraph of Paris-Lille line.
49. Telegraph station at Dilbeek, near Brussels, in 1804, from a contemporary painting. The station was on the Lille-Brussels extension operated during the First Empire.
50. Rue de Rivoli, Paris, about 1830, showing (centre background) Ministry of Marine with its telegraph (Paris-Brest line).
51. St. Malo, showing telegraph of Paris-Brest line on cupola of the cathedral.
52. Present-day view of Notre Dame de l'Espérance tower, Mont-Dol, which formerly carried a telegraph of the Paris-Brest line.
53. A fanciful English engraving, 1798, entitled 'The French Raft', and intended to represent a raft in Brest Harbour for an invasion of England. The artist has mistakenly included an English shutter telegraph.
54. Telegraph on Strasbourg Cathedral, 1825.
55. Old engraving of the tomb of Claude Chappe in Père Lachaise, Paris.

56. Telegraph on lighthouse at Calais.
57. Telegraph on church tower at Ardres (Pas-de-Calais).
58. Rennes Cathedral with telegraph of Avranches-Nantes line on north-west tower.
59. The later central headquarters of the telegraph in the Rue de Grenelle, Paris. The drawing shows the apparatus for four of the five lines of which this was the terminus.
60. St. Sulpice church, Paris, in the 1830s with Lyons line telegraph on north-west tower and that for Bordeaux on south-west tower.
61. Early illustration from *Illustrated London News* showing abstract of the Indian Mail being telegraphed from Marseilles to Paris.
62. Former statue of Chappe in Paris.
63. Telegraph tower at Trou d'Enfer, Bailly, near Versailles, on Paris-Brest line, from a water-colour of 1842.
64. Modern view of Trou d'Enfer station, Bailly.
65. Haut-Barr telegraph, Alsace, as reconstructed to form a monument to the Chappe telegraph.

between pages 160 and 161

66. Telegraph at Naples during Murat's rule.
67. Stone in the yard of the Tower Hall at Susa, west of Turin, with the name of the two operators of the Susa telegraph (Paris-Milan-Venice line).
68. Detail from hand-made map of Amsterdam, 1812, showing Chappe-type telegraph on the Weesperpoort.
69. Netherlands: Lipkens-type telegraph on the Residency at the Hague, 1831.
70. Germany: Johannis-kirche, Magdeburg, showing telegraph on roof.
71. Telegraph on south pavilion of the castle, Koblenz; terminus of line from Berlin.
72. Telegraph inspectors outside Dahlem church telegraph, near Berlin.
73. Hamburg telegraph and code.
74. Bremerhaven telegraph.
75. Cologne-Flittard reconstructed telegraph station.
76. Telegraph station at Old Harbour, Cuxhaven.
77. Denmark: telegraph on Kronborg Castle.
78. Representation of telegraph with Great Belt in background.
79. Portugal: shutter telegraph in use between Cabo Carvoeiro and Berlengas, near Peniche, as late as 1889.
80. Sweden: dimensional drawing of west coast telegraph station, 1838.
81. Sweden: Mosebacke telegraph.
82. Sweden: view showing (left) Osterhällen and Stigsbergsåsen telegraphs in 1855.

x   THE OLD TELEGRAPHS

83. Russia: (left) telegraph on the Winter Palace, St. Petersburg (now Leningrad); (right) sketch of Russian telegraph tower showing lanterns fixed to arms for night operation.
84. Russian telegraph at Otchakov, near the mouth of the Dnieper, during the Crimean War.

between pages 192 and 193

85. South Africa: Grahamstown, Cape Province, about 1850, showing Fort Selwyn telegraph at extreme right, from a painting by Thomas Baines.
86. India: The Alambagh, near Lucknow, in 1857, showing telegraph on roof.
87. Tasmania: telegraph at Mount Nelson.
88. Tasmania: preserved telegraph station in Prince's Park, Hobart.
89. United States of America: New York Harbour, telegraph at Sandy Hook in 1856.
90. Model in New York City Museum showing the telegraph on Staten Island and barrel signals being hoisted.

Acknowledgments

1, 2, 34, National Maritime Museum; 6, Vigers; 16, Luton Museum; 17, Central Library, Great Yarmouth; 18, National Portrait Gallery; 26, Waverley Studios; 35, Liverpool Public Museums; 43, Fire Protection Association; 44, W. Easter; 46, Librarian, Institute of Royal Engineers; 48, 55, 56, 57, 60, PTT-SRCT Photothèque, Paris; 49, Bibliothèque Royale, Brussels; 50, 59, PTT-CNET Photothèque, Paris; 51, Musée de St. Malo; 52, Artaud Père et Fils, Nantes; 58, Musée de Rennes; 63, 64, Direction de L'Architecture, Paris; 65, Association des Amis de l'Histoire de PTT d'Alsace.

FIGURES

1	Map of Admiralty Shutter Telegraph System	10
2	Shutter telegraph and code installed at the Admiralty, 1796	12
3	Fanciful representation of the Nunhead shutter telegraph station	18
4	Shutter telegraph on Southsea Common	23
5	Shutter telegraph routes through London, 1808-14	27
6	Admiralty semaphore code as given in the supplement to the *Encyclopaedia Britannica*, 1824	42
7	The Admiralty semaphore lines, actual and projected	46
8	Semaphore telegraph room at the Admiralty	47
9	The London-Portsmouth semaphore in 1825	48

10	Goddard's sketch of Cooper's Hill semaphore station, 1822	50
11	Haste Hill semaphore station, Haslemere; from a sketch by E. Hassell, 1825	55
12	The former semaphore tower at Portsmouth Dockyard	59
13	No. 2 Battery, Dungeness, with coast semaphore on top (sketch from a drawing by Thomas Goddard, 27 May 1820)	67
14	Watson's first telegraph	69
15	Holyhead-Liverpool semaphore telegraph	74
16	Llysfaen station, Holyhead-Liverpool Telegraph	76
17	Watson's telegraphs, Southern England and Hull-Spurn	82
18	Sketch of Watson's telegraph on Shot Tower at Topping's Wharf, Southwark	87
19	Mulgrave's Channel Islands Telegraph	96
20	Sark semaphore, 1809-14	97
21	The Channel Islands telegraph routes	98
22	Edgeworth's telegraph as applied in Ireland	103
23	Macdonald's British Semaphore	113
24	Conolly's telegraph for stations in India, 1825	117
25	Arrangement of Chappe telegraph in France	125
26	French coast semaphore	131
27	French telegraph system c.1846	145
28	Telegraph used by professors of Jesuit College of Schwyz, transmitting messages over Lake Lucerne from Brunnen to Seelisberg, 1847	153
29	Berlin-Koblenz telegraph route	158
30	Section of operating room of Berlin-Koblenz line station	161
31	Telegraph on Garrison (now St. Pantaleon) Church, Cologne	163
32	Reconstructed telegraph station at Flittard (Cologne)	164
33	Diagrammatic sketch of Signilsskär station, Sweden	168
34	Captain Ohlsen's Norwegian telegraph	176
35	Ramstedt's telegraph used along the Gulf of Finland during the Crimean War	178
36	Russian telegraph routes	180
37	Russian telegraph tower of the type used on the St. Petersburg-Warsaw line	182
38	Mathé's telegraph as applied in Spain from 1846	184
39	Surviving tower near Calcutta	201
40	Map of telegraph stations in Tasmania	205

Acknowledgments

I should have found it impossible to complete my researches without the goodwill of numerous persons who as private individuals or in their official capacity have helped me without stint.

First I owe a debt of gratitude to Mrs. Gweneth Mead, who generously made over to me additional valuable material collected by her husband, the late Commander H. P. Mead; to Captain T. D. Manning, C.B.E., R.N.V.R., Hon. Editor of *The Mariner's Mirror* (the journal of the Society for Nautical Research) for kindly allowing me to use blocks first reproduced in that journal; to Major-General R. N. Batra, Signal Officer in Chief, Army Headquarters, New Delhi, for permission to quote from an unpublished history of the Indian Signal Corps, and to Brigadier H. Chukerbuti for drawing my attention to a valuable article in the *Signalman*, the journal of the Indian Signal Corps.

I wish to thank Miss Florence Comerton for undertaking valued research for me at the Irish State Paper Office in Dublin; Mr. B. Mac Giola Coille of that office for producing relevant documents; Commander M. Godfrey for help in checking references at the Public Record Office in London; Mr. A. W. H. Pearsall of the National Maritime Museum, for general help and advice; Monsieur Jean Germann for information on telegraphy in Switzerland; Mr. Harold E. Gough who as Curator of the Herne Bay Record Society helped me to establish the site of the telegraph station there; Mr. Arthur Percival who as Hon. Secretary of the Faversham Society confirmed the site of a projected station near Faversham; Mr. and Mrs. Philip Chantler who entertained me at their Cobham home when he and I were endeavouring to safeguard the Chatley Heath semaphore tower; and Mr. and Mrs. A. J. Coppock who showed me over the Old Telegraph House at Llysfaen, their former home.

I am also grateful to: Mr. Jonathan Coad of the Department of the Environment for guidance on the history of Chatham Dockyard; to Dr. Maurice Craig who when in the service of the former Ministry of Public Building and Works was instrumental in scheduling the Chatley Heath station; to Mr. George F. Hawkesworth, Mrs. M. M. Hughes and Mrs. Mollie Stoppani for details of the former telegraph houses in which they reside; to Mr. Ewbank Smith for confirming the site of a proposed station near

Farnham; to Captain J. L. Burnett and Captain N. Zambra for help in establishing the site of the Merifield station near Ropley; to General Pierre Pamart for assistance in solving queries on French telegraphs; to Herr Horst Drogge, Herr Dieter Herbarth and Herr Gottfried Korella for supplying material on former German stations; to Mrs. Nan Parnell for help in respect of material belonging to her father, the late Professor R. R. Kirby of the University of the Witwatersrand; and to Mr. Einar Risberg for kindly supplying notes from his history of the Finnish telegraph service. I have also received assistance from the following:

Colonel R. M. Adams (Historical Officer, Royal Signals Institution); Mr. G. Arnold; Monsieur Georges Clairaud; Mr. E. Collins; Mr. Victor Coysh; Mr. Anton Curtecka; The Revd. Charles Davey; Mr. W. Easter; Lt-Col. K. E. Edgeworth; The Revd. the Hon. A. C. V. Elphinstone; Mr. J. N. F. Fairey; Professor Henri Gachot; Mrs. Mary B. Graham; Mrs. Evelyn O. S. Green; Mr. Thomas W. Holmes; The Revd. R. L. Hordern; Mr. Robert S. Horne; Mr. F. Jackson; Monsieur A. Jamaux; Dottore Augusta Lange; Signor Carlo Lozzi; Capt. T. D. Manning; Mr. Walter McGrath; Monsieur J. Maurice; Mr. T. E. Morland; Mr. R. N. R. Peers; Monsieur H. L. T. de Petiville; Mrs. Grace Riley; Mr. Percy Russell; Mr. J. Thomas; The Revd. G. A. R. Thursfield; Vigers; Mr. J. W. Warren; Mr. G. B. A. Williams.

I express my appreciation of their valued co-operation to the following archivists, curators, librarians and officials of national and local government and of organisations:

Angoulême: Services d'Archives de la Charente; Amsterdam: City Archives; Barcelona: Instituto Municipal de Historia; Bedfordshire County Council Archives; Besançon: Musées de Besançon; Bordeaux: Archives Municipales; Archives Départementales de la Gironde; Boston, Mass: Curator of Fine Arts, Public Library; Bremen: Staatsarchiv; Châlons-sur-Marne: Archives de la Marne et de la Province de Champagne; Cologne: Kölnisches Stadtmuseum; Rheinisches Bildarchiv; Stadtkonservator Köln; Copenhagen: Dansk Post- og Telegrafmuseum; Cuxhaven: Archiv der Stadt; Dax: Société de Borda; Dol-en-Bretagne: Association Française Duine; Dorchester: Dorset Natural History and Archaeological Society; Genoa: Direzione Belle Arti del Comune di Genova; The Hague: Historical Section of the Netherlands Postal and Telecommunication Services; Netherlands Postal Museum; Halifax, N.S.: Public Archives of Nova Scotia; Hertfordshire County Council Archives; Iserlohn: Archiv der Stadt; Kent County Council Archives; Koblenz: Mittelrhein Museum der Stadt Koblenz; La Rochelle: Services d'Archives de la Charente-Maritime; Lisbon: Arquivo Histórico Militar; Liverpool: Public Museums; London: National Gallery; Science Museum; South London Art Gallery; Archivist, London Borough of Lambeth; Luton: Museum; Lucerne: Verkehrshaus der Schweiz; Magdeburg: Kulturhistorisches Museum und Forschungsinstitut; Marseilles: Archives Communales; Metz: Services d'Archives de la Moselle; Mont-St.-Michel: Directeur des Annales; Mont-de-Marsan: Services d'Archives du Département des Landes; Munich: Deutsches

Museum; Nantes: Musées Départementaux de Loire-Atlantique; New York: Museum of the City of New York; New York Historical Society; Staten Island Historical Society; Norwich: City Museum; Orleans: Musée Historique de l'Orléanais; Ottawa: Public Archives of Canada; Paris: Direction des Archives de France; Conservateur du Musée Postal; Conservateur du Musée Carnavalet; Bureau des Monuments Historiques; Directeur de l'Institut Géographique National; Poitiers: Services d'Archives de la Vienne; Portsmouth: City Record Office; Victory Museum; Rennes: Musée des Beaux-Arts; St. Helena: Hon. Archivist; St. Malo: Conservateur des Musées et Bibliothèques; Stockholm: Telemuseum; Surrey Record Office; Sydney: Royal Australian Historical Society; Toulouse: Musées de Toulouse; Tours: Conservateur des Musées; Vatican City: Archivio Segreto Vaticano; Warsaw: Post and Telecommunications Museum; Washington, D.C.: National Archives and Records Service; Wiltshire County Council Archives.

Aberystwyth: National Library of Wales; Algiers: Bibliothèque Nationale; Avranches: Bibliothèque Municipale; Boston, Mass, Public Library; Chatham: Public Libraries; Royal Engineers Corps Library; Deal: Kent County Library, Deal Branch; Dijon: Bibliothèque Publique; Draguignan: Bibliothèque et Musée; East Suffolk County Library; Gillingham: Public Libraries; Gravesend: Public Libraries; Great Yarmouth: Borough Libraries; Grimsby: Public Libraries; Halifax, N.S.: Memorial Library; Hamburg: Staats- und Universitäts Bibliothek; Hobart: State Library of Tasmania; Jersey: Public Library; Kent County Library; Kingston-upon-Hull: Public Libraries; Liverpool: Public Libraries; London: Australian Reference Library; British Library; India Library; Ministry of Defence (Navy) Library; and the public libraries of the following London boroughs: Bromley, Camden, Greenwich, Kensington and Chelsea, Lewisham, Southwark, Sutton, Wandsworth, Westminster; Madrid: Biblioteca Nacional; Margate: Public Libraries; Melbourne: Public Library of Victoria; New York: Public Library; Norwich: Public Libraries; Paris: Bibliothèque de la Direction de l'Architecture; Philadelphia: Free Library; Portsmouth: Public Libraries; Plymouth: Public Libraries; Rochester: Central Library; Rouen: Bibliothèque Municipale; Bibliothèque de la Ville; St. Albans: Public Library; San Francisco: Public Library; Sheerness: Kent County Library, Sheerness Branch; Stockholm: Royal Library; Sydney: Mitchell Library; Washington D.C.: Library of Congress; District of Columbia Public Library; Weymouth and Melcombe Regis: Public Library.

Alderney: Clerk of the States; Anglesey: Clerk of the County Council; Bonn: Bundesminister für das Post- und Fernmeldewesen; Cardiff: Welsh Office; Cheshire: Clerk of the County Council; Colwyn Bay: Town Clerk; Copenhagen: General Directorate of Posts and Telegraphs; Deal: Town Clerk; Denbigh County Council: Clerk to the County Council and County Planning Officer; Dijon: Directeur de la Télécommunication de la Région de Dijon; Dunstable: Town Clerk; Flintshire: Clerk of the County Council; Guildford: Engineer, Surveyor and Planning Officer, Guildford Rural

District; Helsinki: General Direction of Posts and Telegraphs of Finland; Hobart: Marine Board; Superintendent of Scenic Reserves; Hoylake: Clerk of the Council; Kingston-upon-Hull: Town Clerk; Lisbon: Chief Conservator, Correios e Telecomunicações de Portugal; Liverpool: Town Clerk; Madrid: Director General de Correos y Telecomunicación; Moscow: Minister of Postal and Electrical Communications; Orpington: Engineer and Surveyor; Oslo: Telegrafstyret; Stockholm: Teleskolan; Surrey: Clerk of the Council; County Valuer and Estates Surveyor (Historical Buildings and Antiquities Office); Sydney: Director, Posts and Telegraphs; Southwark: Borough Architect and Planner; Whitstable: Clerk of the Council.

Not least I have to thank John Murray (Publishers) Ltd. for permission to quote from *Ordeal at Lucknow* (Michael Joyce) and Doubleday and Co. Inc. for permission to quote from *Halifax, Warden of the North* (Thomas H. Raddall).

CHAPTER I

Introduction

This book is a history of the remarkably varied visual mechanical telegraph systems which existed in many parts of the world up to, and in some places well after, the general introduction of the electric telegraph about the middle of the last century. It is the first complete work on the subject in English, and the first in any language to give such wide coverage in detail. Many illustrations, contemporary with the story and present-day, have been assembled, and maps have been included of some of the most important lines whose course is known with certainty.

The book does not, of course, describe every piece of telegraphic machinery ever invented; that would need volumes — Colonel John Macdonald's conceptions alone would fill one.

In the author's view, the usefulness of visual, or optical, telegraph systems has never been fully appreciated, save perhaps in France, where they were most developed and played a notable part in national unification after the Revolution. *Telegraph* and *semaphore* are often regarded as synonymous terms in speaking of pre-electric systems. The book explains why this is an oversimplification, if not an error.

Some of the old telegraphs were ephemeral; others gave decades of good service. All sorts of men — Army men, Navy men, scientists, even clergymen — devised them, and their designs varied from efficient to fantastic. The first English telegraph did not, as some have said, bring London the news of Trafalgar and Waterloo. But the shutters, flaps, discs or arms of the old telegraphs relayed innumerable other tidings, uncommon or routine, welcome or unwelcome, warlike or peaceful. The storming of a stronghold, the outbreak of an insurrection, the coronation of a monarch, the execution of an enemy, the sighting of an anxiously awaited merchantman: these were some of the more exciting of their cabalistic messages. Bad weather was their bane, and they were seldom used at night. But for all their limitations, they performed well. In their day they were a marvel, and they were a worthy forerunner of the ubiquitous electric telegraph that displaced them. And relics of them may still be seen — as the book shows — by those with patience to explore; in 101 places literally from Holyhead to Hobart.

The first British author to enquire seriously into the English visual telegraphs long after they had been given up was Instructor-Captain Oswald

T. Tuck, RN. He had no previous special knowledge of signalling or cognate subjects, but he set to diligently, examining the Admiralty papers in the Public Record Office and ably marshalling details of technical and human interest for his article, 'The Old Telegraph' in *The Fighting Forces*, September 1924.

Tuck fell into one or two errors. Like some other writers before and since he spelt the name of Chappe, the French telegraph inventor, with a final acute accent, and imagined that the English stations were lofty, imposing towers, such as never existed and were only faintly resembled in two isolated instances. For all that, his pioneer work was of the greatest value and every writer on the subject since is in his debt.

The present writer's friend, the late Commander Hilary P. Mead, RN, a signals officer, followed Tuck's lead and much advanced the subject of early telegraphology by his own diligent researches at the Public Record Office, Admiralty Library, National Maritime Museum and elsewhere, and by patient field studies. Some of his scholarly articles in *The Mariner's Mirror* were collected in a limited edition as *The Story of the Semaphore*.

Mead fostered my own interest in the subject, which had first been aroused by a fondness for topography and a desire to find out more about 'Telegraph Hills' marked on maps of southern and eastern England. Mead and I gradually amassed a formidable body of information based on our joint researches. He eagerly accepted my suggestion of a book which would attempt to record what was known of early telegraphs in Britain and abroad. Since his untimely death in 1961 I have continued the quest, discovering new facts and clearing up points which had puzzled us, until I now feel justified in presenting a work which I hope will be a fitting tribute to his notable ground work.

Microwave telecommunication networks of today recall the visual telegraph in their use of a chain of relay stations on high points in 'aerial' communication. M. Georges Rigol, Conservator of the Postal Museum in Paris, has acutely remarked that the towers of such systems follow the same principle as the visual telegraph, the re-transmisssion of signals from station to station in a straight line. The Paris-Lyons telecommunication link and its small number of stations compared with its predecessor of more than a century ago are referred to in Chapter X.

Telegraph systems came into practical use at the end of the 18th century, when the subject virtually begins. Most writers on telegraphy introduce the subject by recalling the efforts of the ancients and of the inventors of the 17th and early 18th centuries to communicate over distances, but as only later achievements can be described as mechanical communication, here it may suffice merely to refer to some of the earlier experimenters.

According to Aeschylus, Agamemnon had a means of communication in 1084 BC by which the news of the fall of Troy was conveyed to his queen Clytemnaestra. The watchman had been waiting 10 years for the message, which then arrived by a line of beacons.

The classical method almost always was to use beacon fires or columns of smoke. To find out the complete story one must study the works of Aristotle, Polybius, Vegetius and other ancient authors.

Polybius tells us that he perfected a torch telegraph that either Cleoxenus or Democlitus had invented. It depended on the display and concealment of torches representing the letters of the alphabet. To send, say, 'Kres [Cretans] 100 have deserted', the person making the signal first held up torches on the left, to show that it was the second tablet which was to be inspected, and afterwards five upon the right, to show that K was the letter for the observer to note. Next he held up four torches on the left, because the second letter, r, was found in the fourth division of tablets, and so on.

The Romans operated signal systems in their empire. Vegetius describes raising and lowering a wooden beam (*trabes*) from a tower. But in general fires or smoke were the means. In Roman Britain there were some signal stations between the Humber and Tyne, as part of a system warning of Saxon raiders, along the North Devon coast, in central Wales, and along Hadrian's Wall and the Solway Firth.

Much as one would like to imagine a mechanical network complementing the imperial highways, one has to reject the notion, even though in L. Sprague de Camp's entertaining novel *Lest Darkness Fall* an American scientist transported to 6th-century Rome sets up a semaphore telegraph company.

Heliography, readopted by the British Army in India and in South Africa, is of respectable antiquity. For instance, in *Theodora and the Emperor* Harold Lamb refers to the sun telegraphs of the Byzantine Empire.

The fire-and-smoke system was practised in the Middle Ages. The Crusader-built towns and forts in Syria and Palestine kept in touch in this way. In 11th-century Roussillon, the Vicomte de Castelnou, who held most of the province, communicated with distant vassals by signal towers from which rose smoke by day and pillars of flame by night. The first signal given was followed by as many more as there were groups of 100 lancers observed.

The best known use of such methods in England is the fire beacon system employed to give warning of the approach of the Spanish Armada. Many of the hills on which these beacons stood are remembered by the name of Beacon Hill. Similarly, the name Telegraph Hill commemorates the use of the newly-invented telegraphic system of communication at the time when England feared an invasion by Napoleon.

An Act of the Scottish Parliament in 1455 laid down that on the approach of English invaders one bale or faggot would be set ablaze. Two on fire would signify that the English were 'coming indeed' and four that they were 'in great force.'

The North American Indians perfected smoke signalling. Not only did they use a varying number of puffs, but by throwing substances on the fire they changed the colours of the smoke. Even in the 1914-18 war, rockets with yellow, green, brown or purple puffs of smoke were used as recognition signals and had the advantage of defying expectation or imitation.

All such examples are best termed signal incidents. They were limited in application and there was always danger of their being misunderstood. The tragedy of Aegeus and Theseus affords a classical instance. Theseus set out to slay the Minotaur. It was arranged that if successful he would change the colour of his sails from black to white when returning. He forgot to do so. His father, Aegeus, concluding that he was dead, was so upset that he cast himself into the sea and was drowned.

The first Englishmen to consider seriously means of rapid communication were John Wilkins (1614-72) and the second Marquess of Worcester (1601-67). In 1641 Wilkins, a proficient mathematician and scientist and a founder of the Royal Society, produced his *Mercury, or the Secret and Swift Messenger, showing how a Man may with Privacy and Speed communicate his Thoughts to a Friend.* But he got no further than torches and beacon fires for conveying intelligence any distance.

In his *Century of Inventions* (full title, *Century of the Names and Scantlings of Such Inventions as at present I can call to mind to have tried and perfected*) the Marquess of Worcester affirmed that he had discovered 'a method by which, at a window, as far as an eye can discover black from white, a man may hold discourse with his correspondent without noise made, or notice taken, being according to occasion given, or means afforded, *ex re nata,* and no need of provision beforehand, though much better if foreseen, and course taken by mutual consent of parties'. The method, he said, was so perfect that correspondence could take place 'by night as well as by day, though as dark as pitch is black'. For all that he is so vague that we are equally in the dark about the method he had in mind. The *Dictionary of National Biography* refers to the book's 'nebulous ideas without any attempt to work them out in practical detail'.

We are on firmer ground with Dr Robert Hooke, a friend of Wilkins, who delivered on 21 May 1684 a 'Discourse to the Royal Society ... showing a way ... how to communicate one's mind at distances in as short a time almost as a man can write what he would have sent.' He published a paper on the subject in the *Philosophical Transactions* for that year. Hooke proposed to have boards of different shapes — square, triangular and others — answering to the letters of the alphabet, hung up in a large square frame divided into four compartments and shown in the order required from behind a screen. Each piece of wood represented a letter according to the compartment in which it was hung.

Hooke describes the distances between stations, mentions the use of telescopes — a recent invention — and suggests how the characters could be varied 10 thousand ways. The same character, he thought, might be seen in Paris the minute after it was shown in London.

The stations, if far apart, should be high and exposed to the sky. Each intermediate station should have two telescopes and three operators. Two operators should suffice at the termini. Characters for daytime use should be made of deal. Those for use at night might be of lights disposed in a certain order and covered or uncovered according to an agreed method. Several other characters could express a complete sentence.

According to R.W.T. Gunther, *Life and Work of Robert Hooke* (1930), Hooke first tried his system in 1672 between the garden of Arundel House and a boat moored off the far shore of the Thames half a mile away. The President of the Royal Society objected that 'the use of it would be often hindered by hazy weather'.

About 16 years after Hooke's paper, Guillaume Amontons (1663-1705) proposed a similar system. He suggested placing men in stations so that a man in one station might with a telescope see a signal made in the next station, immediately make the same signal, and so on. The signals could be letters of the alphabet, or a cipher understood by the most distant operators, not by men repeating them intermediately. The man in the second station should make the signal to the person in the third station the moment he saw the first.

Amontons tried the method 'before several persons of the highest rank' at the court of France. But the experiment was ephemeral. He also experimented with windmills. Letters were fixed to sails and read as they were hoisted aloft, a method said to have been used in Holland.

The 18th-century German scientist Johann Andreas Benignus Bergsträsser tried all kinds of communication, from fire, smoke, explosions, torches and mirrors to a gymnastic signalling experiment using Prussian soldiers, trumpet blasts and artillery fire. He seems to have been too ingenious to perfect any one means. Later inventors and writers are indebted to him for his *Sinthematografie*, a five-volume work of 1784 in which he reviewed all known means of communication ever devised and even gave an account of signalling by using shutters or pivoted arms set at angles and worked by rodding through bevel gears, so anticipating the telegraphs which were to be developed in the next decade or two.

The words 'telegraph' and 'semaphore' both derive from Greek. 'Telegraph' signifies literally: 'write (or describe) at a distance'. It therefore remained appropriate when applied to later systems such as electric, cable or wireless in which, although there is no visual indication, the operator at the receiving end really writes at a distance. 'Semaphore' signifies: 'bear a sign'. It may therefore be logically applied to, say, a finger post, road sign, railway signal or traffic lights. A local, limited indication is implied.

Both terms were coined by the French at the time of the Revolution and its aftermath. The credit for *télégraphe* goes to Claude Chappe, whose important T-type apparatus for long-distance visual communication came into full use in 1793. Chappe had wanted to call it a *tachygraphe* (speed writer) but his friend Miot de Mélito advised *télégraphe* as being less exuberant. The *sémaphore* was the invention of another Frenchman, Depillon, who evolved it in 1801 for ship-to-shore communication by the movement of arms pivoted directly on a mast.

Our forefathers, incidentally, were more exact borrowers from the ancient languages than we sometimes prove. Such hybrid Graeco-Latinisms as 'television', 'telecommunication' or 'teleprinter' would surely have seemed barbarisms to them. What would they have made of the contraction 'tele'?

Laymen have often tried to label every kind of signalling device a semaphore. But at sea, and on railways, semaphore has almost always meant one thing: 'an apparatus for making signals, consisting of an upright post with one or more arms working in a vertical plane,' to use the definition in Murray's *New English Dictionary* (1914).

Popular abuse of the word began with the rise of the electric telegraph. Historians and others then cast about for a convenient way of describing pre-electrical systems. As 'visual' or 'optical' did not occur to them, or at least satisfy them, they called them all semaphores, even systems which had been in use before the word was coined.

To be fair, it must be conceded that long before this the British Admiralty had seemed to blur the original distinction. Admiral Sir Home Popham's arm-type telegraph was inspired by the French coast semaphore and its English copy. To distinguish it from the shutter system which it superseded on land, their Lordships referred to it quite legitimately as a 'semaphore telegraph' or just 'semaphore.'

Confusion is evident when one reads in *Encyclopaedia Britannica* (1929 edn.)

> the semaphore stations, each with its tall mast and signalling arms which linked London with the south coast while Napoleon's Great Army waited at Boulogne, were an advance, for they could spell out any message . . .

The author should have written 'the original telegraph stations, each with its movable shutters set in a frame . . .'

Confusion is compounded in Sir Arthur Bryant's *The Years of Endurance* where we read:

> At the end of February [1798] . . . a report . . . was flashed to London over the new semaphore telegraph which, with its great arms and shutters and chain of towers stretching from the naval ports to the roof of Westminster Abbey, brought news to the Admiralty from the coasts in a few minutes.

For one thing, there was never a combined arm and shutter apparatus. For another, the hilltop huts of the shutter telegraph system then used by the Admiralty were hardly a chain of towers. Finally, one of the towers of the Abbey was used for an isolated experiment only!

The French never lost sight of the distinction between semaphore and telegraph. But then their coast semaphore system developed independently of the already firmly established overland telegraph system and there was never any confusion or mixing of the genres. There was also an acceptable term, *aérien* (aerial), to describe visual, as distinct from electrical, telegraphy.

No sooner had the semaphore been thought of than it was imitated by scores of other inventors, most of them seeking to adapt its simple principle to telegraphy. A semaphore might be devised with three or even four arms on a post. These arms might be displayed on one side of the post only, in which case each would be capable of three changes. If arms were displayed on both sides of the post, the variations would be six in each case. With three separate arms using both sides of the post there were 301 combinations, which should have been more than enough for any telegraph.

Cranks like Colonel John Macdonald considered this total quite insufficient. They multiplied the arms, putting either four on one post, or pairs of arms on the same pivot. It was not uncommon for semaphores to be devised with even six arms and some machines even had four pairs arranged on one post, or divided on two posts as preferred. Semaphores with more than two arms were used for coded telegraphy and their feasibility, if not their simplicity, aggravated the war between the schools of spelling and coding. Such machines employed numbers running into thousands and the meanings had to be taken out of a telegraphic vocabulary, containing perhaps every conceivable word and sentence likely to be necessary and many that were surely not.

The rapid development and extension of the Chappe-type telegraph in France ensured its predominance there, though some French inventors sought to modify or vary the original French semaphore for telegraphic use. In Britain the semaphore became the basic telegraphic form, although the semaphore inventions of Nicolas — later a famous naval authority — Macdonald, who delighted in complication, and Pasley, in his earlier days, were not adopted or applied to telegraphy. Pasley, as will be shown, learned his lessons well and subsequently became a most successful semaphorist.

Although Admiral Popham introduced his sea telegraph (an apparatus of semaphore type for mounting on ships and the predecessor of his similar land telegraph machine) in 1816, it was not until the end of the 1870s that such devices became at all widespread at sea. By that time the Pasley type had swept the board. It is said that W.S. Gilbert originally intended to call *HMS Pinafore, HMS Semaphore*. If so, he may have wished to strike a topical note and the date of production, 1878, may be significant.

The zenith of seaborne semaphoring was at the beginning of the present century when the bridges of men-of-war, British and other, bristled with such signalling machines, while members of the crew communicated liberally with hand flags. This was signalling within the fleet, sometimes at only one or two cable lengths, and nobody really considered sea semaphores suitable for communicating at a great distance. True, for a brief time, there was a huge masthead semaphore with 12 ft. arms, meant for long-distance work, but it was unpopular and generally impracticable.

Finally, it seems possible that railway semaphore signalling was first inspired by a particular visual semaphore telegraph installation. It may not have been coincidence that the first arm-type railway signals, invented by Charles Hutton Gregory, were erected on the London & Croydon Railway near New Cross in 1841, within sight of the imposing masts of Watson's commercial semaphore telegraph on One Tree Hill.

Ever since practical telegraphy began there have been two opposing, or rather antagonistic, schools of thought as to the basic method of procedure. One favoured spelling messages pure and simple. The other favoured coding them and referring to a 'telegraphic dictionary'. A machine that can achieve something like 20 separate and distinct signs is quite sufficient for spelling. But if a detailed code is to be used the apparatus must be capable of several hundred combinations. The British Admiralty policy of restricting its telegraphs to spelling messages persisted successfully for 50 years. The policy annoyed and frustrated the numerous inventors who wanted to impose their own appliances on the Royal Navy.

The Admiralty from the first preferred simple machines, which, during successive periods, achieved 63, 48 and 28 separate signs respectively. They were therefore cheap to make and maintain. Unskilled and ill-educated operators could work them, so that it was necessary only to pay wages of the lower scale. Finally, no code books were needed.

With the rival method, first the apparatus had to be complicated so as to be capable of even 1,000 changes. The telegraphists had to be quicker on the uptake and educated enough to handle a dictionary, and code and decode. They could therefore command higher pay. Lastly, an elaborate vocabulary had to be chosen and worked out by an expert, and printed and distributed. For a system including, say, 60 stations, the cost of the book alone would be considerable. All these things made the second method far more costly.

What advantage was gained by the more expensive means? Most important was its speed. A few signs could stand for a whole message, sentence or phrase, as the following example from Lord's *Code* will show: 4090 = 'We are in great distress, and require immediate assistance of a steamer'.

Much of the Admiralty business was routine, so that the extra time spent in spelling out messages was of no serious account. But there were occasions when the Admiralty departed from the spelling rule. One was on 9 August 1827 when a message announcing the death of Canning was brought to the Duke of Clarence while a review was in progress at Southsea. For some reason it was not in plain. John Barrow, Second Secretary of the Admiralty, who was there, translated it as he knew the numerical key and could read the message straight off. Mead suggests that the message was written out in semaphore characters as received, for instance that the word 'Canning' was made as '3, 1, 25, 25, 21, 25, 15.'

Chappe's system in France, by far the most widespread of any, is an example of an organisation that fell rather between the two extremes. The apparatus was not very expensive or elaborate, the operators were of moderate intelligence and the ancillary code was more in the nature of a shorthand and was not an expensive book to produce as it contained only such common words as *avant, avec, avoir* and *après*.

A point in favour of the spelling method was that it could afford to have a few corrupt signals in transmission, such as might have been fatal in a code. As Pasley pointed out, such a message as the following would be fairly intelligible despite its mistakes: 'she wil saul fsr the Wesr Inbies imrdiatly'. Six errors in a coded message and it would have been impossible to decipher.

A final qualification may perhaps be allowed. Purists may object that four of the overland systems described, that of Lipkens in the Netherlands, that in St. Helena, that of Prince Edward in Nova Scotia and that between Calcutta and Chunar in India, were not strictly mechanical telegraphs, but it is hoped that their ingenuity and associations will be found to warrant their inclusion.

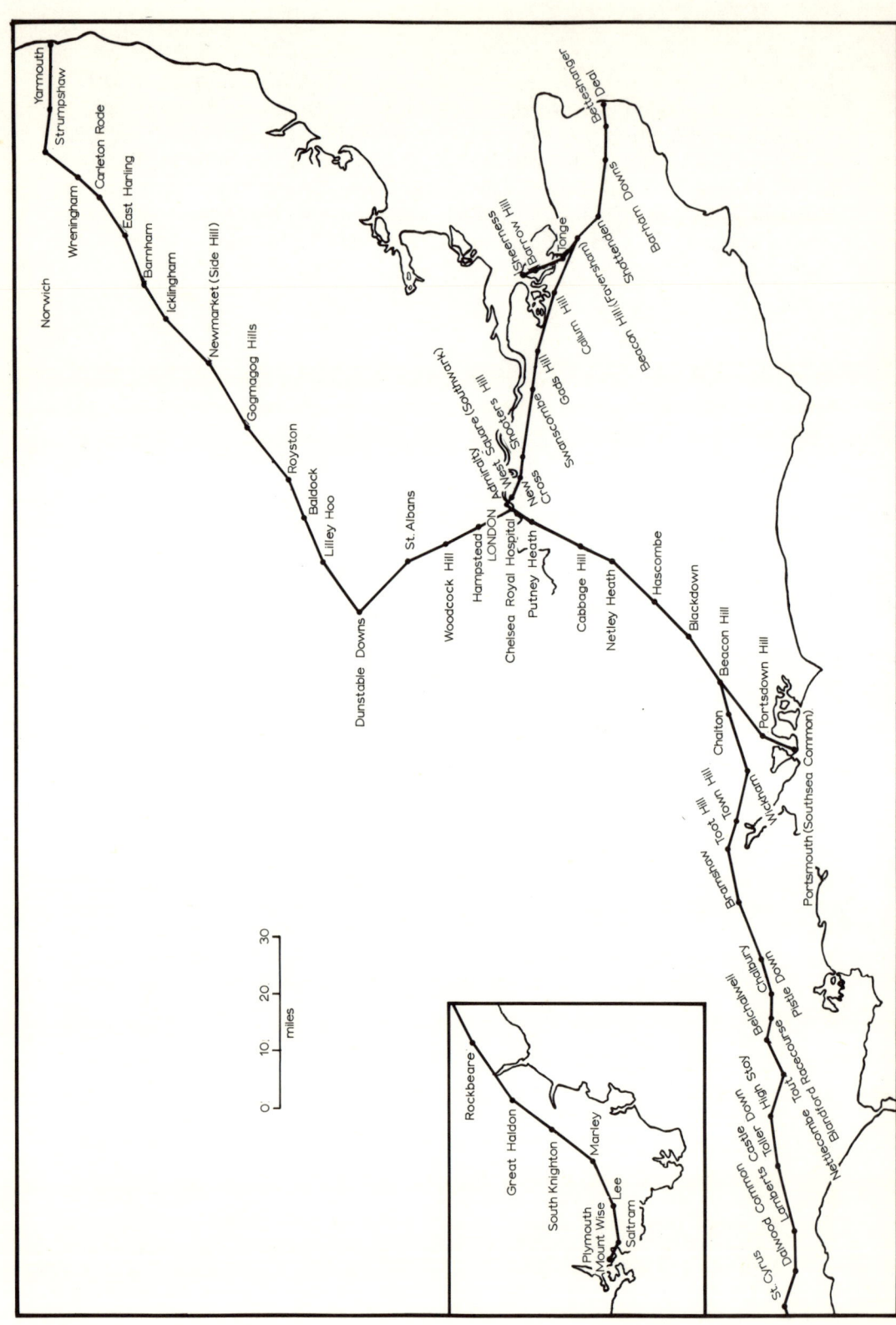

1. Map of Admiralty Shutter Telegraph system.

CHAPTER II

The Admiralty Shutter Telegraph

The fame of the T-type telegraph triumphantly inaugurated by Claude Chappe in France in August 1794 soon crossed the Channel. Charles Dibdin's *Great News, or a Trip to the Antipodes,* first performed in London on 11 October 1794, contained the following:

> If you'll only just promise you'll none of you laugh
> I'll be after explaining the French Telegraphe!
> A machine that's endowed with such wonderful pow'r
> It writes, reads and sends news 50 miles in an hour.
> Then there's watchwords, a spy-glass, an index on hand
> And many things more none of us understand,
> But which, like the nose on your face, will be clear
> When we have as usual improved on them here.
>
> Adieu, penny posts! mails and coaches, adieu!
> Your Occupation's gone, 'tis all over wid you.
> In your place telegraphs on our houses we'll see
> To tell time, conduct lightning, dry shirts and send news.

A description of the Chappe telegraph is said to have been brought to Frankfurt-am-Main by a former member of the Parlément of Bordeaux who had seen it at work at Belleville, outside Paris. Two models of it were made at Frankfurt and sent to the Duke of York, Commander-in-Chief of the British Army. In addition, a drawing and alphabet of the *télégraphe*, said to have been found on a French prisoner, were passed to the Rev. John Gamble, the Duke's chaplain and Chaplain-General to the Forces. At the Duke's behest, Gamble, who had been a mathematics don at Cambridge, investigated and reported on all known means of signalling and at the Duke's expense printed his report on *Observations on Telegraphic Experiments.*

War with revolutionary France pointed the need for rapid communication between London and the naval stations. The success of Chappe's invention prompted Britons to emulate it but the Admiralty looked for a simpler, cheaper kind of apparatus. A series of shutters working in a frame might not be the improvement on Chappe's elegant moving-arm telegraph promised in Dibdin's song but would perhaps provide an instrument robust and serviceable enough for a particular function. It needed only rectangular boards pivoted about their horizontal axes in the frame. Each shutter had but one movement. If it were normally closed, one movement of a crank or

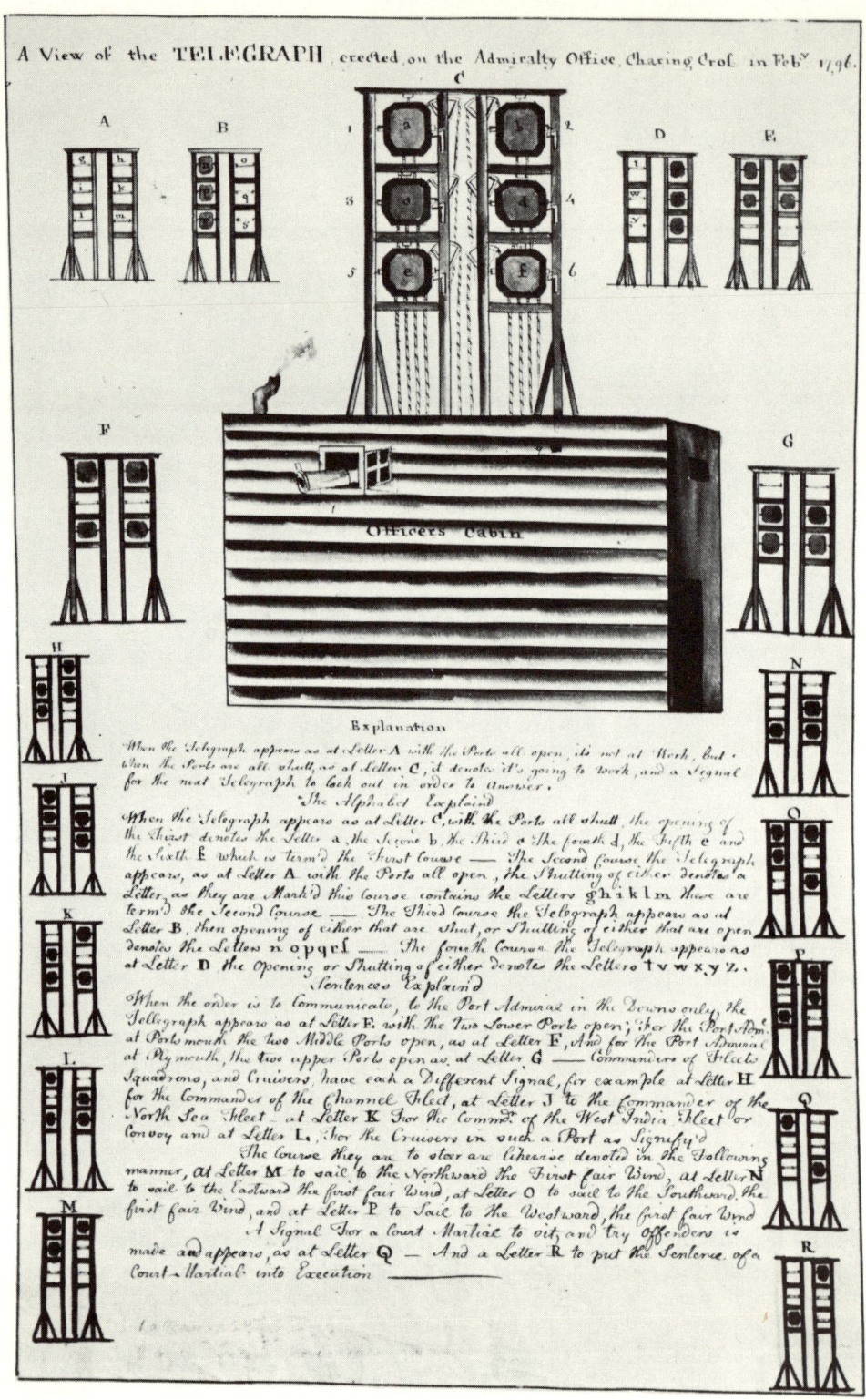

2. Shutter telegraph installed at the Admiralty, 1796, with code details.

quadrant, pulled on by rope or chain, would put it at open. If it were normally open, one movement, too, would set it at closed. The operation has been likened to bell-ringing, and the series of 'pulls' called 'courses', as in campanology.

In April 1795 the Admiralty received a copy of the report and sent Gamble to experiment at Portsdown Hill. With the aid of Portsmouth Dockyard officials, he erected a machine which was ready for trial by 6 August 1795. It was found to be readable from the dockyard with the naked eye. Gamble returned in hope to London on 27 August, only to find a rival claimant for Admiralty favour, Lord George Murray, fourth son of the Duke of Atholl and later Bishop of St. David's and Archdeacon of Man.

Murray had proposed to the Admiralty a machine with six shutters in a 20ft. frame. Gamble had been using five shutters, arranged one above the other, the top and bottom ones being almost square, the middle three oblong. The idea of the different shapes was that signals could be identified at a distance when they otherwise could not have been. Had the five shutters been of the same shape one above the other there would have been nothing to show which was the top one when some of them were open.

Murray's machine had six shutters, of the same shape, arranged two by two vertically. It was essential that they should be separated sideways by a space about half their own width, so as to make exposures more distinctive, and give room for the operating cranks, counterpoises and ropes. The separation is not always shown in pictures.

The larger number of changes, 63, was probably the main consideration that influenced the Admiralty's choice of the Murray system. The 31 changes of Gamble's telegraph would not have sufficed for all purposes, namely the whole alphabet, 10 numerals and several arbitrary phrases.

When told on 27 August, in 'desultory conversation' with one of the Lords of the Admiralty, that Murray had succeeded, Gamble was surprised and justifiably annoyed. As he pointed out, he could easily have given his five-shutter machine an extra movement.

By an Order in Council of 8 August 1795 Murray received £2,000 for his telegraph and in March 1796 the Admiralty entrusted the direction of telegraph lines to him. The *Dictionary of National Biography* records that Murray had a long conversation about telegraphs with George III on 18 December 1795. It is believed that a machine was tried out on Wimbledon Common.

Murray's shutters were stated to have been 3ft. square. Sometimes the corners were cut off and the boards were consequently octagonal. This is how they are often depicted. In Murray's own model, now in the National Maritime Museum, the shutters are certainly octagonal. On the other hand, other views show square boards.

Having decided on the Murray system, the Admiralty had to consider where to apply it. It lost little time. On 25 September 1795 George Roebuck, a surveyor, was appointed to select sites for stations on lines to link London with Deal, Sheerness and Portsmouth. He began work before the end of the year and had the lines in working order before the end of 1796. This was a fine achievement for he had no precedent. The labour of seeking out likely elevated spots, some in remote districts, in all weathers, and then testing them and treating with landowners, some of them no doubt ill disposed, was considerable.

The task of building and equipping the lines devolved on the Navy Board, the supply division of the Navy, operated from Somerset House.

In the *Royal Kalendar* of 14 April 1798, under 'Admiralty Office, Marine Department,' Roebuck is listed as Inspector of Telegraphs, and therefore must have superseded Murray by then. He received £300 a year. Some years after, Roebuck was referring to Murray, who died in 1803, as 'my late friend.'

It was agreed to pay Roebuck £215 per station on the Deal route, the amount including the erection of the building with its two small apartments, two 12-guinea telescopes, a time-piece worth eight guineas and a stove with an iron funnel pipe. Roebuck tried to get £240 per station on the Portsmouth line but had to settle for £230.

A letter of 2 December 1795 shows Murray's expenses to have included a payment of £34 to a Mr. Browne for surveying and settling with owners and his own disbursement of £171 14s. 6d. for travel and engaging staff. Beds, bedding, tables and chairs were estimated to cost £150. The whole establishment was expected to cost, at lowest calculation, £2,938 2s. 6d. a year to run.

To understand the complications the shutter system sometimes involved, we must look at the construction of a station. There was a two-roomed hut or shack, comprising an operating room and a general living room-cum-kitchen. On the roof was fixed the shutter frame, of stout timbers, about 20ft. high, and supported by heavy wooden baulks embedded in the ground in front of and behind the hut. In some pictures and in Murray's model, the supporting baulks end on the eaves of the roof. All the work had to be substantial to withstand the winds that must have whistled through the contrivance on bare, bleak hilltops.

The construction was a fixture. It could not be trained round like some other telegraphs to point in various directions. Inevitably there was an angle at which the aspect of the frames and their apertures could not clearly be made out. At 90° nothing could be seen. At such a location the only seeming solution would have been to place two machines at right angles to each other. (This may have been done on Dunstable Downs; see page 28). But the Admiralty would have jibbed at employing two crews at one station. Where the angle was less, the difficulty could be got over by 'dividing the angle' and setting up the frame so that the openings could be seen partly askew from both corresponding stations.

The operating space must have been restricted and cramped at the best of times. Conditions were worst when the telegraph was set up on the roof of an existing building, as at the Admiralty in Whitehall; at a house in West Square, Lambeth; at Chelsea Hospital (two places); on the town clock tower at St. Albans; and on the South Gate at Great Yarmouth.

On the roof of the Admiralty the congestion must have been serious. There were soon two main frames set up together but at right angles to each other, the one facing SE. towards West Square, the other SW. towards Chelsea. Later, a smaller, subsidiary frame was installed for signalling to Chelsea Hospital on the Yarmouth line. When all three frames were being worked at the same time, the confusion and din must have been overpowering!

The full complement of a shutter station has been given as a Lieutenant, Royal Navy (in charge), a Midshipman, and two assistants, probably 'jack-tars' or 'foremastmen'. Many Midshipmen in those days of slow promotion were hardly youngsters – there is a classic instance of one aged 35. The duties of the crew are partly explained in, of all things, a lady's fan of the period. It shows the movements of the shutters above the hut, and the persons within it, the 'disabled' officer writing down the message in a ledger, the Midshipman at the telescope, and the Foremastman attending to the ropes. The other assistant is not shown, as he is probably supposed to be looking through his glass in the opposite direction. The fan is dated 1 July 1796.

The *Navy List* of corresponding date does not give telegraph appointments, but it lists Lieutenants at coast signal stations, and states that their pay was 7s. 6d. a day plus half-pay of their rank. That of the Midshipman was 2s. a day plus half-pay of a fourth-rate and of the one seaman, 2s a day. All had lodging, coals and candles free.

Although a similar complement may have been allowed at the most important shutter telegraph stations, such as the terminals, it is not certain that there were any officers out in the country, and the crew there probably consisted of not more than three men per station. At the stations on the line from the Admiralty as far as Beacon Hill, South Downs, the crew was ultimately to be four men because of the double volume of traffic coming up from both Portsmouth and Plymouth. There were finally 64 shutter telegraph posts in England. Had each been manned by a Lieutenant – even if 'disabled' – and a Midshipman, their Lordships would hardly have approved of so many officers on subsidiary duties in wartime.

At the Admiralty one 'foreman' superintended the Portsmouth and Plymouth line, another the Deal and Sheerness line. The first received 28s. a week, the other 24s. 6d. and both had 1s. 6d. a week for lodging. A night's lodging was no great problem if the post was near a town or village but finding convenient accommodation in the remoter parts of, say, Dorset or Devon, may have been difficult. It would be interesting to know what the natives thought of the 'furriners' and their cabalistic trade.

Fortunately, the 'glass men' did not have to have eyes glued all the time but kept watch every 5 min. The operators soon acquired great skill. An average message between London and Portsmouth might take about 15 min. to pass. As with the later semaphore telegraph between the same places, the shutter spelled the words but often contracted them, leaving out some vowels. The preparative signal could be sent from London to Deal or Portsmouth and acknowledged back in 2 min. A similar pre-arranged signal is said to have passed from London to Plymouth and back, 500 miles at least, in the incredible time of 3 min.! Such exploits compensated for the many days when bad weather delayed or precluded transmission.

On windy days or in very cold weather the shutters might stick and have to be freed. At such times the chimney would no doubt reek and conditions in the operating room and the 'living' room-cum-galley cannot have been ideal. But on a fine day the life must have had some of the advantages and few of the drawbacks of life on board one of the 'wooden walls'.

The printed textual description on the lady's fan gives details of the operations. The allocation of the 63 changes seems to have been definite, as several sets have come down to us. As well as the 26 letters of the alphabet (sometimes omitting J and U), there were the 10 numerals, while the remaining arbitrary extras included the following:

All shutters open, Not at work, or Message ended
All shutters closed, About to work, or Ready

Page	Word ended	Fog
And	Sail	Arrived
The	French	Line-of-Battleship
Deal	Dutch	Frigate
Downs	Russia	Convoy
Sheerness	Admiral	Fleet
Portsmouth	Captain	Transport
Nore	India	East
North Seas		West

Port Admiral in the Downs
    at Portsmouth
    at Plymouth

Commander of the Channel Fleet
    North Sea Fleet
    West India Fleet

To sail, the first fair wind, to the northward
                                eastward
                                southward
                                westward

Court-Martial to sit
Sentence of court-martial to be put into execution

The working of the crew can also be seen in a rather crude contemporary sketch of the Nunhead (Plow Garlic Hill) station in Deptford Library. Two men were needed for the shutter ropes, and one each at the forward and rear telescopes — 12-guinea instruments supplied by Dollond. The best drawing of shutter telegraph men at work is perhaps in a former advertisement of Cable & Wireless Limited, in a series *Communications Old and New*. It looks to have been adapted and improved from the Plow Garlic sketch.

The simple business of pulling the shutters open sufficed generally, and counterpoise weights ensured that they shut when released. Each of the six shutters may have been subdivided again into several slats, on the principle of the venetian blind, which with the help of fairly simple apparatus could be made to open and close together. There is no direct evidence of such a contrivance having been used, though one or two contemporary sketches suggest something of the sort.

It will be most convenient to consider each of the four shutter lines in turn, and discuss their stations as they occur.

The Deal Line

Admiralty. Murray reported that the Deal line was finished on 27 January 1796. By the end of that year, when the Portsmouth line had also been completed, there were two frames on the roof of the Admiralty, but little pictorial evidence is available except a rather indistinct print of Horse Guards Parade dated 1 March 1805, with what is evidently meant for the two telegraphs on the corner of the roof of the Admiralty, as seen from the rear of the building. *The Picture of London for 1808,* states:

> On the top of the Admiralty are erected two telegraphs, one communicating with Deal, and the other with Portsmouth, the inside of which may be seen, on proper application, or very small interest with the porters, or persons who work the machines.

This evidently meant a handsome tip, for one can hardly imagine the authorities encouraging strangers on the premises.

West Square. The first station towards Deal was at the present No. 36 West Square, Lambeth, 1¼ miles from Whitehall. It is one of a terrace of Georgian houses on the SE. side of the square, lying just off St. George's Road, at the rear of the Imperial War Museum. The house was taken over and adapted by the Admiralty and long remained a kind of annexe to the Whitehall station, being used as a residence for the Superintendent of Telegraphs. It remained Admiralty property for several years after all visual telegraphs had been given up. Charles Hawes Jay, Superintendent from 26 June 1828 (Commander 1841, Captain 1857), was allowed to go on living at West Square, where he died in 1864, aged 71. Lieutenant Stephen Perdriau (Perdrieau in the *Navy List*) was in charge at West Square in 1817. On 1 January 1935 a Mr. M. H. Lapidge wrote that he visited the house when it was occupied by a cousin of his, the daughter of Perdriau, and that he remembered seeing the remains of the mast which went up through the house and had been cut off where it passed through the roof. (The reference was certainly to the semaphore apparatus which succeeded the shutter at West Square).

It is satisfactory that the square has been declared a conservation area and that, at the time of writing, its houses are being restored by the landowners. No. 36 is the only one of the houses to have both a ground floor bow window and an additional square, as distinct from mansard, storey. An early print, reproduced as Plate 5, shows a wooden tower, lacking the shutter apparatus, erected on No. 36, which then shared a gabled pediment to the front with No 37.

3. Fanciful representation of the Nunhead shutter telegraph station.

Nunhead (Plow Garlic). Just over three miles from West Square was the station on **Plow Garlic Hill**, at 'Nun Head', near New Cross. The site is maintained as Telegraph Hill Gardens by the Greater London Council.

Shooter's Hill. Some 4½ miles beyond Nunhead was the site of the next station on this prominent point. The place was not as ideal as might be thought, as it was subject to fog. The site was not at the summit of the 446 ft. hill but on its S. shoulder, formerly called Hazlewood Fields and now in the War Memorial Hospital Grounds. The only relevant print surviving shows highwaymen being hanged on Blackheath, with what is meant to be the telegraph in the background.

Swanscombe. There was a gap of rather more than 10½ miles to the Swanscombe station, on a site easily recognised as being the only convenient piece of rising ground in the line of sight and lying just south of the main A2 road (Watling Street).

Gad's Hill. Another 6¾ miles brought the line to Gad's Hill, made famous by Charles Dickens's later residence nearby. From the *Sir John Falstaff* Inn on the main Gravesend–Rochester road, Telegraph Road leads steeply northwards and the position of the station, even if it cannot now be recognised, can well be guessed, at a height of 300 ft., clearly overlooking the country between Swanscombe and Chatham.

Callum (Calham) Hill. Some 10 miles beyond Gad's Hill lay the next station, on this rise of about 200 ft., near Lower Halstow. The line of sight just skirts the north of Chatham Dockyard. Mead, in his *Mariner's Mirror* articles collected as *The Story of the Semaphore*, supposes a fork from Gad's Hill direct into the dockyard, but no firm evidence of such a branch has come to light. By that period the silting of the Medway much hampered the passage of large men-of-war, and Chatham had become less of a fleet base than a building and repair centre. Rapid communication between Chatham and London was not therefore as vital as with such wartime operational bases as Deal and Sheerness.

Beacon Hill (near Faversham). Mead erroneously assumed that the Beacon Hill station named on the Deal line was sited on the hill of that name above Luton, Chatham, near the present Royal Naval Hospital. From the sequence of stations given in a letter from the Navy Board to the Secretary of the Admiralty dated 5 December 1795 it is clear that the Beacon Hill about two miles west of Faversham is meant. The exact site would seem to be indicated by 'Telegraph Bank' marked on the Kent OS and about 20 chains SSW. of Watling Street off a lane passing Beacon Farm, west of Ospringe. It is eight miles ESE. of Callum Hill.

It was here that the Sheerness branch diverged. The Sheerness transmission had to go on a pronounced 'back leg', messages being almost sent back on themselves for some distance, so narrow was the angle of convergence between the main and branch lines. (By coincidence, as we shall see, another Beacon Hill was destined to be a junction, that of the Portsmouth and Plymouth lines.)

Tonge. The first station on the Sheerness branch was at 'a field near Tong *[sic]*', four miles north-west of Beacon Hill. The Kent OS marks 'Telegraph Hill', a modest eminence about 30 chains south of Tonge Green, about 2½ miles north-east of Sittingbourne and only a little north of the Callum Hill-Beacon Hill sight line. (Mead placed Tonge on the main line and took it to be the junction for Sheerness).

Barrow Hill. About five miles from Tonge was the only other intermediate station on the Sheerness branch, on this 100 ft. rise east of Queenborough-in-Sheppey. The original list of stations gives the site as 'Furze Hill', which adjoins on the east. It seems surprising from a glance at the map and a visit to the locality that messages could not be telegraphed direct between Callum Hill and Barrow Hill. It can only be assumed that fogs in the Swale or some other local condition dictated the choice of a circuitous route.

Sheerness. The final 2.3 miles brought the branch to its terminus at Sheerness dockyard. The apparatus is represented as being at Garrison Point, rather than in the dockyard itself, in the background of a print showing the execution of Richard Parker, ringleader of the Mutiny at the Nore, on 30 June 1797. A letter dated 1 July 1797 included in John Evans, *Excursions to Windsor* (1817), and quoted in Coles Finch, *The Medway River and Valley,* says:

The instant he [Parker] was visible to the garrison from the yardarm, the telegraph was put into motion to announce it to the Admiralty; and from the clearness of the atmosphere and quickness of working, the advice must have been received in seven minutes.

Shottenden Hill. The Deal route continued from Beacon Hill to Shottenden Hill (5 miles), where a windmill stood on ground 462 ft. high, formerly a well-known landmark and seamark. The station was sometimes known as Old Wives' Lees, from its postal address, some miles away. It was the main subject of a letter dated 2 April 1796 to the editor of the *Gentleman's Magazine,* from one Z. Cozens, who described a ramble a few weeks earlier into the interior of Kent. The telegraph, he said, was placed on 'Shottenton-hill' in the parish of Selling, a considerable eminence. He added that the Romans used the hill as an encampment and included sketches of the site, showing the Roman ditch around the land containing the windmill and the telegraph.

Cozens's drawing of the telegraph hut shows three square shutters open and three closed. He credits them with 720 positions! The 'low building' has two rooms for the personnel, the first of whom observes the signals at the first station, the second makes the necessary movements, the third notices when they are repeated at the next station and the fourth, the 'superior', enters the observations in a journal.

Barham Downs station was about ½-mile NNW. of Womenswold, where there is a cottage said by some to be the old building, but hardly resembling the type. The span from Shottenden was the longest on the Kent line, 11.2 miles, though not the greatest on any shutter route; it was made possible here by the height of both sites.

Betteshanger, 5.6 miles beyond Barham Downs, is marked by the present Telegraph Farm on the Roman road, ¾ mile S. of the village.

Deal. The site of the terminal station at Deal, 4.3 miles from Betteshanger, was in the Navy Yard, abutting on the shore. It is likely that the telegraph stood on or very close to the beach, as at Southsea. Such a position is low-lying for telegraphy but doubtless the sea itself provided a suitable background. *The Kentish Traveller's Companion* (1799) states that 'At the south end of Beach Street, close to the sea, stands a telegraph, being the first of the series ... conveying intelligence from the Downs to the Admiralty office in Westminster.' (*The Telegraph* public house and Telegraph Road at the back of the town apparently have no direct connection with the shutter telegraph and probably commemorate the later electric telegraph).

The Portsmouth Line

When the Portsmouth line was planned, Murray asked for a house in Lower Grosvenor Place for use as the first station, giving the poor visibility in London as the reason for needing one so close to the Admiralty terminus. It was found to be redundant.

Chelsea. The governors of the Royal Hospital, Chelsea, were prevailed on to allow a hut and shutters to be installed on top of the E. wing near the Governor's residence. *The Diary and Letters of Madame D'Arblay* (Fanny Burney) show that in 1798 she was taken over Chelsea Hospital where 'Good Mister punning Townshend showed me the telegraph...' Many prints depict the Hospital, mostly from across the river, with the telegraph on the roof. The best seems to be that by S. Owen, engraved and published by W. Cooke in July 1809. No. 18 Jubilee Place, less than half-a-mile from the Hospital, was used to lodge the telegraph crew. The distance from the Admiralty to the Chelsea station was 1.75 miles.

Putney Heath station was sited just over 4 miles from Chelsea. It stood in what is now the triangle formed by Kingston, Wildcroft and Telegraph roads, approximately marked by the present *Telegraph* inn, which bears a fine signboard showing the shutter telegraph. It is recorded that on 21 September 1809 the officer and crew watched the duel in which Lord Castlereagh wounded George Canning.

The station inspired several pages of useful remarks and observations in Sir Richard Phillips, *A Morning's Walk from London to Kew* (1817 and 1820). Phillips visited the station and spoke to the operators. He gives the total number of stations correctly, but erroneously states that the Plymouth line diverged in the New Forest. The average distance between stations is given as about eight miles, 'yet some are 12 or 14 miles'. (No span, in fact, was ever 14 miles). The distances, adds the author, are often increased by circuits for want of commanding heights. 'After 20 years' experience' — rather an exaggeration — signals could be transmitted on 200 days of the year throughout daylight hours, but on 60 days they could pass only during part of the day, and on about 100 days few of the stations could see one another.

Putney was generally made useless during E. winds when the smoke of London filled the valley of the Thames between it and Chelsea. The least favourable time for transmission, Phillips relates, was an hour or two before and after the sun's passing of the meridian. There was no excuse for bad timekeeping as each station had an eight-guinea clock.

Cabbage Hill. The line of sight first crossed Wimbledon Common, then a largely treeless heath, making for Cabbage Hill, now more generally known as Telegraph Hill, a slight eminence easily recognised in open country one mile south of Chessington Zoo, close to the Surbiton-Leatherhead road and marked approximately by a modern reservoir. A growth of blackthorn trees in the rough form of a rectangle is supposed to be a relic of the hedge planted round the station in 1796.

The sites of the following stations, at *Netley Heath* (Hackhurst Downs) on the North Downs above Gomshall (8.4 miles), *Hascombe* (8.4 miles) and *Blackdown* (7.8 miles) are sadly indefinite, mostly because thick woodland has grown up in the intervening 150 years, and the task of searchers is disappointing indeed when no view in either direction is to be had. The Netley Heath site was at the junction of several tracks near the scarp of the downs. Telegraph Hill, the Hascombe site, lies at about 600 ft., just N. of the Deer Park and east of Loxhill.

The commanding ridge of Blackdown has its summit 916 ft. above sea level, the highest point of Sussex. The telegraph stood some distance below the summit, on a projecting knob on the hillside near Blackdown Cottage, subsequently known locally as 'Tally Knob.' When in March 1816 orders were given to surrender the shutter sites to the owners the Blackdown station was referred to as 'Blackdown under the Knob, near Barfold under the Beacon.' The present house called Telegraph Cottage is at about 600 ft., about ½ mile ENE. of Blackdown House.

Beacon Hill. A span of about 10 miles brought the line to the next station, on Beacon Hill above South Harting, another lofty, windswept position, 795 ft. up on the South Downs. The site is a little south of Casey's Farm on the N. edge of the downs. There are few records of this post, though it was of much importance when it became the junction for the Plymouth line.

Portsdown Hill. The next span was one of the longest in the south of England, 11.5 miles to Portsdown Hill. The post was a few yards S. of the cross roads at the top of the down, where the view to N. and S. on a clear day is superb. A cottage called Cliffdene occupies the site.

Portsmouth (Southsea). The first communication between Portsdown Hill and Portsmouth was successfully carried out, as we have stated, in 1795, by Gamble experimenting with his five-shutter machine. The situation at Portsmouth needs explanation. At that time Portsmouth was still contained within the walls of fortifications pierced by only a few heavy gateways, and posterns. The Navy Office was in High Street, opposite the *George Hotel* and formerly known as Admiralty House, the official residence of the Port Admiral. In front of the bastions and ravelins on the SE. side stretched the glacis, an expanse of smooth open ground, sloping and terminating in the ditch, which was theoretically susceptible to uninterrupted fire from the walls and forts.

Murray's terminal shutter was erected on the ground, where the glacis and the rest of Southsea Common merged. It stood five miles from the Portsdown station. There was a postern gate with a footway across the Spur Ravelin (subsequently Pembroke Road), by which a messenger no doubt could go quickly to and from the telegraph and the Navy Office in High Street.

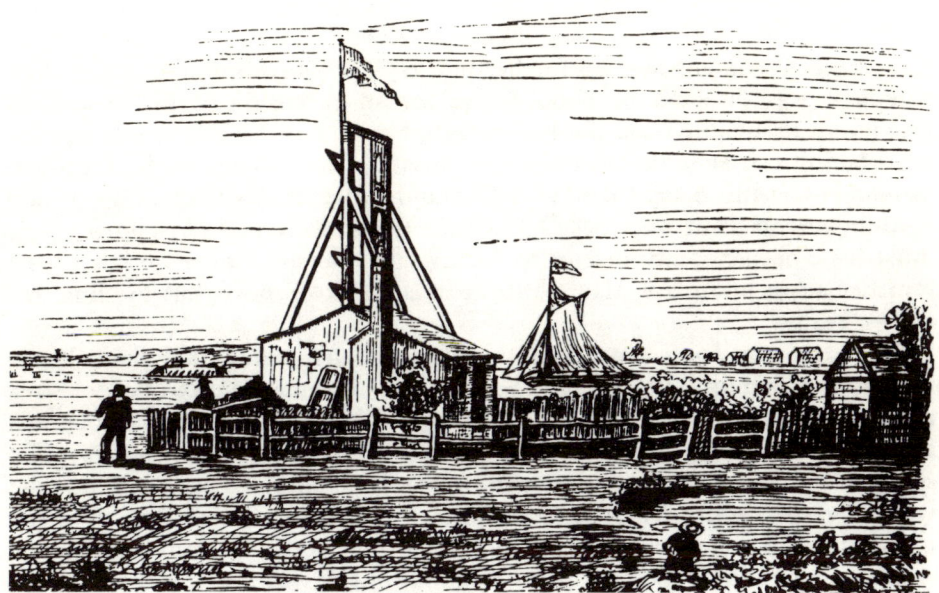
4. Shutter telegraph on Southsea Common.

As late as 5 October 1823 the hut on the glacis was still standing, and although it was long in disuse and ordered to be pulled down, the Telegraph Lieutenant persuaded the army authorities to let it remain for the present, on account of his large family, furniture and garden produce. The Admiralty approved.

In August 1807, Roebuck spoke of raising the telegraph because the smoke of the town was 'extremely detrimental' and when more houses were built it would be almost impossible to see over them. He examined the Montague Ravelin, the northernmost on that site, which was much nearer the Navy Office, but the station was moved nearer the seafront close to the Spur Ravelin, again on the glacis. One cannot believe that at the end of the 18th century the Army troubled much about keeping a glacis clear, but the site was nevertheless objected to and the shutter was finally sited by the Little Morass, close to the position of the present Clarence Pier.

A painting by Richard Livesay of a review of the Worcestershire Regiment on Southsea Common on 14 October 1800 has a fine representation of the telegraph in the background. Still further beyond are seen men-of-war under sail proceeding towards Spithead. Another good picture of the post on Southsea beach was presented to Portsmouth Corporation in 1945; the name of the artist is lacking. The shutters are clearly shown, and are portrayed as square, without the corners cut off. The station appears to be very close to the shore, as vessels are passing near in the background. W. G. Gates, *The Illustrated History of Portsmouth* (1900) has a drawing of the telegraph, certainly from the same source, save that a man with a wheelbarrow in the foreground is omitted.

The Plymouth Line

The Peace of Amiens held up further telegraph progress but in October 1805 Roebuck was directed to choose sites for an extension of the system to Plymouth, branching from the Portsmouth line at Beacon Hill. It was a much more ambitious scheme than the two existing lines. Roebuck had a tough assignment, with many sites to select and prove in difficult country and winter weather, but he worked fast and the sites were soon approved. It must have been no easy matter to arrange the transport of materials to the remoter sites, but on 4 May 1806 he was able to report that the line was almost ready.

Roebuck was soon practising the stations and stated that from Portsmouth he called up Plymouth (via Beacon Hill, of course) and 20 minutes after making 'Message ended' began to receive the reply! He was now concerned about sudden fogs, and said that a message from Plymouth becoming stopped by fog, after it had got within one day's post of London, should be written down and sent to the Admiralty by letter. There was then the question of messages meeting to and from the west. Naturally, preference would be given to a message from London, but it was decided that when a Plymouth message had passed 22 stations on its way up, it should have preference or priority. It then would virtually have got as far as Beacon Hill.

At Beacon Hill there was an angle of about 28° between the lines of forward transmission to Portsmouth and to Plymouth, not too great for the shutters and their apertures to be clearly seen at the two receiving stations.

Chalton. The first post on the Plymouth line was 5.8 miles from Beacon Hill on a down at Chalton, 476 ft. high, next to a windmill overlooking the Portsmouth road.

Wickham. a span of 7.7 miles led to a station on a slight rise near Wickham, after which the route zigzagged, first WNW., then due west and finally back to WSW. and SW., but the angles between stations remained acceptable. The site was 1 mile ENE. of the village and E. of Rooksbury Park, at about 300 ft.

Town Hill. 'Tacking' first brought the line to Town Hill (9 miles), about two miles ESE. of South Stoneham church.

Toot Hill, 5.7 miles, is a prominent hill near Nursling, SE. of Romsey and now on the edge of the built-up area of Southampton. Telegraph Wood is a reminder.

Bramshaw, 9.5 miles on, is a fine open site in a wide heathy part of the New Forest, a well-known spot recognised today as Bramshaw Telegraph, with an excellent view to westward.

Across Dorset and Devon there are many high hills suitable for telegraphy, their only drawback being their remoteness from villages and sources of supplies. One can but speculate on the impression which the strange machines made on the rustics of Wessex, for Hardy merely mentions the telegraph in *The Dynasts*.

As little detail is known about most of the succeeding stations they are listed below, with the distances between them so far as can be ascertained. On the Ordnance Survey printed in 1810 the telegraph sites are shown up to the borders of Dorset and Devon but not beyond. The Devon Survey may have been completed before 1806, when the posts were installed. From Bramshaw it was 9.2 miles to *Pistle Hill*, 2.7 miles SE. of Cranborne Church. The site is marked today by Telegraph Plantation. The 1870-71 OS, revised 1895, marks 'remains of telegraph'. 4.8 miles to *Chalbury* (or Pentridge), N. of Wimborne Minster. The site is close to the church. 8.6 miles to *Blandford Racecourse*, 1.2 miles E. of Pimperne Church. The site is known as Telegraph Clump, a small wood about 400 ft. above sea level. 5.2 miles to *Belchalwell*, (Bell Hill). This 846 ft. hill is about four miles S. of Sturminster Newton. 5.2 miles to *Nettlecombe Tout*, an 850 ft. hill about two miles S. of Mappowder. 5.8 miles to *High Stoy*, one mile NNW. of Minterne Magna. 'Telegraph Hill' is 871 ft. 7.8 miles to *Toller Down*, 812 ft, 2.5 miles W. of Rampisham. 9.7 miles to *Lambert's Castle*, 842 ft., 4.5 miles E. of Axminster. 9.4 miles to *Dalwood Common* ; 'Telegraph Cottage' stands at about 700 ft. on the southern tip of Stockland Hill. 4.7 miles to *St. Cyrus (St. Cyres) Hill*, 828 ft., one mile NNW. of Honiton. 7 miles to *Rockbeare Hill*, 504 ft. 10 miles to *Great Haldon*, 766 ft., on the edge of racecourse. Formerly Kenn Beacon. Marked 'Telegraph Hill'. 9.2 miles to *South Knighton*, 640 ft., four miles NNE. of Ashburton. Marked 'Telegraph Hill'. 9 miles to *Marley*, 629 ft., half-mile S. of Marley House. 8 miles to *Ivybridge*, 471 ft. Marked 'Telegraph Hill' (Telegraph Cross is half-mile W. by N. of Westlake village). 5.4 miles to *Saltram*. Hardwick Hill, 374 ft., on the W. side of the road between Plympton and Plymstock and near the Gables Hotel. 4.8 miles to *Mount Wise* (Plymouth Dock, later Devonport).

More will be said about Chalbury and Saltram stations. At Dalwood the present 'two-up', 'two-down' stone cottage contains nothing of the original.

A good reliable drawing of the shutter telegraph in use appears in the *Proceedings of the Dorset Natural History and Antiquarian Field Club* (1890). vol. XI. It shows the hut at High Stoy with high-pitched roof, and a smaller lean-to outhouse, water butt and chimney. The shutter frame is supported by struts secured to the eaves and not embedded in the ground. Three shutters are shown closed; they are black with a large white oval spot in the centre. The author of the accompanying article gives the complement of a shutter station as an officer and two men and says that a stock of firewood was kept on hand for immediate kindling as a beacon fire in case of invasion. He quotes a letter from a Cranborne doctor who as a boy was taken to High Stoy station, where the officer improvised a short message to his neighbour at Toller for the lad's edification.

In 1808 Henry Ward of Blandford received 10 guineas from the Society for the Encouragement of Arts, Manufactures and Commerce for an ingenious crank he invented to overcome the difficulty of moving shutters in windy weather. The device held the shutter firmly in place in either the open or shut position. It is not known whether the invention was officially adopted at Blandford or elsewhere.

Great Haldon station was burnt down on 23 September 1806. *The Times* of 10 October reported that it had been rebuilt.

The *Naval Chronicle* (1806) contained a report from Plymouth dated 12 July that the new telegraphs within the Higher Lines Dock and on the heights above Saltram were nearly completed.

It is sometimes stated that a branch was built from the Plymouth line to Wyke Regis or some other point near Weymouth for use when George III visited the resort. There seems to be no foundation for such a statement. The telegraph was for routine Admiralty business only.

Through Dorset and Devon the line kept generally to a WSW. course, with variations from WNW. to SW. There was an awkward angle at Nettlecombe Tout — SW. from the rear post and WNW. forward to High Stoy — but as the distances were only 5.2 and 5.8 miles respectively, the difficulties of the angles were probably minimised.

By the time the Plymouth line was decided on John Barrow (later Sir John Barrow, Bt.) had become Second Secretary at the Admiralty, a post he held, with a brief intermission in 1806, from 1804 to 1845. As Christopher Lloyd says in *Mr. Barrow of the Admiralty* Barrow ran the Admiralty office and 'supervised the extensive correspondence with naval officers all over the world, and with agents of other boards.'

The day-to-day running of the telegraph was but one of the many responsibilities of this zealous civil servant, to whom must go much of the credit for the efficiency of the Navy's bureaucratic machine, complicated as it was, at a time of great expansion and commitment.

John Wilson Croker, M.P., journalist and friend of Wellington, was appointed First Secretary in 1809 and held office until 1830. He and Barrow became lifelong friends and both contributed much to the famous *Quarterly Review*. Croker was concerned primarily with policy and with representing Admiralty interests in Parliament. Decisions on establishing extensions to the telegraph network would have been his responsibility.

The Yarmouth Line

With the eastward marches of Napoleon the defence of the North Sea and the Baltic became increasingly important to Britain. As early as 1801 the Admiralty had considered a telegraphic link with Yarmouth, the most easterly naval station, but the Peace of Amiens soon intervened. By 1807 the need had become urgent and Roebuck was instructed to survey a line. At that time the fleet was making much use of Yarmouth Roads and there was a Port Admiral at the base.

Originally a direct NE. exit from London, possibly making use of coast signal posts, had been suggested, but London smoke and the incidence of fogs in Essex posed problems which led Roebuck to a clever solution. He decided to take the line out on quite another course, first almost along the line of Watling Street towards the NNW. at a tangent to the required direction, until the Chilterns were gained. There, reaching its summit on Dunstable Downs, it turned a right-angle to go forward on an ENE. course along the Chilterns and the Gogmagog Hills. Nor did Roebuck go direct from Whitehall but made the Chelsea Hospital the diverging point, doubtless because of the height so gained and the desire to get clear of smoke and intervening obstacles in the built-up area. To communicate with the next point, at Hampstead, a second shutter was erected at Chelsea, *The Times*, 25 March 1808, stating that it was much larger than usual. It was placed on the W. square of the inner quadrangle of the hospital, as distinct from the existing Portsmouth line shutter on the E. wing. At the same time an additional, smaller shutter was erected at the Admiralty specially for the

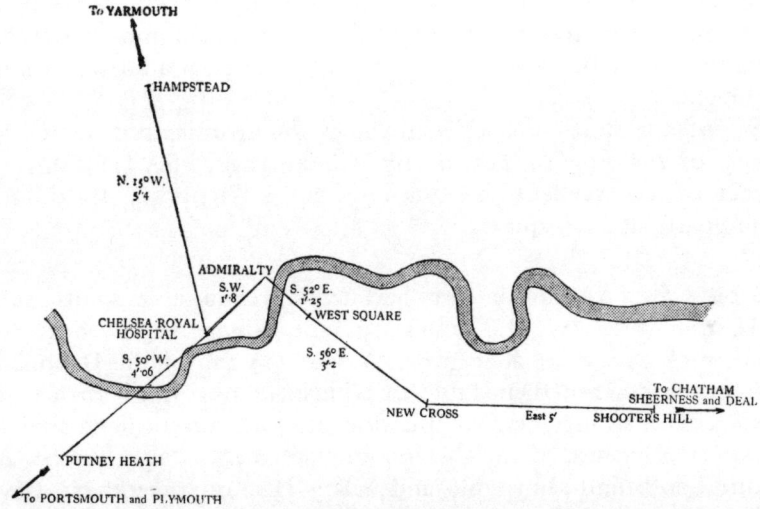

5. Shutter telegraph routes through London, 1808-14.

Yarmouth transmission, making three there. The accompanying sketch map of the shutter routes in inner London shows the divergence at Chelsea. It seems possible that the observers at the next station on the Yarmouth line, at Hampstead, had to read Chelsea 'in rear', as it were.

Hampstead. The site, originally called One Tree Hill, lay between Hampstead Church and Childs Hill, and 5.4 miles from Chelsea. Long since built over, it is commemorated in Telegraph Hill, the name of a road close to the intersection of Platt's Lane and West Heath Road. G. W. Potter's *Random Recollections of Hampstead* (1907) says that there was a hut still standing on the slope of the hill and that it contained the framework for the shutters. T. J. Barratt's *Annals of Hampstead* (1912) has a fairly accurate description of the telegraph and the appendix reproduces a contemporary sketch of it by J. J. Park entitled *Telegraph Hill 1808,* showing the then newly erected machine.

Woodcock Hill, 6.5 miles N., is part of the ridge pierced by the Elstree railway tunnel. The site is half-a-mile SSE. of Elstree station, though not very recognisable.

St. Albans. Eight miles N. of Woodcock Hill was the next station, on the roof of the medieval clock tower, at the end of the market place. The Corporation Minutes record that the Government was granted permission on 4 November 1807 to erect a telegraph on the tower, standing 310 ft. above sea level, which was erected between 1402 and 1410 to hold the curfew bell. Although the shutter was placed askew, from one corner of the narrow flat roof diagonally to the other, the position must have been cramped for the operators, who were presumably crowded into the room immediately below. There was the consolation that they were in the heart of a busy town, not on a windy height.

The St. Albans shutter is well shown in the frontispiece of *St. Alban's (The Story of the English Towns)* by William Page, FSA (1920). It is also the subject of an excellent painting by George Shepherd, dated 1812, in which the boards appear square.

Dunstable. The Yarmouth line had some remarkable spans, some of almost 12 miles. It was 11.3 miles from St. Albans to the next station, which Roebuck placed at a height of 796 ft. on top of the Downs, in the angle of the roads from Dunstable to Whipsnade and Kensworth Common respectively. The arrangement at this fine site has caused much speculation among experts. Because of the distances between Dunstable and St. Albans on the one hand, and Dunstable and Lilley Hoo on the other, they have doubted whether the transmissions would have been sufficiently clear had the angle at Dunstable, where the line took a right angle from NW. to NE., been divided, and the frame placed at 45°. They have conjectured that there must have been a second frame on the downs, at right angles to the other, involving double the work and an augmented crew.

Admiralty records do not mention two machines or extra crew. The author believes that, because of the elevation at Dunstable, it would have been possible for one machine to have sufficed.

The Luton Borough Museum contains an excellent water-colour, also by George Shepherd, purporting to show the Dunstable station. Shepherd painted it in 1819, five years after the shutter had been abandoned, and may have worked from memory. Although the hut is exactly like others known, he allowed his fancy rein by showing no fewer than 18 shutters, most of which would overlap and frustrate one another. His picture certainly shows no second frame.

Lilley Hoo. A nine-mile span led to the site at Lilley Hoo, a 602 ft. hill, also called subsequently Telegraph Hill. The diary of farmer John Carrington of Bramfield records: 'July 1808. Next Sunday 10th over Lilly Hoo wheare is now building a Talligraff on the Hooe.' (Bramfield is 14 miles away).*

As little appears to have been recorded of the following intermediate stations in the counties of Hertford, Cambridge, Suffolk and Norfolk, they are listed. It was 9.4 miles from Lilley Hoo to *Baldock* (Hickman's Hill, 486 ft., two miles SE. of town). 8.1 miles to *Royston* (379 ft., one mile SE. of town). 11.7 miles to *Gogmagog Hills* (one mile NE. of Babraham church). 11.7 miles to *Newmarket* (King's Chair, Side Hill, 200 ft., 1.25 miles E. of town). 9.8 miles to *Icklingham* (1.25 miles NE. of church. Marked 'Telegraph Plantation'). 6.7 miles to *Barnham* (one mile W. of Euston church). 8.8. miles to *East Harling* (Marked 'Telegraph Hill'). 7.3 miles to *Carleton Rode,* 227 ft., 1.5 miles NW. of church. (Marked 'Telegraph Farm'). 5.4. miles to *Wreningham* (near church). 8.2 miles to *Norwich* (Thorpe). 6.6 miles to *Strumpshaw* (S. of church). 10.8 miles to *Great Yarmouth* (South Gate).

The Newmarket site is not precisely established. The name of its vicinity, King's Chair, derives from the fact that it was a favourite resort of Charles II, who discovered it when walking on Newmarket Heath.

A modern reservoir marks the site of the Norwich station in the suburb of Thorpe; Telegraph Lane leads from Thorpe Road. Unfortunately, no painter of the Norwich School seems to have considered the telegraph in their midst worthy of his brush and no representation by a lesser artist seems to have to come to light. The station was rather awkwardly placed as the lines of transmission were at an angle of about 120° (SW. and E. by S.).

The only Norwich records of value come from the contemporary *Norfolk Chronicle* and Charles Mackie's *Norfolk Annals* (1901). On 19 December 1807 it is recorded that 'a telegraph communicating with Yarmouth is now erecting and nearly completed on Rymer Heath, between the 7th and 8th milestones on the Thetford Road from Bury. A telegraph or signal station is on the point of being erected upon the hills leading from Norwich to Thorpe.' The mention of Rymer Heath is evidently in error for the site at Barnham, which is not at the point quoted.

The site at Great Yarmouth is well authenticated, as prints exist showing the apparatus erected over the South Gate. One is reproduced in *The Perlustration of Great Yarmouth,* by C. J. Palmer (1872-75), which also refers to the end private residence known as Telegraph House adjoining the Marine Parade having been built for the officer in charge.

*See W. Branch Johnson, *Memorandoms for ... The Carrington Diary* (1973), p. 160.

On 25 June 1808 Roebuck reported that the Yarmouth line was ready and that the practising messages had passed perfectly correctly. The first official message was from the Admiral at Yarmouth who telegraphed on 24 August 1808, '*Calypso* ready for sea.'

By the time the Yarmouth line was completed, inspections of the system were occupying Roebuck for half the year. He was relieved of much routine work by deputing the principal man in the Admiralty telegraph office to act as clerk when not operating shutters. For the extra duty the man received £40 more a year.

It is appropriate, while considering Norwich and Yarmouth, to note that, according to White's *Norfolk Directory* (1845), a telegraph was erected on 29 September 1803 on the top of Norwich Castle to communicate with Yarmouth, through intermediate stations at Strumpshaw and Filby Church. It was presumably part of local anti-invasion measures and may have been short-lived, if indeed it ever functioned. For in March 1805 the *Norfolk Chronicle* reported that it understood that a Norwich—Yarmouth telegraph 'for mercantile purposes' was contemplated, and in October Norwich Corporation agreed to pay £10 a year to support it for three years 'if it shall continue to be worked that time'. Possibly it operated for just that period before being overtaken by the Admiralty line over the same course. It would be interesting to know the type of apparatus selected for both these 1803 and 1805 pioneer local ventures.

There is a Telegraph Hill between Ringland and Honingham, 8 miles NW. of Norwich and another WSW. of Weybourne. Both sites appear on Bryant's 1823 map of Norfolk. The one-inch OS of 1838 also marks 'Telegraph Hill' about a mile SSW. of Holt. They can hardly have had any relevance to the Admiralty line to Yarmouth and any connection with local schemes mentioned seems pointless. The Weybourne site was possibly that of a coast signal station (Kelling). (See Chapter IV).

A Falmouth Line?

There has been speculation as to the existence of a shutter line to Falmouth. The idea has been fostered by references in the *Letters of Lord Barham*, vol 3, (1911). In a memorandum of 28 November 1805, when he was First Lord of the Admiralty, Barham wrote, '... to which place [Falmouth] the telegraph, now ordered to be established between London and Plymouth, is proposed to be extended.' Again, in a letter of 25 December 1805 he stated, 'We have ordered a telegraph from Plymouth to Falmouth which must prove very convenient.'

In a memorandum 'For the King and Cabinet', 11 January 1806, Barham observes, 'The telegraph now erecting between London and Plymouth, is ordered to be carried on to Falmouth, the importance of which will be very apparent...' The only trace that has been found in the records is an Admiralty Minute of 4 December 1805, directing the Navy Board to take the necessary measures. The 40 or 50 miles of extension would have required seven or eight stations in rather difficult country. No action was apparently taken.

With the Peace of Paris signed on 30 May 1814, the Admiralty deemed the telegraph superfluous. On 4 May 1814 the Admiralty wrote to the Navy Board, 'Whereas we have this day directed Mr. Roebuck ... to discontinue the line of telegraphs from London to Yarmouth, to sell the materials of the buildings ...'

On 6 July their Lordships wrote that they had

> directed Mr. Roebuck immediately to discontinue the line of telegraphs from London to Sheerness and Deal, and from London to Portsmouth and Plymouth the beginning of September next and have directed him to return the telescopes to this office.

There had been no formal agreement about the land for the shutter telegraph huts, the ground being held at a yearly rent varying between 5s. and 15 guineas. The estimated annual cost of operating the 15 stations of the Kent line, plus rent and repair was £2,950.

Some of the huts were apparently taken down at once and the wood sold locally. Many a local builder or farmer would have been glad of such stout Admiralty planking and baulks. But a number were allowed to remain, for use as residences for superannuated Lieutenants and men. Some became ramshackle in a few years. We learn that in October 1821 the Bramshaw hut was much shaken by gales. The occupant wrote on 28 December 1821,

> I beg leave to inform you that the late heavy gales of wind has done considerable Damage to the Slatering at this Telegraph has likewise blown part of the chimney down which fortunately I prevented from falling on the roof the cealling is Also shattered in the house, and that a little repairs is required to be done to the weatherboarding round the Telegraph and coal-house — The privy has been Blown over and the sills and weatherboarding Damagd ...

A carpenter suggested removing the shutters to prevent more trouble.

In April 1826 it was decided that the station would not be required for the new Plymouth semaphore line and Reynolds, the then occupant, was told that his allowance would cease at once, but there was no objection to his continuing to reside in the hut, provided that he would keep it in habitable repair at his own expense. He lived there until 1834, when the Commissioner of Woods and Forests resumed the site.

From a receipt dated 5 January 1821 relating to Barham station we can form an impression of the full equipment of a shutter station. Lieutenant Gibbon took over the following items from Lieutenant Belsey:

> Two mattrasses [sic] very much worn
> Five blankets very much worn
> Three coverlets very much worn
> A set of telegraph shutters
> Quantity of ironwork — 40 lb.
> Four chairs, one deal table, one pair steps
> One water tub, two grates, decayed telegraph frame
> Two ladders, two double cabins, one cupboard

Records show that the stations at High Stoy and Toller, in particular, were finally given up at this time.

Nevertheless, at least two of the Plymouth line stations, Chalbury and Saltram, lasted until the 1960s. Chalbury, enlarged but still recognisable, was a farm worker's dwelling until 1967, when it was bought by the Bournemouth & District Water Company. It was pulled down in 1968 because it had become derelict. A reservoir extension will ultimately cover the site. The Saltram house happily survives, in private ownership, and also recognisable, as part of a larger building, and there is little doubt that it is the only shutter structure still standing. Its roof still carries stout beams.

The English shutter telegraph, despite its rather crude composition, had an aura of romance. To Britons who knew little or nothing of Chappe and the French telegraph it was one of the wonders of the age that messages could be sent so far and so rapidly. It was inevitable, then, that it should have been credited with more romantic exploits than it achieved. It has been said that the telegraph brought the news of Trafalgar and Waterloo to London. As to Waterloo, the fable is easily demolished. No telegraphs operated in this country in 1815. The shutters had ceased; the semaphores were not yet in service.

If the news of Trafalgar had been brought into Portsmouth, the telegraph there might have been set in motion unofficially. But the Admiralty was strict about such things. For instance, it discovered in September 1805 that the telegraphists at Portsmouth were using an improper private code. They declared that they had nothing whatever to do with the Commander-in-Chief, much less with his Flag-Lieutenant, and took orders from Roebuck. It transpired that Roebuck, 'the engineer officer in charge of telegraphs', had a private code with which he was accustomed to exercise his men, without interfering with official signals. One would imagine that he was within his rights but the Admiralty reprimanded him and ordered him to confiscate his code.

The Trafalgar despatches were in fact brought to London by road from Falmouth. Lieutenant John Lapenotiere, commanding the schooner *Pickle*, made his landfall and came ashore at Falmouth on 4 November 1805. There was in any case no telegraph there. He travelled to London by post-chaise, taking 38 hours for the 270 miles and changing horses 19 times, and delivered the despatches with his own hand to the Secretary of the Admiralty.

A favourite story is that during the Peninsular War a message 'Wellington defeated...' began to be sent to London but was stopped by fog. The message would have continued, '.... the French at Buçaco (or Salamanca)'. It was first quoted in the *Encyclopedia Britannica* (1824), the originator being the Admiral at Plymouth. The story gained more and more weight of disaster, one writer stating in 1827 that the message was being relayed from Portsmouth to London and that 'suspense and alarm prevailed throughout the day.' Still later accounts mention 'great consternation in London and a panic on the Stock Exchange' or that 'funds were violently agitated' because of the disastrous part-message received.

The story tellers forgot that messages over the Admiralty telegraphs were private and confidential and that the authorities were not likely to 'release' such a story to the public before it was substantiated. But at least they are right in adducing it as a reason for the importance of putting the significant part of a message first, in case of such interruption.

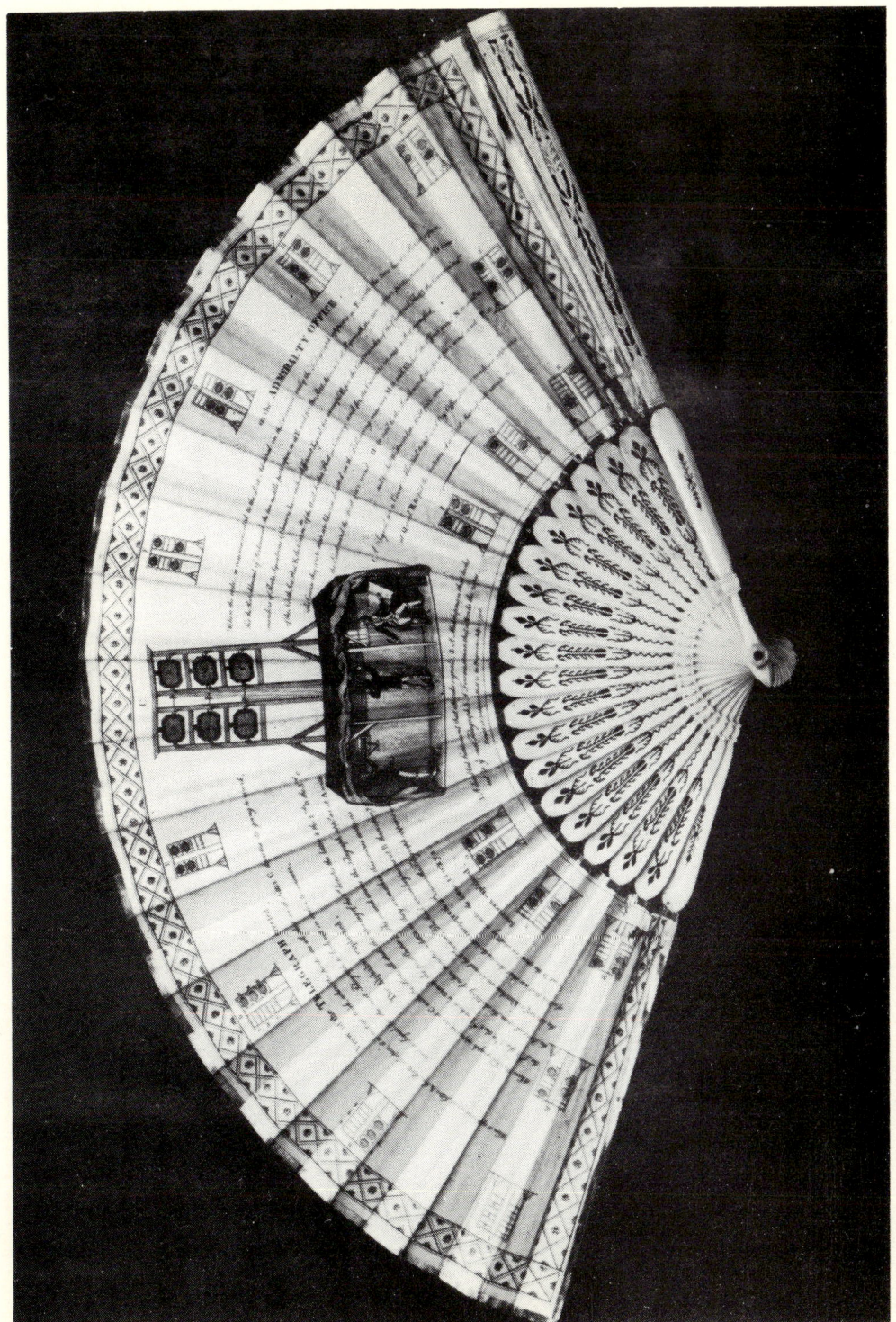

Plate 1: Lady's fan of the late 1790s, showing the operation of the shutter telegraph.

Plate 3, above: Cable and wireless Advertisement depicting the shutter telegraph and its operation.

Plate 2, left: Lord George Murray's model of his shutter telegraph in the National Maritime Museum, Greenwich.

Plate 4: The Admiralty and Horseguards, showing shutter telegraph.

Plate 5: No. 36 West Square, Lambeth: the telegraph tower, 1810.

Plate 6: No. 36 West Square, Lambeth: recent view.

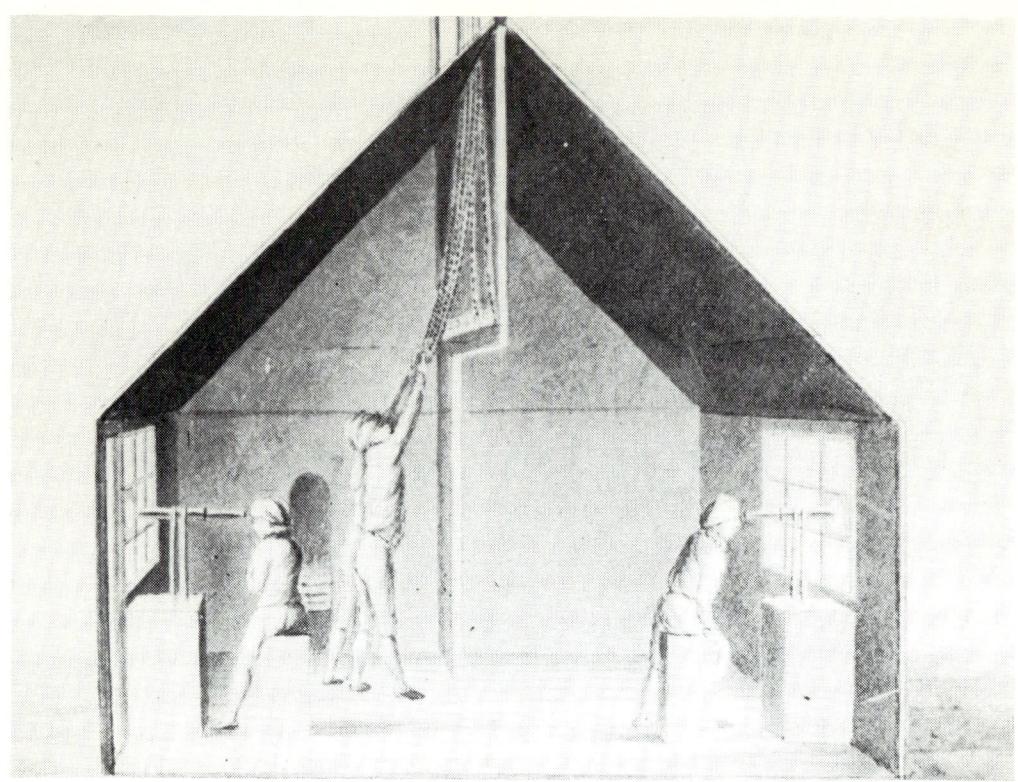
Plate 7: Shutter telegraph operating room, Nunhead.

Plate 8: Highwaymen being hanged at Blackheath, with representation of Shooter's Hill in the background.

Plate 9: Execution of Richard Parker at Sheerness, 30 June 1797, with the shutter telegraph on Garrison Point.

Plate 10: Chelsea Hospital.

Plate 11: Putney Heath in the early 1800s.

Plate 12: 'The Telegraph', Putney Heath.

Plate 13: Shutter telegraph at High Stoy, Dorset.

Plate 14: Shutter telegraph, Hampstead, 1808, from a water-colour by J. J. Park.

Plate 15: Clock tower, St. Albans, with shutter telegraph.

Plate 17: Shutter telegraph on South Gate, Yarmouth.

Plate 16: Shutter telegraph on Dunstable Downs, from a water-colour by George Shepherd.

Plate 18: Rear-Admiral Sir Home Riggs Popham, 1762-1820, from a portrait by M. Brown.

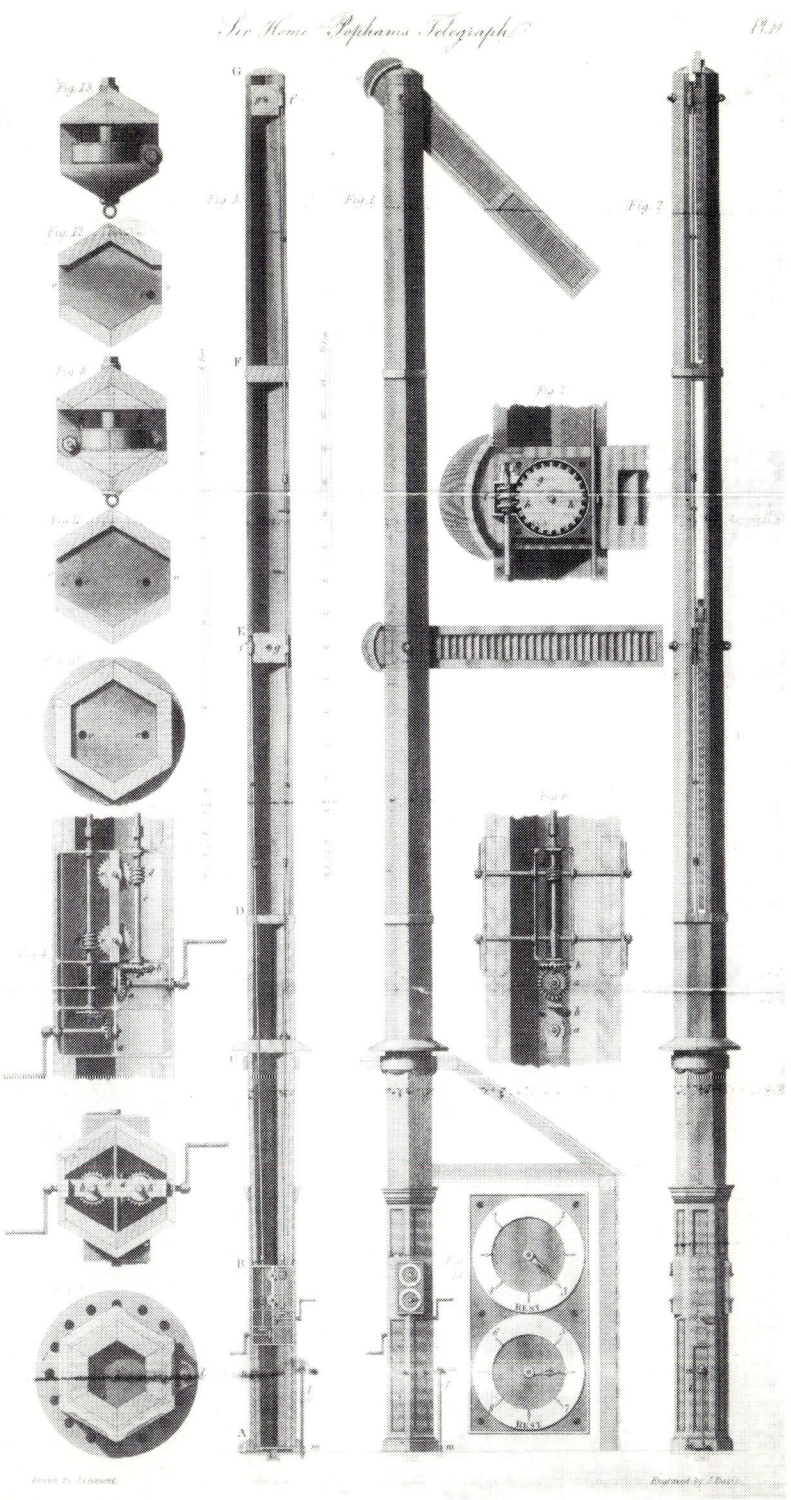

Plate 19: Sir Home Popham's Telegraph, London-Portsmouth line.

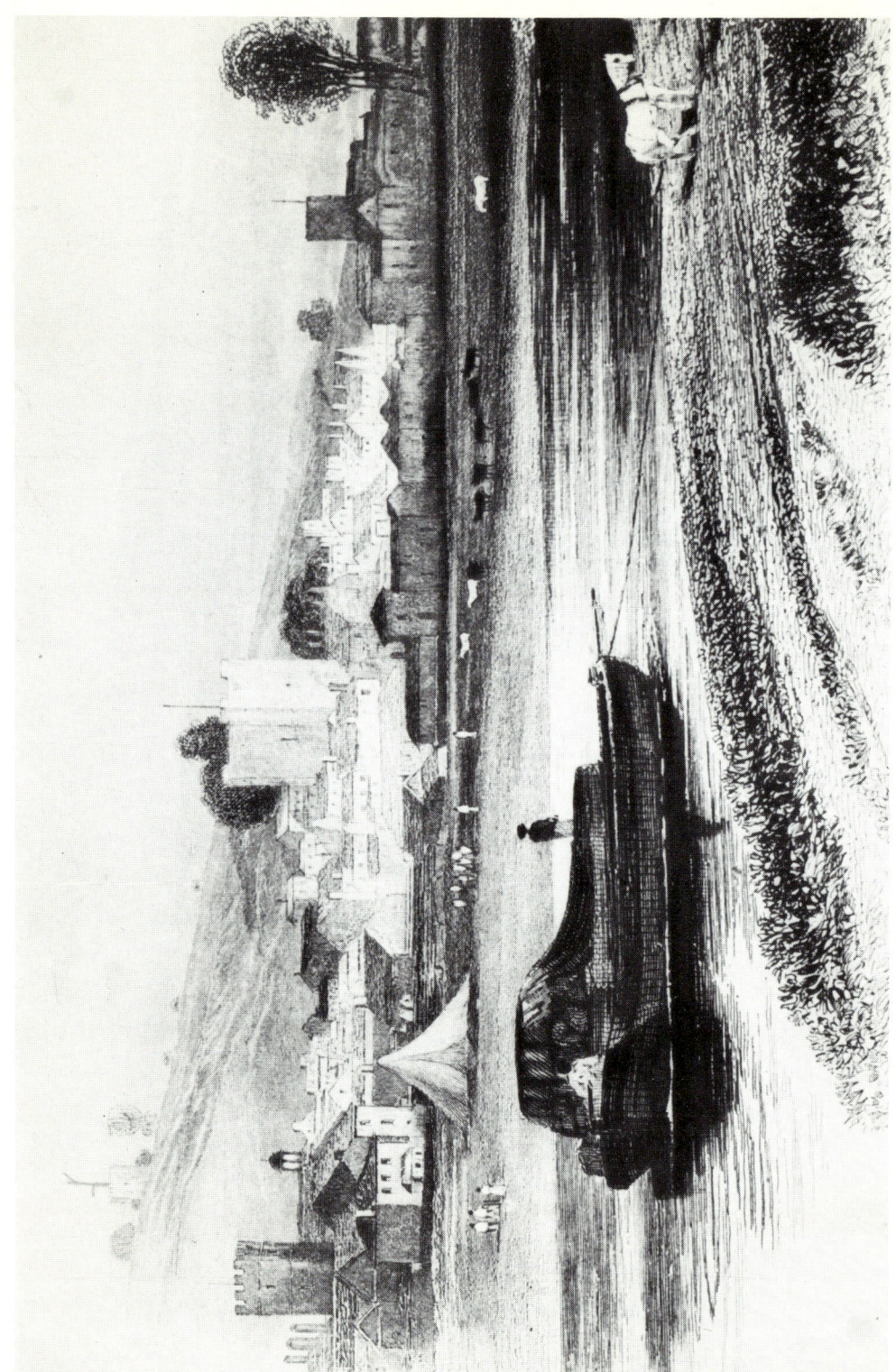

Plate 20: Guildford in the 1820s, showing Pewley Hill semaphore at top left.

Plate 22: Farley Chamberlayne: derelict semaphore telegraph station.

Plate 24: Worplesdon: semaphore telegraph tower (subsequently demolished) and church.

Plate 21: Surviving semaphore telegraph house at Sherfield English.

Plate 23: The Admiralty, showing semaphore telegraph on roof.

Plate 25: The Duke of York's School, Chelsea, showing semaphore telegraph on roof.

Plate 26: Semaphore telegraph, Putney Heath, in 1844.

Plate 27: Modern view of Cooper's Hill, Hinchley Wood, semaphore station.

Plate 28: Chatley Heath (Cobham) semaphore telegraph tower before dereliction.

Plate 29: Chatley Heath: operating mechanism in first floor room of tower.

Plate 30, above: Semaphore telegraph station, Pewley, Guildford, from a watercolour by E. Hassell, 1826.

Plate 31, left: View of Pewley Hill semaphore station in 1939. The cupola is one of several later additions.

Plate 32, above: Older Hill semaphore telegraph station.

Plate 33, left: Compton, Sussex, semaphore telegraph station.

Plate 34: Portsmouth semaphore station, in the High Street, on square tower. The arms are at rest.

CHAPTER III

The Admiralty Semaphore Telegraph

In 1813 *The Satirist* carried these verses:

> Our telegraphs, just as they are, let us keep,
> They forward good news from afar;
> And still may send better — that Boney's asleep
> And ended oppression and war.
> Electrical telegraphs all must deplore,
> Their service would merely be mocking;
> Unfit to afford us intelligence more
> Than such as would really be shocking.

The famous Admiralty dictum that 'telegraphs of any kind are now totally unnecessary, and no other than the one now in use will be adopted' occurred in a letter of 5 August 1816 from John Barrow to Sir Francis Ronalds, the celebrated electrician and meteorologist. Ronalds had written to Lord Melville, First Lord of the Admiralty, offering an electric telegraph of his invention. Ronalds records that he felt hardly any disappointment or resentment, 'because every one knows that telegraphs have long been great bores at the Admiralty.'

During Napoleon's Hundred Days there had been no time to resurrect the shutter lines, but it was intended to reinstate the service if necessary by using a new land telegraph devised by Sir Home Popham and derived from his telegraph for use on ships. Only 11 days after the Battle of Waterloo an Act was passed 'for Establishing Signal and Telegraph Stations' (55 Geo. III, c.128, 29 June 1815). It authorised that land could be legally acquired and obstructions prevented or removed between any two stations, and it laid down the law as to persons who declined to treat, and their right to appeal, and so on. The preceding day the Admiralty had directed the re-establishment of stations permanently on Popham's semaphore plan and Thomas Goddard, purser of the Royal yacht *Royal George*, was instructed to seek sites.

By August 1816 the Admiralty was experimenting with Popham's semaphores on a trial line between London and Chatham, the system referred to as the 'one now in use.' As to telegraphs being a great bore, no fewer than 100 plans of telegraphs had been sent in to their Lordships, who, as Colonel Macdonald also said, were sick of the subject.

Sir Home Riggs Popham was born in 1762 at Tetuan, where his father was British consul. He was the 21st child and his mother died when he was born.

Educated at Westminster and Cambridge he had a distinguished naval career, but was court-martialled at Buenos Aires after an excess of zeal. His political actions earned him the hostility of Lord St. Vincent and other Whigs. He died at Cheltenham on 11 September 1820, a Rear-Admiral and a K.C.B., K.C.G., K.M. and F.R.S. His tomb in Sunninghill church, Berkshire, is most elaborate. In addition to a long and fulsome eulogy, it includes, cut in the stone, a globe, compass, sextant, telescope flags and a semaphore.

Popham produced the first edition of his *Telegraphic Signals* in 1800. It became an official Admiralty publication in 1816. He submitted a description of his land semaphore to the Society of Arts which was published in the Society's *Transactions* in 1816.

Popham's invention soon proved ideal for its purpose. At Nunhead a new semaphore was tested against a shutter machine which had been left standing. Lieutenant George Pace, who had been appointed Superintendent of Telegraphs at the Admiralty, observed Nunhead from West Square and decided in favour of the semaphore. He declared that he could distinguish the movements of semaphore arms with the naked eye, whereas he had never been able to discern those of the shutter without a telescope.

The best account of what took place is contained in the article 'Telegraph' in the *Encyclopaedia Britannica, 1824 Supplement*, written by Barrow above the pseudonym 'K'. Barrow discredits Macdonald, who was still obsessed by multi-shutter telegraphs, and shows by the evidence of optics and so on, and from the nature of things, that a semaphore arm would always stand out much more clearly than a shutter in a frame with five others.

Orders were given on 6 November 1815 for an experimental line to be laid out from London to Sheerness. *The Times* of 9 May 1816 reported:

> The telegraph frames at the top of the Admiralty are to be removed, and the improved semaphore consisting of a hollow mast from whence two arms project in various directions, will be erected in their stead. The utility of this invention is to be tried, by way of experiment, in a few days, from London to Sheerness, and the number of stations, it is said, will not exceed nine; several are erected.

On 15 May *The Times* stated that the old apparatus communicating with Deal and Sheerness had been taken down and the new semaphore for Sheerness would be erected in a few days. On 28 May 1816 Popham was told that the line would not go beyond Chatham and on 27 July the Navy Board reported a proposal, which it adopted, from Sir Charles Rowley for semaphores to be mounted on ships to be moored between Chatham and Sheerness.

All mechanical apparatus was supplied by Henry Maudslay, of Lambeth, and the telescopes were ordered from Dolland. Popham received £2,000 for his invention. Officers and men selected to operate the line were appointed with effect from 3 July 1816 and the line was ready for working at the end of that month.

From New Cross over Shooter's Hill to Swanscombe the shutter line had traversed a region more susceptible to fogs than was tolerable. The new line therefore took a more southerly course out of London, from New Cross to

Red Hill (White Horse Hill), Chislehurst, thence to Rowe Hill (half-a-mile N. of Huxtable on the 300 ft. mark), and Betsham, near Swanscombe, rejoining the old route at Gad's Hill.

Trials were satisfactory, deciding once and for all the Admiralty's preference in the matter of telegraphs. It was agreed to set up permanent lines on the Popham system from Whitehall to Portsmouth (subsequently also to Plymouth) and to Chatham, Sheerness, Deal and Dover. A new link with Yarmouth was not considered.

Meticulous regulations for operating and maintenance were drawn up.

The machinery was to be kept clean and always well oiled. The station house was to be swept once a day, and washed twice a week in summer and once a week in winter. 6d. a day was allowed for coals and candles. One man had to sleep in the house to prevent its being plundered.

> As the Machinery of the Telegraph is made with the greatest exactness, the officers are never to force the winches; and as the post turns on a pivot, it is recommended that, before the men leave the Telegraph at night, they be directed to turn the front of the post to the wind, that in case of rain it may not beat into the grooves where the arms traverse; and to set up the shroud to windward, to prevent the mast yielding to the force of the wind.
>
> The stations will be frequently visited without any previous warning...

Each station was to keep a journal, signed by the officer and sent to the Secretary of the Admiralty at the end of each week. It had to state the time every communication was received, with the position of the arms so that by comparison the origin of any mistake which might occur could be detected.

Attendance was to be between 10 a.m. and 3 p.m. and the glasses (telescopes) were not to be left for more than 2 minutes at a time.

If a fog arose when the telegraph was not operating, the station which could not see its corresponding stations was to make its own sign to indicate that communication had ceased. If a fog occurred during transmission the station that was clear was to take the whole message and send back, 'Received,' and its own number. It was to send on the message as soon as the fog lifted.

Some officers seem to have indicated the fog signal when they felt like a respite from duty, a practice against which they were sternly warned!

Officers were instructed never to lower the arms until the sign was correctly exhibited at the next station. If a mistake was made they were to make the correction sign and, when this was acknowledged, to exhibit the correct sign and continue with the message. When a movement was made by one arm only, that arm was to be completely housed before any other movement of arms took place.

Lieutenants received 3s. a day plus half pay and men received 2s. 4d.

In 1818 we find Lieutenant Harrison, of Betsham station, complaining that the station was much shaken by winds. He was living in the former shutter station house at Swanscombe, not far away. Harrison, whose complaints led to the Betsham station roof and other parts being covered with canvas, refers to the 'Mast stept on the upper floor', which suggests that the stations were of more elaborate construction than might have been supposed for an experimental line. They were almost certainly of wood, as an estimate for painting Rowe Hill station in 1816 includes the 'whole of the weatherboarding.'

Incidentally, when the officer took charge at Red Hill station on 7 May 1816 he listed the inventory as: telegraph house, mast, two arms, complete machinery, iron stove and funnel, set of fire irons, stools, two telescopes, one canvas cot, one hair bed, one blanket, one cot frame and small items from the Shooter's Hill shutter station, where he lodged.

On 19 January 1820, Lieutenant Atkins, of Rowe Hill, wrote to the Navy Board about sleeping accommodation for his man Samuel Gladman, 'who sleeps in the Rowe Hill Telegraph, agreeable to your Hon. Board's instructions for its protection...'

In 1820 the Admiralty decided to make the London-Chatham line permanent and extend it to Sheerness, and Deal and Dover. Between London and Sheerness the freehold was to be obtained under the Act if the sites could not be had by treaty.

Goddard re-surveyed the London – Chatham section and chose sites that varied only slightly from Roebuck's of 25 years earlier. His final list was: *Admiralty; West Square; Nunhead; Red Hill; Rowe Hill; Swanscombe; Hartshill; Gad's Hill; Chatham Lines; Callum Hill; Beacon Hill* (near Faversham); *Old Wives Lees; Swaddling Down* [sic] ; *Barham; Malmains;* and *Deal* on the main line, and *Barrow Hill* and *Sheerness Garrison*, and *Coldred* and *Dover Citadel* on the forks.

It is not certain whether Goddard proposed to re-adopt the Shottenden site or choose a new site nearer Old Wives Lees, the postal address for Shottenden. The selection of a new site at Swaddling Down – obviously Swarling – suggests that shutter experience had dictated the need for an intermediate station between Shottenden and Barham. The omission of Tonge implies that direct communciation by semaphore between Callum Hill and Sheppey was accounted satisfactory whatever may have been the shortcomings with the shutter.

On 5 September 1820 he wrote to Croker that he would arrange to purchase the Nunhead site and all others between London and Sheerness.

Swanscombe was recommended for re-adoption in place of Betsham as it communicated equally well in the London direction and better with the new intermediate site at Hartshill to the east, which Goddard advocated to improve communication in view of the traffic expected when the Deal line was working in conjunction with the Deal-Beachy Head coast semaphore.

Goddard found the shutter site at Gad's Hill acceptable but at Red Hill he moved operations to an adjacent field which commanded a better view of Nunhead and Rowe Hill. The new site was bought for £100.

With the coming of peace the function of the telegraph had changed and a direct link with Chatham Dockyard was now sought. The decision gave Goddard trouble as smoke hampered communication between it and Gad's Hill. He proposed a new site on the Citadel, communicating with Callum

Hill. If that did not serve it could be removed to a site in Prospect Row, Brompton, communicating first with a station on Darling Hill and then with Callum Hill. A later choice, made on 9 June 1821, was for a site on the King's Bastion in Chatham Lines; this seems to have been final.

Some of the sites in Kent were decidedly expensive. Goddard had to close for £130 for that at Callum Hill, and Lord Harris demanded £200 for Barrow Hill. Beacon Hill cost £190 and Coldred Hill £200. New dockyard works at Sheerness had not yet upset the proposed site there, which had also been that of the shutter station.

The Beacon Hill purchase granted the right to cut down a chestnut plantation if necessary. On 24 April 1829, by which time no brick had been laid anywhere on the projected Kent line, the proprietors at Beacon Hill were told that they might occupy the ground to prevent the land from becoming a nuisance in its wild state, but would have to sacrifice crops if and when it became necessary to build the station.

The Beacon Hill station would have been particularly commanding. A five-floor octagonal tower was designed for the station and was the model for two actually built, at Chatley and Worplesdon, Surrey.

The line was to fork at Barham. Goddard's first choice for the next station towards Deal was Betshanger (Betteshanger), as on the shutter route. He finally fixed on Malmains Down as cheaper and equally suitable.

The Deal terminus was to be on the site of the former shutter station in the Navy Yard and was to serve the needs of both the new line and the coast semaphore. Goddard refers to the Deal semaphore on the 'lofty building' lately erected there on the site of the old telegraph house.

The Dover terminal site at the Citadel on the Western Heights was proposed with a view to communicating with Coldred, the coast semaphore and naval ships in the offing. Goddard suggested that two brick buildings on the ramparts, formerly used as cookhouses, might be adapted.

Although all sites were marked out, nothing more was done.

The West Square station at the London end of the proposed line was nevertheless kept up as a lieutenant's residence. On taking it over in 1820 Lieutenant Nops reported that he had paid £4 for the stores. He did not question the sum as the widow of the former lieutenant, Perdriau, was very distressed, with her husband lying dead in one room and a child dead in another.

Lieutenant Pace was in residence there by 1822. He received permission for work to be done as the house was 'very much out of repair.'

Their Lordships were broad-minded enough to order the construction of a Chappe-type telegraph and set it up alongside the semaphore at Nunhead on 21 December 1820. Ignace Chappe (Chappe l'Aîné), eldest brother of the famous Claude, says in his *Histoire de la Télégraphie* (1824) that the British ambassador approached the French government in 1819 for a Chappe telegraph. It was supplied but without instructions. Ignace Chappe ridiculed the idea that the English would profit by it, as experience with the Chappe system could be found only in France.

On 1 March 1820 Goddard sought permission to paint the Nunhead Chappe telegraph red (the colour of the Popham apparatus there), except for the indicator arms which were to be black, so that it could be more easily distinguished, especially in bad weather.

At length the Admiralty decided to give up the Chatham experimental line and concentrate all resources on the lines to Portsmouth — already under construction albeit slowly, for it had been agreed on 3 August 1818 — and to Plymouth.

Instructions were issued in February 1822 for the telegraph between the Admiralty and Chatham to cease on 25 March and its equipment to be transferred to the Portsmouth line. Lieutenant J. B. Robertson, of Nunhead, wrote that he had been ordered to deliver up 'the two Semaphores at Nun Head with machinery' to the charge of the Navy Board. They were still there in April when Robertson reported that the owner of the site had no objection to taking charge of them until they were removed.

On 19 February 1818 Goddard had been instructed to survey likely sites between London and Portsmouth. The portable semaphores in Deptford Yard needed repair so he could not begin until 8 March, when he set out with two riggers to help him.

Like Roebuck, Goddard was a meticulous man. He was, too, a good letter writer. His letters well describe how, sometimes drenched and scratched by briars, he climbed numerous promising hills, mounting a ladder held by his assistants, to scan the neighbouring heights.

Goddard's difficulties were not lessened by the attitude of some of the landowners whose properties it was proposed to cross. The 1815 Act could be invoked but the Admiralty preferred not to offend property and privilege. For example, after the route had been settled, Lord King of Ockham forbade a tenant to cut down branches of a tree that obscured the line of sight near Cobham in Surrey. In such cases the only remedy was to raise the semaphore and its arms higher than had been planned.

Goddard tested alternative routes out of London. Working up to Town, he was at Sutton in May 1818. He was finding it difficult to plot a course from there to Putney without interfering with the 'tasteful ground' of Mrs. Merrick at the SW. end of Wimbledon Common or with Lord Spencer's park to the E. He then tried a route via Mitcham and Brixton to Lambeth. He climbed Lambeth church tower and describes how he watched the semaphores at the Admiralty and Nunhead in correspondence.

The 'Sutton' survey produced of a list of possible stations out of London: *Lambeth Church; Brixton Hill; Green Hill; Epsom Downs; Ranmoor* (Ranmore); *Winterfol* (Winterfold); and *Hascombe* to *Haste Hill*, Haslemere, picking up there the route of the alternative route via Kingston Hill, Cobham and Guildford.

Green Hill refers to the ridge just north of Sutton, south of the present Rose Hill cross roads. (There is a 'Telegraph Track' across the Little Woodcote Estate on gently rising ground between Carshalton and Woodmansterne but it seems to be too far E. to have any connection with Goddard's scheme).

The Telegraph public house on the brow of Brixton Hill is certainly in the right position to mark Goddard's probable site but it would be surprising if it had been named in recollection of a tentative survey. The Epsom Downs, Ranmore Winterfold and Hascombe sites have eluded precise identification.

The choice of Lambeth church is interesting. Except for Gamble's trials, recorded later, it would have been the only sacred edifice in the country so used. Telegraphing from its tower would have been a cumbersome business and it is probable that Goddard would have substituted a special structure close by, as he was to do at Worplesdon.

The Admiralty asked Goddard to finalise his surveys and give, *inter alia*, the heights of stations deemed necessary. He seems to have left open the choice of route. In the event the Admiralty chose his Putney-Kingston-Guildford course.

Goddard wanted the line to run from Kingston Hill to Pewley Hill, Guildford, via Fairmile and Hatchford (between Cobham and Ockham). A site at Fairmile had been bought for £35 when the objections of the Hatchford owner compelled a deviation. Eventually the whole route was settled, Goddard receiving authority to mark out the line on 29 June 1820. All the land, comprising an acre at each site, had been acquired by 21 April 1821.

A notice in the *London Gazette*, 7 April 1821, invited tenders for building a semaphore house at each of the following sites: Putney Heath; Kingston Hill; Claygate Hill; Pointers; Pewley Hill; Bannicle Hill; and Haste Hill, all in Surrey, Older Hill (Holder Hill, or Olderhill); Beacon Hill; and Compton Down, all in Sussex; and Post Down (Camp Down) in Hampshire.

Each tenderer was to engage in the sum of £1,000 for due performance of the contract. Henry Maudslay was to supply 14 sets of telegraph apparatus.

Thomas Corfe, junior, of Putney, successfully tendered and was given 18 April 1822 as completion date. The average price for the sites was £47 3s. 9d., and average ground rent £2 7s. 2¼d. Older Hill and Beacon Hill sites were cheap at £20 each, but Pewley cost £120.

As with the shutter line, a station was proposed at Chelsea, where Goddard decided on the Duke of York's School instead of the Hospital.

To enter Portsmouth he thought it necessary to build an intermediate station at Lumps Fort, Southsea. Writing on 27 September 1820 he reported that the Ordnance ground at the fort was ready for building but suggested a postponement until it was possible to work the terminal station on the Glacis at Portsmouth. If the smoke of the town did not obscure the signals it might be possible to cut out Lumps Fort.

The Portsmouth and Plymouth semaphore lines were designed to last. Gone were miserly notions of wooden shacks with two primitive rooms. Instead there were to be well-designed dwellings fit for permanent habitation.

The stations were built to three patterns, depending on topography. For convenience they will be referred to as the bungalow, house (three-floor) and tower types.

If the building were on a high hilltop it would be unnecessary to raise the machine still more, so that in the most elevated situations we generally find stations of bungalow type. On medium elevations it was found best to erect houses of three floors with the semaphore mast on the roof. For low-lying ground or in wooded country where the arms had to be raised as high as possible, towers of several storeys were envisaged.

There were thus substantial quarters with good accommodation for the officer and his family, and usually for the assistant as well. At Beacon Hill, on the South Downs, however, the officer had eight children, so the rating, who had lost a leg in battle, had to billet out!

Corfe seems to have been a reasonably satisfactory contractor, though on 28 June 1821 he was complaining of a brick shortage. In some respects Goddard helped him, for most of the sites were quite accessible from the Portsmouth Road and the Haslemere — Chichester Road, in contrast to the remoteness of some of the old shutter station sites.

On 11 February 1822 Goddard reported that all the buildings as far as Portsmouth were ready for use. On 13 February the Navy Board ordered the machines used on the Chatham experimental line to be transferred to the new stations. The Portsmouth line began regular working in 1824.

All this activity had not passed unnoticed by the captious William Cobbett. In *Rural Rides* he wrote of what he had seen near Hambledon, in Surrey, in 1823:

> On one of these hills is one of those precious jobs called Semaphores. For what reason is this pretty name given to a sort of telegraph house, stuck up at publick expense upon a high hill?

Again, he refers to a 'queer-looking building on the skirts of Hindhead, called a "semiphore" or something of the sort.'

Cobbett was sure it was a 'job', the Georgian and Victorian word for a 'racket.' He considered that 'alarm-post' would have been too nasty a name for it and appears to have had no notion that it was intended to keep up a routine, rapid communication with Portsmouth.

The stations Cobbett saw were evidently Bannicle Hill and Haste Hill. Incidentally, Greenwood's map of Sussex of 1823 also marks the Sussex stations as 'Semiphores'.

The new line ran from the *Admiralty* to: *Chelsea; Putney Heath; Kingston Hill (Coombe); Cooper's Hill; Pointers (Chatley Heath); Pewley Hill; Bannicle Hill,* near Witley, *Haste Hill,* near Haslemere, *Older Hill,* NW of Midhurst; *Beacon Hill; Compton; Camp Down* (Post Down, Bedhampton); *Lumps Fort;* and finally to the *Portsmouth* terminus in the High Street.

These sites will be examined in more detail, but meanwhile the layout of the buildings and their apparatus call for description.

In the house-type and tower stations the operating room was on the first storey. Into the floor of that room a substantial beam was inserted diagonally to carry the base of the mast. The general pattern of the house-type stations was three sub-basement rooms, three rooms above them at just above ground level and two more rooms, one above the other, in the central 'tower' section, the lower being the operating room. The flat roof, reached by ladder and a trap door, was leaded and had a low parapet all round.

The bungalows were laid out with the operating room alongside the living quarters, though not in all examples in line with them. The fronts of the stations all faced towards Portsmouth.

The Popham semaphore was a well-designed and well-made machine, full details of which fortunately are to be had. A fine engraved folding plate accompanying the *Transactions of the Society of Arts* vol. 34, shows all parts of the machinery with proper scales.

The mast had a hexagonal section and was formed of fir boards 10 in. wide, and was 1 ft. 8 in. in diameter. Above the roof it stood nearly 30 ft. high, and the arms were pivoted 12 ft. up and at the top, with 16½ ft. between the pivots. Thus the arms could not touch or overlap, for they were 8 ft. long and 15 in. wide, and when at rest were housed inside the mast. Each arm could take up six positions, enabling in all 48 separate signals to be made.

The whole apparatus could be trained round in its collar in the roof of the station, and could be fixed at every two points of the compass by means of a ring at its base containing 16 holes, with a bolt to go through it into a hole in the floor. The length of the mast inside the building naturally varied, but in a bungalow it was 12 ft.

On the outside of the 'shaft' (mast) two rows of wooden blocks were fixed alternately to serve as steps, so that with the help of a rope a man could go up and attend to and oil the mechanism of the arms. These steps are not shown in the engraving, but in several contemporary sketches they can clearly be seen.

The arms were actuated by bevel wheels and rods leading from winch handles in the operating room. They could convey 48 different symbols which were allotted to the letters of the alphabet and the numerals 1 to 0, with a few arbitrary expressions such as 'and,' 'Commander-in-Chief,' 'fog' and 'closing down'.

The crew of an ordinary semaphore post was usually only two. The Lieutenant, Royal Navy, generally a veteran, was in charge. The 'man' or assistant, usually an old sailor or even a superannuated warrant officer, was generally chosen by the Lieutenant, possibly because he was an old shipmate or known to be a good 'glass-man'.

Admiralty Semaphore now in use.

No. of Signal by 1 and 2 Arms.	Signification.	No. of Signal by 1 and 2 Arms.	Signification.	No. of Signal by 1 and 2 Arms.	Signification.
1	1	15	G	43	X
2	2	16	H	44	Y
3	3	21	I	45	Z
4	4	22	K	46	
5	5	23	L	51	
6	6	24	M	52	
1	A	25	N	53	
2	B	26	O	54	
3	C	31	P	55	
4	D	32	Qu	56	
5	E	33	R	61	
6	F	34	S	62	
11	7	35	T	63	
12	8	36	U	64	
13	9	41	V	65	
14	0	42	W	66	

In all—48 separate and distinct signals; being the whole which the two arms are capable of making, as under; in which the two arms actually exhibited (in black lines) represent the number 16 or H, according to the table or key, as above arranged.

We have here, in addition to the alphabet and the numeral digits, 13 signs over, applicable to the names of stations, preparative, finish, stop-signals, &c.

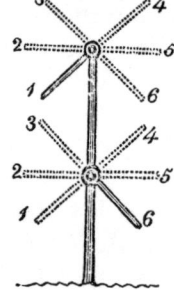

6. Admiralty semaphore code as given in the supplement to the *Encyclopaedia Britannica*, 1824.

At a few busier posts, such as the Admiralty and Portsmouth termini, the crew numbered two assistants. Had the Plymouth line materialised, Chatley Heath, as the junction, would doubtless have had at least two assistants. Records exist giving the names of all the Lieutenants throughout the lifetime of the semaphore service.

The semaphores, with their efficient machinery, mounted on specially-designed, well-sited buildings were a great advance on the shutters. But both when building and after occupation the houses were the subject of incessant correspondence. On 2 April 1822 Lieutenant James Robertson asked for early possession of the new Cooper's Hill station as the 'peasantry' had forced open the window shutters out of curiosity to see inside the unusual new house. On 15 May 1825 we find the officer at Kingston Hill asking permission for a fence to be put up as the house had been broken into on three successive nights.

Lieutenant Edward Harries, posted to Chatley Heath, seems to have had frequent cause for complaint. In May 1826 the Navy Board agreed to pay him up to £20 to buy a seat in perpetuity in Cobham Church for the officer at Chatley; meanwhile the vicar had allowed the Harrieses to use his own family pew. Some more of Harries's grumbles will be examined when the Chatley station is described in detail.

There was a contretemps at Haste Hill at the outset when Lieutenant Carpenter was arrested in 1822 for debt. Croker directed that one of the 'old telegraph men' be sent there as a temporary measure.

The next consideration was the extension of the semaphore to Plymouth, a project never to be completed. *The Times*, 15 April 1822, stated:

> An immediate Survey is to take place for the most eligible situation to erect telegraphs between London and Plymouth. A communication from the Admiralty to Portsmouth takes place directly.

On 13 July 1825 Barrow instructed the Navy Board to have 30 semaphores ready for the Plymouth line. The former shutter sites were not resumed, a new course being preferred, its first section lying close to the present A31 road.

With customary thoroughness Goddard proposed as many as six variant routes, all with sites chosen, as he said, 'after a winter examination to prove the currency thereof.' They show how seriously he took his difficult task. At first he thought of branching from the Portsmouth line at Kingston Hill and going via Claremont and Queen Anne's Hills to Worplesdon but soon substituted Chatley, well fitted to serve as a junction, as the point of divergence. The choice was probably at the instigation of Croker, always anxious to keep down the amount of new construction. Goddard's report to the Admiralty was minuted to the Navy Board on 11 April 1826,

> directing them to lose no time in obtaining the scites [sic] and making all other preparations for the establishment of the telegraph line to Plymouth on these sections, beginning at Worplesdon ...

The following stations were built: *Worplesdon,* on glebe land in the churchyard of St. Mary's. Standing at about 180 ft above sea level only a few yards from the church tower, it was itself a tower, of apparently six floors, otherwise like that at Chatley, and the only other specially-built telegraph 'tower' in the country. W. of Worplesdon, Goddard first proposed stations at both Wick (near Ash) and Swingate (just south of Farnham) but finally substituted one at *Poyle Hill,* at the W. end of the Hog's Back. It has been impossible to trace the exact site or type of building. It communicated on the west with *River Hill, Binstead,* a house still standing and occupied. Next came *Farringdon Common,* near Four Marks. A bungalow, it survives, occupied and in excellent condition. At one time it was used as an isolation hospital. It is also one of the most accessible surviving buildings, being situated in a metalled road called Telegraph Lane, leading S. from the A31 at Four Marks.

The next two stations, at *Merifield* (Merryfield) and *Chestford* (Cheesefoot Head) have not survived. The first was a house, the second a bungalow. Merifield station lay near the present Merryfield Manor Farm between Ropley and Tisted at about 620 ft. A prominent water tower marks the approximate site. The Cheesefoot station site is remembered by the name Telegraph Hill (533 ft), though it cannot be identified precisely. The hill adjoins the A272 three miles ESE. of Winchester.

West of Winchester are two surviving structures. The first is at *Farley Chamberlayne,* where a bungalow, empty and derelict, adjoins the church, and the second near *Sherfield English,* where a well-preserved, occupied house stands at Mount Pleasant, close to the Hants-Wilts border. The present occupant states that on the lead of the flat roof have been scratched dates, initials and crudely sketched ships. No doubt these graffiti date from the Admiralty ownership of the house — time must have hung heavily for a telegraph man with no regular communication to keep up.

A station is known to have been completed at *Woodfield Green,* the present Woodfalls, near Redlynch, south of Downton. It stood on the east side of the road, close to the plantation called Tinney's Firs.

Beyond Sherfield English Goddard had much trouble in selecting workable sites. Like Roebuck before him, he must have had considerable fortitude to ride and tramp through the wilds in winter, setting up sights in windswept places, often baffled by topography or thwarted by landowners.

Goddard finally recommended the following route in continuation:
Woodfield Green, Rushmore, Badbury Rings, Bere Regis, Piddletown Heath, Steepleton, Coombe, Pilcombe Hill, Whitelands, Beer Head, South Down (Sidmouth), West Down, Holcombe, Roccombe, Beacon Hill, Diptford South, Mount Widdicombe, Coyton, Dunstan's Hill, Mount Wise (Plymouth).

The Rushmore site is hard to fix with certainty but was probably about 2 miles NE of Cranborne. That at Bere Regis was almost certainly Woodbury Hill, just to the east. Piddletown (now Puddletown) Heath — Hardy's Great Heath — extends to the south of Puddletown — the site may have been close to Yellowham Hill on the present A354. Steepleton is Winterborne Steepleton, 4 miles W. of Dorchester. Coombe can be approximately identified with the high land at High Coombe, 4½ miles ESE. of Bridport. Pilcombe almost certainly stands for Filcombe, inland from Golden Cap. Whitelands is Whitlands, 2½ miles WSW. of Lyme Regis. West Down adjoins Budleigh Salterton. Holcombe is 1½ miles N. of Teignmouth. Roccombe is undoubtedly Higher Rocombe, 3 miles NNW. of Torquay centre. Beacon Hill (640 ft.) is 2 miles WNW. of Paignton. Diptford lies 5 miles SW. of Totnes. Mt Widdicombe is SE. of Ugborough and Coyton 2 miles SW. of Ivybridge. Dunstan's Hill is 1 mile N. of Plymstock.

The Worplesdon site was purchased for £82 17s. 6d., with £10 to the rector. Badbury, Beer Head and West Down were to be held on lease. The rental for the three sites, plus Gad's Hill on the Chatham line, was given as £39. Coyton was the dearest site on the new line, at £240.

There were some long spans between stations, the maximum being Rushmore — Badbury (9 miles), Cheesefoot — Farley (8¼ miles) and Woodfield — Rushmore and Beacon Hill — Diptford (8 miles each).

As on the Portsmouth line, access had been studied and Goddard's sites were rarely far from a highway or the coast, so that in case of interruption other means could be used to transmit a message.

It is a pity that the Plymouth line was never completed. Telegraphy along the Dorset and Devon heights near the coast would have looked impressive and the present-day investigator would have had pleasant sites to inspect. Unfortunately by April 1827 interest was waning. On 24 April Barrow ordered the suspension of station building that year and added that only two or three houses were to be erected in each subsequent year. At that rate Plymouth would not have been reached until at least 1834!

As his letter of 10 April 1826 to Croker makes clear, Goddard, writing from Goldsworthy Place, Rotherhithe, had trouble in getting through Dorset and avoiding the Earl of Sandwich's land at Drakenorth, near Powerstock, the earl objecting to any of it being taken.

Enclosing six variants with his letter, Goddard said:

Finding myself foiled... in respect to Dean Hill, and Long Critchell, and bearing in mind an observation which you were pleased to make when I last had the honor to accompany you on the examination of ground, namely "the great saving of expense to be effected by the abridgment of stations," I proceeded in my endeavour to carry the line more to the south from Farley, as well as for this purpose as to endeavour by such means to avoid Drakenorth ground .... the proprietor of which has signified to me by writing in the strongest terms, his great objections to its being taken, but which could only be avoided by the success of going in the direction of *Dorchester,* where numerous obstacles seemed to present themselves by the intervention of the hills between that place and *Bridport,* which are continually lifting in the direction from the former towards the latter place; however by persevering in moving to and fro in every possible direction, and by placing marks on sundry scites [*sic*], I at length succeeded in fixing upon *Steepleton* and *Coombe,* the former to work with *Piddletown Heath* and the latter with *Pilcombe Hill,* near Chideock, thereby accomplishing my object, at least so far as regards Drakenorth, as per the accompanying list of routes, of which that under the letter *B* must, under all circumstances, be considered as having a decided preference.

After stating the pros and cons of his six variants, Goddard went on to point out that the shutter line to Plymouth had numbered 30 stations, including the termini, and that his had only five more and at the same time embraced a communciation with Torbay. The increase, he observed, was proportionately no more than on the Portsmouth line. He noted that the semaphore was well defined at its present temporary site at Mount Wise. It did not command a general view of the Hamoaze but could be permanently and eligibly placed for all purposes upon the brick building (a guard house) within the keep by removing the roof and raising the substantial walls sufficiently high to form the semaphore room and other rooms on top.

Their Lordships concurred that Goddard's Route B, with 35 stations — later reduced to 34 by the substitution of Poyle for both Wick and Swingate in W. Surrey — was to be preferred. Routes E and F would have needed only 33 stations but were ruled out because of the impediment of smoke from Farnham and other places between 'Guildford Hill North' and Binstead. The semaphore at Chatley was found to be well defined to the church tower at Worplesdon as was the church tower to the lower room at Chatley, 7¾ miles away.

By the time the buildings at Farley Chamberlayne, Sherfield English and Woodfield Green had been completed in September 1831 costs were proving much higher than had been foreseen and the project was abandoned. The house at Merifield had cost £1,394 and the bungalows at Farringdon and Cheesefoot, £987 and £850 respectively.

The history of the Popham semaphore telegraphs is happily supported by drawings, prints and scattered references in text, as well as by the evidence of such of the stations as remain. Of the 21 buildings in the country, outside the termini, it is fortunate that examples of all three types have survived.

The Portsmouth line stations will now be described in more detail. It is a pity that while some are well documented, others are devoid, or almost so, of record.

Admiralty
The *Mechanics' Magazine,* 24 September 1825, contains a drawing entitled 'The Semaphore, now in use between London and Portsmouth.' It depicts the inside and outside of a station, obviously made of wood and with three people in the transmitting room. It is reasonable to infer that it is meant for the Admiralty station itself. One person is working the winch handles, another is gazing through his telescope at the distant station (Chelsea), and the third is at a desk, evidently at paper work.

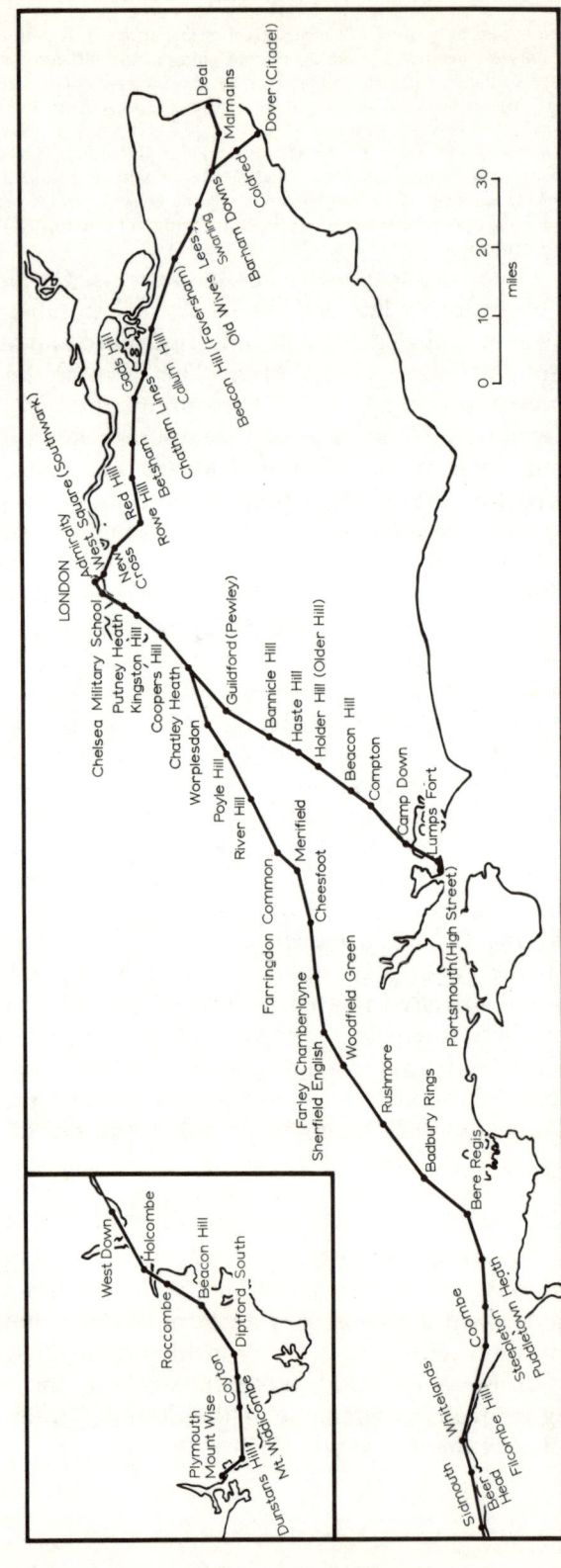

7. The Admiralty semaphore lines, actual and projected.

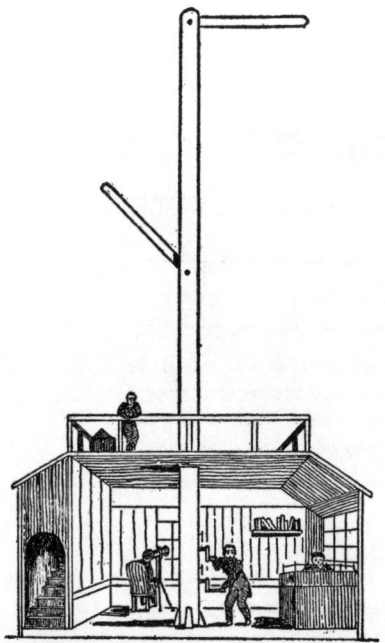

Many small engravings exist showing the front of the Admiralty in Whitehall. Nearly all occur in 'guide' books of the time, such as *England and Wales Delineated.* Some show the arms of the semaphore vaguely branching out from the mast, others with the apparatus at rest.

The Admiralty semaphore does not lack literary references. Marryat's *Peter Simple* tells of an officer who, after a row at the Admiralty, hastens out into Whitehall and is relieved to observe that it is too foggy for the telegraph to work, and therefore a message cannot reach Portsmouth ahead of him while he drives there post-haste.

8. Semaphore telegraph room at the Admiralty.

Leigh Hunt quaintly observes in *The Town:*
Where the poor archbishop sank down in horror at the sight of King Charles's execution, telegraphs now ply their dumb and far-seen discourses, like spirits in the guise of mechanism, and tell news of the spread of liberty and knowledge all over the world.

The *Nautical Magazine,* April 1840, contained a letter to the following effect:
Mr. Editor — The other day on my return from my station aloft, I was glad to find that you had taken the hint I gave you one cloudy day . . . I am, Sir, it is true, a moving fixture, yet I am not without a mind, as indeed you may well conceive when you reflect on the important communications to the Fleet, etc., constantly pointed out by me; and this mind of mine, I can assure you, Sir, is very fond of being comfortable, and when the day is over is glad to retire to the fire-side, and over some hot rum punch, which pleases best after such a cold day as this has been, study the news, how those succeeded in the destinations to which I have sent them, and how far it is probable I shall have to direct others (weather always permitting), and so forth.

The sun having set in the haze, and my fingers very cold, from which, as you may perceive, they have not yet recovered, having stirred the fire and my punch, I took up your March part of the *Nautical,* with which I am well pleased . . .

I am, Sir, your servant, The Semaphore on top of the Admiralty.

In 1934 the Admiralty Librarian suggested that the writer of this letter was Charles Hawse Jay, Superintendent of Admiralty Semaphores. Jay, incidentally is supposed to have been the prototype of a fictional Lieutenant Squib in Frederick Locker-Lampson's *My Confidences,* holding the same appointment, and described as tall and thin, elderly and brush-headed, and having a long nose and about 10 bouncing daughters.

An amusing description occurs in Charles Knight's *London* (1843), which is worth repeating as a contemporary appreciation:

Mechanics' Magazine,

MUSEUM, REGISTER, JOURNAL, AND GAZETTE.

No. 109.]   SATURDAY, SEPTEMBER 24, 1825.   [Price 3*d*.

"To industrious study is to be ascribed the invention and perfection of all those arts whereby human life is civilized, and the world cultivated with numberless accommodations, ornaments, and beauties. All the comely, the stately, the pleasant, and useful works, which we view with delight, or enjoy with comfort, industry did contrive them—industry did frame them."—*Barrow*.

THE SEMAPHORE,

NOW IN USE BETWEEN LONDON AND PORTSMOUTH.

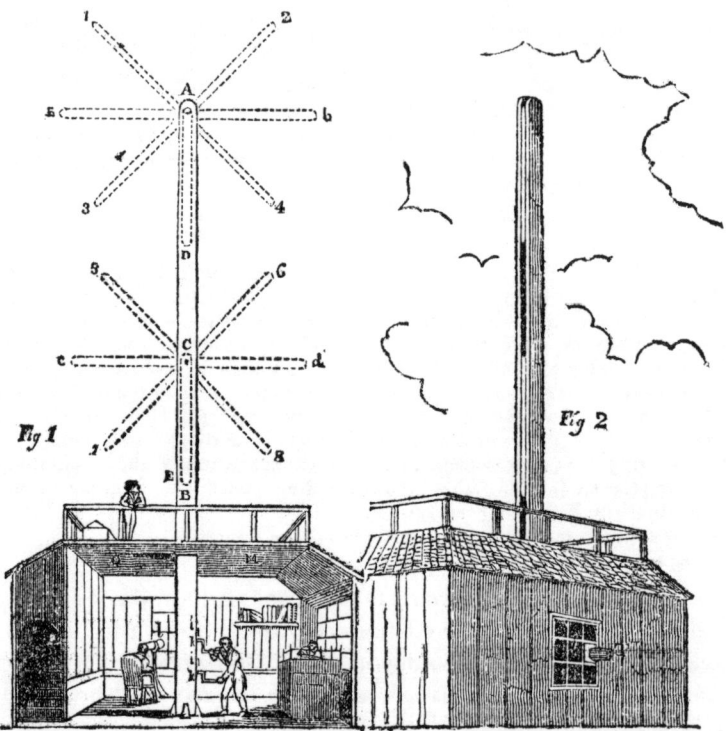

9. The London-Portsmouth semaphore in 1825.

> If we take up our station on the esplanade in St. James's Park, the eye is caught by a huge upright beam erected on the roof of the Admiralty, with straight arms extending from it laterally at different angles. They may be seen altering their positions, remaining a few moments at rest, and then changing again. The giant upon whom the stranger gazes with uncomprehending curiosity is whispering to his huge brother on Putney Heath [in fact, Chelsea] who will repeat the intelligence to his neighbour behind Richmond, and he to the next in order, so that by their unconscious agency the heads of the Navy in London give and receive intelligence to and from the great naval stations hundreds of miles off [sic] as quickly as they can communicate with a storehouse at the other end of the metropolis. The semaphore is, as any man may see, but a block of wood, and, heaven knows, no beauty, yet in the hands of man it becomes instinct with wondrous power...

This is picturesque but exaggerated — there was but one 'great naval station', Portsmouth, hardly 'hundreds of miles off.'

Chelsea

Many similar engraved book illustrations exist of the Duke of York's School, Chelsea, showing the machine on the roof. At first it was over the centre of the building but in 1844 a 'handsome' new church was built in Pimlico on the Marquis of Westminster's estate, and its steeple got in the line of sight of the telegraph. This was the 'society' church of St. Michael's, Chester Square. In consequence, the whole of the machinery was to be taken down and re-erected on the roof of the Hospital, where the shutter had stood, the Marquis having to bear the sole expense of the removal and refitting in the new position. As *The Times* reported, 'some fresh difficulty presented itself' and the semaphore was simply removed to the southern end of the original building, meaning the School.

Putney Heath

The first station outside London was at Putney Heath, the only intermediate site common to both shutter and semaphore. The present *Telegraph Inn*, at about 180ft., is close to the spot on which the two telegraphs stood. This inn possesses a contemporary water-colour of the semaphore, depicting the mast, and arms working, and, as it was of the bungalow type, the operating room to one side of the front door.

In October 1822, while the semaphore line was still in its probationary stage, Charles Pollard was the officer in charge. He wrote to the Navy Board many times to point out the need for improvements before the telegraph was in proper working order. When it blew hard the arms became jammed and could not be moved until there was a lull; consequently, wrong signals were apt to be received at Chelsea. He also wanted the arms at Chelsea to be painted partly white because the background was so black. At this time the shutter hut was still standing close by but the intention was to pull it down. The Navy Board approved Pollard's request to have it or buy it.

Ogilvy's *Pilgrimage in Surrey* says:

'... in later years one of the semaphore telegraph towers [sic] stood near the obelisk [on Putney Heath]; it is curious to read in *The Times* report of the opening of this line from London to Portsmouth, that the movement of the arms was "hopefully expected to assist greatly in the dispersion of fogs".'

Kingston Hill (Coombe)

In Warren Road, a private road traversing the exclusive Coombe Warren Estate — the area is the Robin Hill of Galsworthy's *The Forsyte Saga* — stands Telegraph Cottage, a house occupied by Eisenhower when Supreme Commander, Allied Forces in Europe, in 1943-44. The house, with a fine garden, almost certainly occupies the site of the semaphore bungalow, superseding it in the early 1900s but seemingly incorporating little or nothing of it. Bryant's map of Surrey, 1823, shows 'Telegraph' on the opposite side of the road, on land now part of a golf course, but later OS. sheets mark the more likely site at 178ft. The district was once famous for its springs, which Wolsey tapped to provide pure water for Hampton Court. The telegraph house had a 30ft. well in addition to its cellar; all other stations save Chatley Heath were equipped with pumps.

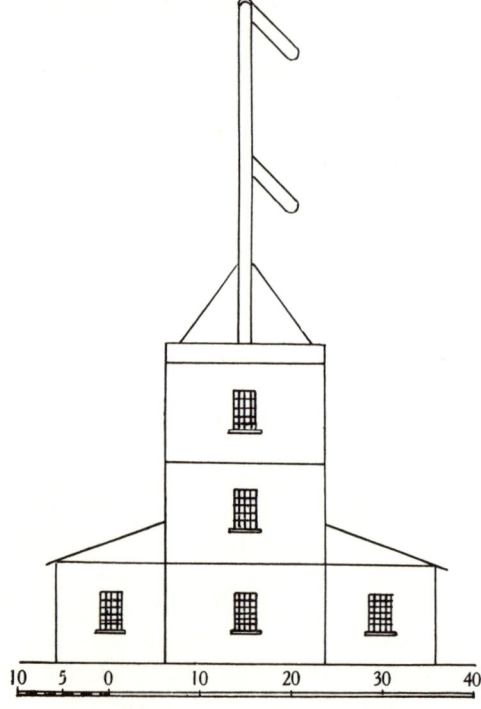

10. Goddard's sketch of Cooper's Hill semaphore station, 1822.

Cooper's Hill

Cooper's Hill station, on the 160ft.-high hill (long since known as Telegraph Hill) lying behind the modern high-class suburb of Hinchley Wood, on the Kingston By-Pass, is the nearest surviving telegraph house to London. Owned by the local authority and tenanted, it is in good condition. Water and electricity are laid on, but there is no main drainage.

The three-storey house is easily reached from Hinchley Wood station by ascending a wooded path from Manor Road South. From the Claygate direction a track called Telegraph Lane gives access.

The *London and Southampton Railway Companion* of 1840 refers to the house:

> ... at the eleventh milepost [from Nine Elms] is a well-known spot, designated Cockcrow Hill; it is an eminence of considerable height, crowned by some large trees, and commands an extensive view of rich and diversified scenery... looking forward, the telegraph, which forms a portion of the line of communication between the Admiralty and Portsmouth, is seen upon a hill on the left...

It may have been the clump on Cockcrow Hill, at Long Ditton, which necessitated raising the mast at Cooper's Hill by 8ft. to allow the arms to show above them.

Chatley Heath

The station at Chatley Heath (otherwise, Hatchford, or Pointers) tops a 100ft. hill S. of the junction of Portsmouth Road and Chatley Road, between Cobham and Ripley. The great octagonal tower, with its five floors and basement, is imposing but the trees which have grown up in the last century or so have all but screened it. The nearest metalled road and other habitation are about half a mile away, and access is by tracks and footpaths.

The tower is a close copy of that proposed to have been built at Beacon Hill on the projected Deal/Dover line.

Like other semaphore buildings, Chatley tower is of red brick rendered with stucco; in this case much of the stucco has dropped off. Because of its height, 60ft. 6in., the construction is massive. At the base the brick courses are 2ft. 3in. thick and at the top 1ft. 1½in. The wooden staircase, with 75 steps, is bricked off from the rooms, so that each room has six sides. A ladder leads from the top floor to a trap door in the roof. From the flat roof, supported, like the other floors, by heavy beams, and edged with a parapet, the view extends E. to the edge of Richmond Park and W. over woods and Wisley airfield to the North Downs. To the N., Windsor Castle can sometimes be seen.

After the semaphore was given up in 1847, Chatley, like the other stations that have come down to us, seems to have had an uneventful century or so in the hands of successive civilian occupants. But in 1963 the local authority gave the then occupants, tenants of a local landowner, notice to quit. It considered the building no longer fit for permanent habitation as it lacked piped water and sewerage.

As the tower was on the telephone, its occupants' immediate needs could always be met. When water from its wells was insufficient, the fire brigade was ready to help out with emergency supplies.

Once empty, the lonely building soon attracted the attention of vandals, who broke in, smashed all the window panes — some window spaces had been blank from the beginning — and pushed over the chimney stack from the roof, wrecking the small outbuilding below, with its main floor, used as a kitchen, and a semi-basement.

The writer and friends tried to stop further depredation and interest authority in the fate of the unique structure. The then Ministry of Public Building and Works scheduled it, partly for the reason that the operating gear was still in position.

Subsequently, Surrey County Council sought to take over 70 acres of Chatley Heath, including the acre on which the tower stands, to add to its Ockham and Wisley Commons. The purchase was completed in August 1965, since when the tower has been the council's property. The council has carried out commendable restoration, partly to make the tower weather- and intruder-proof. All the ground floor windows of the tower and the outbuilding have been filled in and the immediate surroundings have been cleared of rubble and growth. Floors have been renewed, stairs repaired, and a new roof put on.

Possible uses for the tower have been debated but at the time of writing no decision had been taken. Whatever the outcome, the action of the council has, one hopes, saved this fascinating relic.

The rodding, in its substantial vertical wooden casing, passing through the floors between the operating room and the roof, and the bevel gearing in the operating room remain.

A Surrey County Council notice on the door includes the information that some items of operating equipment are in the keeping of the council's Historic Buildings and Antiquities office.

The Chatley tower is central to the theme of Warwick Deeping's novel *The Woman at the Door* (1937). Deeping evidently spent some time studying the place, but was either wrongly informed or invented erroneous detail. He said it had been built in 1795 and had been manned by a Lieutenant, a Midshipman and two sailors, and he gave the names of the eight shutter stations to support information that signalling from Chatley was by shutter. A 'chap' on the roof was said to have read signals!

In Deeping's book the tower is burnt to the ground to destroy any finger prints that the heroine might have left while taking refuge there with the hero after shooting her husband.

On 11 August 1822 Lieutenant Harries wrote to the Navy Board to ask for defects in the new Chatley house to be set right. Harries had five children, boys and girls, his wife and a maidservant. Consequently he had only one sitting room, the other three being necessarily used as bedrooms. The sitting room was stone-flagged and therefore cold and uncomfortable. He asked for a wooden flooring for this room and the operating room and a wooden partition to separate his daughters and the servant in an upper room. He also wanted the entrance doorway to be moved to prevent rain and sand blowing in.

Holl, a Navy Board Surveyor, and assistant, apparently, to Goddard, approving the requests, also sought permission to indent for a 'common press bedstead,' to shut up with a pair of folding doors, for the use of the assistant. As the well at Chatley was 80ft. deep and it was laborious to draw the water, Holl asked for a wheel and pinion to be fixed on the barrel, as also at Kingston Hill and Cooper's Hill stations. Also, some scamped work by the contractor at Chatley was to be made good.

Chatley station, although officially in Cobham parish, was also under the patronage of Ockham. The Ockham parish register records the baptism of the following infants:

1825 March 6		son of Thomas Greenaway, tallegraph mariner
1831 July 17		daughter of the same, of Ockham semophor, mariner
1834 December 21		son of Thomas Phillips, mariner of Ockham semaphor

Greenaway, incidentally, was probably the second assistant to be posted to Chatley. The first was discharged after Harries had complained in November 1822 that he was repeatedly insolent and 'unbecoming.'

In January 1823 Harries complained that snow made contact difficult with their postal village, Cobham, 1½ miles away. He asked for a donkey and a pair of 'panniards.'

Because it was so exposed the semaphore mast at Chatley yielded considerably to the wind. Harries considered it should have four quarter stays of 'roap' or iron to steady it, as it was forced out of the perpendicular and the arms could be worked only with difficulty. The mast was soon puttied and given two good coats of brick red oil paint.

Pewley Hill

The Pewley Hill telegraph, at about 350ft., is one of the most conspicuous of the survivors and is easily seen from the railway station and other parts of Guildford. The extra structure erected on the roof after the station's operational days were over carries a cupola, which glints on sunny days, making the house still more easily visible. It is also one of the most accessible survivors. Occupied and known as Semaphore House, it stands at the top of Semaphore Road in the E. of the town and is flanked on its E., or Pewley Down side, by reservoirs.

The centenary of the station's closing prompted an interesting article in the *Surrey Times,* 27 December 1947, by the late Mrs. J.K. Green. It recalled that the Lieutenant in 1823 was James Smith, later transferred to Worplesdon. Lieutenant James Poad, a later incumbent, was posted to Pewley in December 1841 and remained until the closure. The last assistant at Pewley was George West, who became a gardener, entered Abbot's Hospital, Guildford, as a brother in 1895, and died in 1907.

The Guildford city records can produce several water colours showing the station in its original state, long before suburban Guildford crept up to it.

Bannicle Hill (Banacle Hill)

This station stood on a sandy ridge about a quarter of a mile N. of Sandhills, itself a little W. of the present Witley railway station. The site was about 300 yd. due E. of the present Hill House. At about 400ft. it is quite a commanding position, the ground falling away sharply enough to give good views N. and S. even today.

Unfortunately, no trace of the house has apparently survived.

Tradition relates that mounted men were employed to carry messages between Bannicle Hill and the next station S., Haste Hill, Haslemere, when local haze or fog blotted out signals. The practice is most unlikely in view of the Admiralty's economic operating policy, unless the area was notorious for such hazards.

Haste Hill

After so disappointing a site as Bannicle Hill, it is pleasant to turn to the next station, sited at 650ft. on Haste Hill, just E. of Haslemere and NNW. of Blackdown.

James and Edward Hassell, father and son, painted some of the semaphore houses, including Putney and Haste Hill. Their sketches clearly show what the bungalow-type station looked like in its working days, with the semaphore erected above the operating room.

The station was built on part of several enclosures called Jackman's Fields. The land was farmed by one George Heath, a cripple, with nine children. In 1821 while the house was being built, the Admiralty paid him compensation because the waggons bringing up bricks and other materials ruined his peas and other produce.

In 1849 the Admiralty sold the house and land to James Fielding, son of the former proprietor. They were later acquired by a Miss M. Parson, who enlarged the building in the 1860s, named it Highfield and lived there with her sister. It was further enlarged in the 1880s by a Mr. Stewart Hodgson and again by the Marquess of Sligo who bought it in 1894. Lord Sligo renamed it Whitwell Hatch in 1911, reverting to an old local name.

The building is now the Whitwell Hatch Hotel. It is said that parts of the semaphore bungalow are incorporated in the walls of the hall and dining room, but the large late-Victorian mansion bears little resemblance to the original. Also, as at other sites, the great growth of trees and shrubs in the intervening years has changed the aspect almost out of recognition.

Older Hill

A partial survivor — the operating room has gone — this bungalow-type station stands at 676ft. on Telegraph Hill at the NW. end of a high escarpment, SW. of Fernhurst. It can be reached by following a track leading from the Haslemere — Midhurst road westwards from a point south of the bend at Henley Common. The house, which is occupied, commands a fine view over Woolbeding Common and the valley of the Rother to the South Downs.

Beacon Hill

Beacon Hill was the official title of both the shutter and semaphore stations high up on the South Downs above South Harting, though the Beacon is the name preferred locally for this 795ft. hill.

The semaphore site, with its bungalow, was some 1,000ft. S. of the shutter site, at about 650 instead of about 795ft. It lay to the E. of the present Petersfield-Chichester main road and between Little Round Down and Milpond Bottom.

Between 1822 and 1825 Lieutenant Charles Leaver was in charge. He often complained to the Navy Board, mainly about the effects of the exposed site, which he said was bleaker than any other on the line. Not the smallest thing lay 'betwixt it and the sea.' Even the Portsdown (Camp Down) station, he alleged, was more sheltered as it was much nearer the Isle of Wight.

As workmen were still on the site, Leaver asked for, and got, inside shutters for the SW. and NW. rooms. The house was threatened with flooding in rainstorms and so he demanded a porch. Next, he grumbled, the house was exposed to the vulgar gaze and he wanted white roller blinds for the sitting room, kitchen, each bedroom and the operating room. The Navy Board allowed him brown Holland blinds.

11. Haste Hill semaphore station, Haslemere; from a sketch by E. Hassell, 1825.

Another trouble was the sheep which ate his vegetables. Even when a ditch and a bank with a hedge were made passing carts and waggons shook the earth into the ditch.

In *My Life and Adventures* (1923), Earl Russell, elder brother of Bertrand Russell, says that he bought the freehold of the bungalow for £450 in 1901. He relates how in the years following he gradually pulled down the house and its annexe — then used as kitchen — and built the present mansion, Beacon Hill House, in which little or nothing of the original survives. However, Lord Russell preserved the memory of the former use of the place by erecting a tower room with windows which give fine views in all directions.

A former vicar of Harting recalled that the grandfather of a local man then living saved the timber of the semaphore mast and used it to make window frames for a house in the village. The hexagonal wrought-iron collar which went round the mast to stay it was saved and is now in the Victory Museum at Portsmouth.

Compton

S. of Beacon Hill the line went to a down at Compton, sometimes called Hobbes Down. This station, at 534ft., lies about half-mile E. of Compton village and is another bungalow survival. The 2.3 mile gap separating it from Beacon Hill station was one of the shortest on any English telegraph. For an unexplained reason the operating room was apparently built at right angles to, instead of in line with, the rest of the building. The arrangement seems to obscure the view to the S., at least to the observer of today.

The papers relating to Compton station in the Public Record Office are particularly copious. They give a vivid insight into the conditions and problems of the old telegraph service.

Lieutenant John Harrison, whom we have met at Betsham, was the first officer to be posted to Compton. He first thought of taking with him John Dickens, who, although he had not served in the Navy, had been for some time at both Swanscombe and Betsham and worked well. But he finally recommended Andrew White, a superannuated boatswain who had served nearly 40 years in the Navy with unblemished character. White's appointment, in April 1822, shows how glad even an ex-warrant officer was to get any job, however humble.

Harrison could not move immediately as his wife was close to confinement. He asked leave for White to take advance possession of Compton, while he and his family stayed on at Swanscombe without cost to the government. Meanwhile Croker had put forward a nominee of his own called Thorpe. Harrison had to accept him but the Navy Board upheld White's appointment.

Poor White had to be discharged the following July with cold and 'inflammation.' Thorpe had meanwhile gone to sea, so Harrison called in John Muston, a former naval sailmaker and a 'steady, sober and good man.'

By 21 May 1822 Harrison was installed at Compton and a series of complaints began. As at other places, the principal trouble was the water supply. There was an underground tank, holding '7½ tuns or about 15 butts.' It had no pump and Harrison declared the tank was not big enough anyway. The nearest well was on the premises of the landowner. Harrison asked that they might have a right of supply from there, but it was little use as the water could be ill spared and was only fed from a land spring. So, the tank was to be filled through 'spouts' from the buildings. Harrison reported in June that the semaphore room was not fitted with spouts, so that they lost a valuable quantity of rain. Holl passed for approval the 'charge as reasonable' when Harrison sent him an estimate from the Harting carpenter, as follows:

43 feet run of yellow deal spouts at 7d. a foot, painting two coats and fixing	1	5	1
10 iron brackets at 6d.		5	0
Stout painted water butt	1	2	0
Oak stand for same		5	0
Carriage		4	5

On 21 July, the water supply was again reported to be very bad, the spouts defective, and the water not going into the tank. Harrison pulled the spouts up where they went to earth and found they were choked with rubbish and rubble. The work had been 'performed by such drunken negligent men, that they left their work at every opportunity for the public houses.'

Harrison also complained that the station was more than nine miles from the nearest market-town, Chichester; that mutton and veal only were obtainable in the village once a week, and he had no cool place in which to keep provisions. His 'numerous family of infant children' were often obliged to live on salt provisions. As beer and other liquids froze in winter he asked to have a cellar.

The grievances are most confused. We learn that much snow entered the semaphore room between the tops of the walls and the woodwork of the roof. Writing in August, Harrison says 'I very seriously apprehend it will be quite at the risk of our lives to inhabit the station.'

On 23 January 1823 Harrison complained he was

> under the necessity of making a bedroom for myself and family of the kitchen we suffer great inconvenience and discomfort this severe weather from the searching wind that drives up the joints of the stone floor..., and also that the Man is labouring under severe cold in his limbs in consequence of sleeping exposed to the weather that drives down the mast, through the walls, windows and pavement of the semaphore room, which in our very exposed situation, renders it severely cold and injurious to our health both to perform the duty and to sleep in.

And so on.

In justice to Harrison, it does seem that the contractor had botched some of the work at Compton, possibly by sub-letting unofficially. At all events Holl agreed he was much to blame and that his payments would be withheld. The defects must have been made good eventually — at all events the house remains, still inhabited and well-maintained, after almost a century and a half of winters.

Camp Down

The shoulder of a hill had been found by experience often to be better placed optically for signalling than the summit. The discovery is evident from the choice of the site of this station, on the E. slope of Portsdown Hill, at about 250ft., half-a-mile due W. of Bedhampton. It bore the same relation to Portsmouth as had the shutter post higher up on Portsdown itself, about two miles to the W.

The station, shown on the six-inch OS. of 1859-66, was demolished to make way for Farlington Redoubt, one of the Palmerston 'folly' defences of Portsmouth built in 1861. Portsdown Hill Road, on the S. side of which the station had stood, was deviated southwards to improve the line of fire from Fort Purbrook, also built at that time.

Lieutenant Williamson, who took charge on 4 May 1822, had a family of young children. He at once complained of the shortage of water and was allowed to have a cask on wheels, made by a wheelwright at a cost of

£3 10s. 6d. On 26 June he wrote again to the Navy Board to say that it would cost 10s. to hire a man with a horse to pull the cask. He therefore asked to buy a pony for £4 but the request was refused as it would create a precedent.

Williamson was luckier with a request for more privacy. The Navy Board allowed him to obtain an estimate from a 'husbandry labourer' to dig a ditch, form a bank and plant it with whitethorn, up to a cost of 3s. 6d. a rod.

Lumps Fort

As has been said, the smoke of Portsmouth was expected to make it difficult to signal direct from Camp Down to the terminus at High Street, and it was thought necessary to provide an extra telegraph to the eastward, at Lumps Fort, close to Southsea front and one of the town's earlier defences.

The site was immediately E. of the present Canoe Lake, in the angle of Lumps Lane (now Eastern Villas Road) and the seafront. There was already a rather ramshackle bungalow serving as a coast signal post and it was pressed into service for the telegraph. It was so near the sea that in bad weather spray beat over the glacis of the fort and often entered at the living room windows.

Goold, the officer at Lumps Fort station in April 1822, complained that as there was no proper road access supplies had to be dragged over a rough beach track in a four-horse wagon.

The station was reported to be under reconstruction in 1825, probably to bring it more into line with the standard of the others.

No trace remains of the station and little of the fort save the walls, which surround attractive public gardens.

It is likely that the station was cut out when visibility was good. A signal from Camp Down taken in at Lumps Fort and relayed to High Street had to be passed through an angle of about 100°. Popham's land telegraph was certainly built to be trained round in service, but there is no record that this was regularly done. If the mast had stays they would have to be loosened and cast off, wasting much time.

Portsmouth

A place known as The Platform, near the end of High Street, once mounted a battery of guns commanding the entrance to the harbour. It remains a popular vantage point for sightseers, with its fine views of shipping in Spithead and the Isle of Wight in the background.

Before 1822 a subsidiary semaphore had been erected at this site for signalling to passing ships. When the Popham line was introduced the High Street site became the main Portsmouth signalling centre. A telegraph house was built on the foundations of the famous Square Tower, the oldest building in Portsmouth and called the Square Bulwark in Elizabethan times and latterly the Magazine or Victualling Magazine.

12. The former semaphore tower at Portsmouth Dockyard.

In July 1822 the building on the Platform was taken down and everything transferred to the Magazine site, under the superintendence of Lieutenant Smyth. As the Port Admiral's office was also in High Street, the arrangement was ideal.

In 1833 the Port Admiral's office was moved into the Dockyard. An imposing tower was built over the centre of the Sail Loft and Rigging House and a new semaphore erected there, the High Street machine being retained as a reserve.

Both stations were given up when the line ceased operations in 1847. The Sail Loft and Rigging House, with the tower, were destroyed by fire on the night of 20 December 1913 but rebuilt.

The completed section of the Plymouth semaphore has contributed little to anecdote and local history, with the possible exception of Worplesdon. George Darley, an Irish poet, stayed at Worplesdon in 1846. Writing of the church he says:

> Guess its companion! A huge eight-sided edifice taller than the church tower, covered with yellow white-wash, and every alternate window blind, a pole on the top like a candle stuck on a loaf — and this is the semaphore, a kind of gigantic stand for a government telegraph [sic].

An amusing description but it would have fitted Chatley almost as well. It certainly seems to prove that the Plymouth line stations retained their semaphores to the end even though they can never have been used in regular service.

It is not certain how long the Plymouth line stations were fully manned. A list in O'Byrne's *Naval Biographical Dictionary* (1849) shows that Lieutenant Henry Belsey had been appointed to Cheesfoot station in 1837 and was still alive when the semaphore service ended in 1847.

On an average the semaphore telegraph functioned throughout the working day on about 200 days of the year and for part of the day on about 60 more. The worst time for transmission was around midday in hot weather when the heat haze impaired visibility. Working from Whitehall was sometimes hampered or stopped by the state of the atmosphere, smoke or even the vapour arising from the lake in St. James's Park. Such conditions were present on 133 days in the 1839-40 financial year. Over the three years ended 5 April 1842 telegraphy was impossible for 323 days from London and 65 from Portsmouth. At such times a courier took important messages from the Admiralty to either Chelsea or Putney. Operating expenses for the period were £5642 9s. 3d., pay of officers and men included.

In favourable conditions the time signal might take only 1 min. to pass along the line, when, of course, all operators were at the ready for it. A short service message might take 15 min. to pass all the way. The transmission rate of ten words a minute might be reckoned a fair average.

Each station would keep a sign displayed until the next station onward repeated it correctly. The next sign would then be shown, and so on. With every man alert and good visibility, only seconds would be needed to relay each sign. A watch was kept up every five minutes between 10 am and 5 pm in summer and 10 am and 3 pm in winter.

The period between 1824 and 1847 was quiet enough for naval operations and it is not surprising that in 1839 there were abortive discussions on the possible use of the London – Portsmouth telegraph for commercial purposes.

In 1842 *The Railway Times* reported that negotiations were well advanced for laying down the electric telegraph alongside the London & South Western Railway – 'Her Majesty's Government take a very lively interest in the matter.' At length the Admiralty had been persuaded of the usefulness of the new-fangled electric telegraph. In 1844 it arranged with the London & South Western that the company should provide an electric telegraph alongside its line from Nine Elms terminus, London, to Gosport via Bishopstoke (now Eastleigh). The inventors, Sir William Fothergill Cooke and Sir Charles Wheatstone, contracted to work the system for £1,500 a year.

The electric wires were duly laid alongside the tracks and into Clarence Victualling Yard, Gosport, whence a marine cable, one of the first put into service, was laid from there under the harbour to King's Stairs. Trials were made as early as 1845. 'On Saturday last,' reported *The Builder* of 1 February of that year, the electric telegraph 'lisped its first sentence. A shareholder at Nine Elms asked the keeper at Portsmouth – "How's the Wind?" He was answered on the instant – "South by West!"

The cost, according to *Besley's Devonshire Chronicle and Exeter News* of 21 January 1845, was about £24,000 and borne equally by the company and the Board of Admiralty.

In March 1847 the electric telegraph was taken into the semaphore room in the Sail Loft tower, ready for the changeover, timed for midnight on 31 December 1847. On 13 September 1847 the semaphore crews were notified that their employment would expire and their pay cease at the end of the year but that they might continue to occupy the houses for the time being.

In preparation for the change the Admiralty had laid down some rules for the forthcoming electric service. The hours of attendance were to be until 4.30 pm in November and February and until 4 pm in December and January. Men employed at Clarence Yard and in the semaphore room were to be on duty at 9.30 am except on Sundays, when the service was to be maintained for one hour only, and receive 14s. a week.

At the time their appointments terminated, officers had still been receiving 3s. a day, in addition to the half pay of their rank, and their assistants 2s. a day. To this should be added free lodging, so far as the semaphore building would allow, and an allowance for fuel if the site were very exposed or the winter particularly severe.

Commander Jay, as Superintendent, was ordered on 15 December to visit all the stations and see the machinery properly secured prior to the discharge of the officers. He reverted to the half pay of his rank on 1 January 1848 but was allowed to retain the old Superintendent's residence in West Square.

The Times of 31 December 1847 devoted much space to the change. 'The case of a semaphore officer is a hard one,' it declared.

> The semaphore has been the home of many a veteran lieutenant, the last berth to be given — the very last to be asked or accepted as long as a spark of hope remained of obtaining anything better, but now even this resource is no longer available. The following table shows the names, dates of commission of lieutenants, at what semaphore employed, and the date of appointment thereto, of those thus engaged, by which it will be seen that there is reason and justice for this complaint:-
>
> Lt. William V. Lee, station Chelsea, date of commission Oct 27, 1827, date of appointment June 20, 1844
>
> Lt. Lardner Dennys (Putney) Feb 6, 1812; March 16, 1847
> Lt. Wm. H. Dixon (Kingston) May 25, 1807; Nov 15, 1841
> Lt. Thomas Wright (Esher, Coopers Hill) Feb 17, 1815; June 20, 1844
> Lt. Edward Robinson (Cobham) Apr 15, 1813, May 5, 1847
> Lt. James Poad (Pewley Hill) Aug 24, 1812; Nov 25, 1841
> Lt. George Blurton (Godalming) Apr 15, 1813; July 28, 1838
> Lt. Joseph Williams (Haslemere) Nov 17, 1807; Mar 22, 1847
> Lt. Thomas Hills (Midhurst) Nov 7, 1806; Nov 25, 1841
> Lt. Henry Garrett (P'field) Aug 21, 1809; Dec 8, 1841
> Lt. George Fraser (P'field — Compton Down) Nov 1, 1821; Nov 30, 1839
> Lt. John Wildey (Portsmouth) Apr 10, 1807; Mar 10, 1842
>
> Four of the above are Trafalgar men; one was in Hotham's action of 1795; one was a mate in Sir Richard Strachan's action in 1805; one was a lieutenant of the *Denmark* in the Walcheren expedition; one lost a leg at Navarino, and all the others have distinguished themselves in their country's service.
>
> We are all aware of the proneness to derangement of the electric mode of communication, and its services have not been available on some of the most important and interesting occasions. What, for instance, would be easier than for any evil-disposed person to cut the wires of the electric telegraph in an obscure place on this line? The means of communication would be at once annihilated, and a considerable space of time lost in repairing the damage. Such mischief as this could not be perpetuated [*sic*] upon the semaphore; there the Govt. had an exclusive and safe, as well as speedy means of communicating their commands or instructions at their own disposal. It will now be dependent on the fallibility of the electric mode, for the officers and men of the semaphores, once discharged, these heretofore valuable appliances to the sources of speedy intelligence cannot be worked again without a complete reorganisation of the system of instruction, as it takes years to thoroughly comprehend the vocabulary of figures by means of which all the Govt. notifications are made. Many of the assistants to lieutenants in the semaphores have been upwards of a quarter of a century acquiring and practising their duties. It will, therefore, be no easy task to resuscitate a system such as this, if once allowed to lapse into desuetude. At all events, whatever may be the resolution of the Admiralty with regard to the disposal of the stations

and the telegraphs, we trust those who have hitherto looked to them as their last resource will receive some 'consideration' for the loss of their appointments, especially those who, from the shortness of their tenure, or other ill-luck, may more particularly have deserved compensation.

The reference to Godalming means the Bannicle station, Midhurst stands for Older Hill, Petersfield for Beacon Hill. It is interesting that there is no reference to either Camp Down or Lumps Fort. Lee was the officer who had lost a leg at Navarino and Garratt was one of those who had been at Trafalgar. Poad had been a midshipman in *Victory* at Trafalgar at the age of 16; he was a commander when he retired in 1855, three years before he died. Several officers had also served at other telegraph stations.

Most other officers appointed to the semaphore stations had distinguished themselves on active service, many with scars to prove it. None was in sadder case than Lieutenant William John Snow, who died at Putney semaphore only 11 days after being posted there at the age of 38.

Snow entered the Navy as a Midshipman at the age of 16. A year later he was wounded while serving in *L'Achille* at Trafalgar. Later, while prisoner in an enemy ship in the Baltic, he staged a successful rescue and brought all his comrades back home in their floating prison. Still later, he was cast away in a Danish prize vessel off the coast of Norway but managed to bring the dismantled and waterlogged ship into Scarborough, though he was suffering from frostbite.

While serving in *Guerrière* he helped to capture, recapture or destroy 16 vessels and while in *Poictiers,* another 23 vessels. As master's mate he received severe chest wounds in the action between the *Guerrière* and the American frigate *Constitution.*

He was promoted Lieutenant but his wounds compelled him to live on half pay with his family in Wiltshire until the Lords of the Admiralty appointed him at Putney. The expense and fatigue of the transfer brought on a fatal illness when he arrived at the station. He was buried in Putney churchyard.

It is pleasant to know that his eldest son gained a presentation to the Royal Naval School at Greenwich and that Putney residents contributed to a fund to buy a cottage in Jersey for his wife and other three children and to support them until her widow's pension from the Navy became payable.

According to *The Times* 28 March 1849 the electric telegraph 'will be complete in a few days into the Port Admiral's Office...' One signalman was to remain to take care of the semaphore tower and keep a look out to seaward. *The Times* hoped that Lieut. Williams would receive some reward or promotion on discharge after his 50 years' service, 42 as Lieutenant. In fact he was appointed for service in the ordinary and promoted Commander in 1851; he died in 1853.

Despite the misgivings of *The Times,* the new electric telegraph was not sabotaged even in 1848, the year of European unrest. By 1850 the newspaper had no reservations. Its 25 September 1850 issue stated:

> The great advantage of the electric telegraph over the old system is well known, as by the present one despatches can be sent off and received at all times, night and day, and no matter what the weather, whereas the former telegraphs could work only by day and that in fine weather...

It added that orders could now be transmitted in a moment not only to Portsmouth but also to Woolwich, Chatham, Sheerness, Plymouth, Devonport and Pembroke.

The houses and their acre of land were sold. The Duke of Cambridge paid £100 for the Kingston Hill ground, £30 for the house and £11 for fixtures. The Chatley tower and land went for £280.

CHAPTER IV

The Admiralty Coast Signals

As early as April 1785 a board of naval and military officers was appointed to report on the defences of Portsmouth and Plymouth. The members recommended the erection of signal houses on headlands which would be of 'effectual advantage in conveying intelligence of the approach of an enemy and for the protection of commerce.'

The first signal posts were established in 1794 all along the S. and E. coasts of England, between Land's End and Yarmouth. There were finally 74 stations, some on prominent headlands like Start Point, St Alban's Head and Beachy Head. They were intended to communicate with men-of-war in the offing by means of a system of balls and flags and not necessarily to intercommunicate. All were erected under the superintendence of Nicholas Vass, Master House Carpenter of Portsmouth Yard, and were completed in 1795. In the ensuing years some posts were re-sited and new posts built.

Most stations were huts. Their original full complement was a Lieutenant, paid 7s. 6d. a day, exclusive of half pay, a Midshipman at 2s. a day and the pay of a fourth rate, and two seamen at 2s. a day.

Signals were made by showing a rectangular flag, blue pendant and four black balls (hoops covered with canvas). For example, a pendant and flag at the masthead and two balls at the gaff end told a ship in the offing that she was off St. Anthony's Head. A pendant at the masthead and three balls at the gaff end signified No. 13 – the enemy are landing to the westward.

The signal post comprised an old topmast of 50ft. with cap, crosstrees and fid to secure the flagstaff; a 30ft. flagstaff with truck and sheaves, and a 30ft. gaff yard with an eye, bolt and hoop at each end.

A printed sheet entitled *Signals denoting the Principal Signal Stations along the Coast of Great Britain and Ireland* (1803) showed 28 stations in all. The 1804 reprint lists 35.

A spurt of activity in 1804 was responsible for more stations, on the Norfolk coast and in prominent positions on the NE. coast, and also including those at St Martin's (Scilly), between St Abb's Head and Leith, at Milford Haven, at Formby, and between Liverpool and Holyhead. The Liverpool-Holyhead line was instigated by Prince William (later William IV). Its eight stations were Liverpool (St. Domingo), Bidston Hill, Point of Air, Cave Hill, Great Orme's Head, Table Hill, Point Lynas and Holyhead.

In 1804 also 15 posts were erected on the Irish coast and 66 more were ordered to be built to form a chain from the Pigeon House on the Dublin Quays round the E., S. and W. coasts to Malin Head.

(Incidentally, Vass submitted a report dated 10 October 1803 listing possible sites for a line for military use from Shoebury to the Tower of London, via Southend, Canvey, Lower Hope, Tilbury Fort, Purfleet Hills, opposite Galleon's Point and the Brunswick Dock Meeting House. It was agreed, however, that it would be better to station a hulk at Great Nore and communicate via Sheerness and the Admiralty shutter telegraph to the Admiralty).

In Chapter X it will be shown that in 1801 a new form of signalling apparatus was devised in France for communicating between ship and shore and was adopted enthusiastically by the French Ministry of Marine. This was the *sémaphore*, not at first thought of as a telegraph in the true sense.

The invention did not go unnoticed by the British Admiralty. In 1810 it began to replace the slow and complicated flag and ball system on parts of the English coast by signals on the French pattern. As in France there was no thought of a general coastwise telegraph, though it sometimes happened that posts were sufficiently in sight of each other for them to communicate.

Even in Victorian times, when the Coastguard had taken control of the stations, added more and manned them with efficient retired bluejackets, it was by no means established that they should be in visual communication, one with another. To this rule there was a notable exception, a continuous coast line set up in 1812 in open imitation of the French semaphore, even to the number of signal arms and mode of working. It was then or soon after that the word 'semaphore' entered the English language, later to become synonymous with visual mechanical telegraph. The semaphore system became the approved pattern for future visual telegraph systems in England and a number of other countries.

The 1812 line, superseding part of the original coast system, ran from Deal to the Nore and then to Yarmouth, but had no direct communication with the shutter telegraph. Where there was a break in the coastline, as at the mouth of the Thames and the entrance to the Colne and Blackwater rivers, special vessels were moored to ensure continuity. The positions of each arm of the semaphore were numbered 1 to 6 and the signals were therefore limited to a few numerary groups conveying brief and arbitrary messages, such as 'The enemy have sailed from the Texel.'

The stations between East Hill (Deal) and Yarmouth Dean (naval hospital roof) were:

Sandwich Flats; St. Lawrence; St. Peter's; Birchington; Bishopstone (Reculver); Whitstable; Shellness (Sheppey); Warden Point (Sheppey); Flagship at the Nore *(HMS Namur)*; Wakering Stairs; Priestwood; Cotesend, Foulness; Southminster Marshes; Tillingham Grange; Sandbeach; St. Peter's Chapel, Bradwell; Entrance to Colne and Blackwater rivers *(HMS Warning)*; East Mersea; Beacon Hill, near St. Osyth; Cotewick Farm; Clacton Wash; Little Holland Wall; Burnt House, Walton; The

Naze; Harwich; Landguard; Felixstowe; Bawdsey; Hollesley Bay (Orford Haven); Orford Castle; Aldeburgh; Thorp Common; Beacon Hill, Dunwich Common; Easton Cliff, Southwold; Kessingland; Gunton; Hopton Common

From Norfolk the line of signal stations was carried right up the east coast to Calton Hill, Edinburgh. Between St. Abb's Head and Edinburgh, certainly, they are sited suitably to have served also as a local telegraph. Colour is lent to such a theory by the fact that the site at Dow Law appears on present-day maps as Telegraph Hill, a 572 ft. hill E. of Cockburnspath in Berwickshire. Thence to Edinburgh stations were erected at Black Castle Hill, Dunbar Pier, North Berwick Law, Garleton Hills and Port Seton. Certainly the course is well inland in places. In December 1816 there is a record that Leith, a possible new additional station, was to be equipped like the stations between London and Chatham, that is with Popham's two-arm semaphore.

By Admiralty Minute dated 6 May 1812, all Lieutenants of signal stations between the Nore and Orford were reappointed to the substitute semaphore posts. To quote Mead's *The Story of the Semaphore*:

> As the lieutenants 'drop off' — an unfortunate expression, perhaps suggestive of their great age — there was to be only one officer to each two stations within a day's walk of each other. Otherwise the crew of a station was one midshipman and one man.

Wherever a river intervened and an intermediate station was needed on one side of it, the officer was to have charge of it as well as his own.

Mead adds that 10 May 1812 was the date fixed for the semaphores between Wakering Stairs and Orford Haven to come into service. He also cites a false signal passed along the line in September 1812 which resulted in Rear-Admiral Lord Beauclerk proceeding to Texel, only to find that the enemy had not stirred. The officer of the offending station was merely told to be more careful in future!

At 1 July 1814 there were also another 80 non-continuous stations along the south coast between Deal and Land's End, including five in the Isle of Wight, as well as the 19 by then in service in Ireland. All still used the original system.

The particular expansion of the coast system in 1811-12 brought with it new code and signalling systems, under the threat even then, if the signal tables are a guide, of invasion.

In 1814 a new station was erected on Newford Down (Scilly) and worked by semaphore from the start. Mead records that in 1816 the station was not considered to be fulfilling its purpose. Sir John Duckworth, Commander-in-Chief at Plymouth, confirmed the fact and added that the officer was often drunk, even if he did not connive at smuggling as had been alleged. The station was put down, even though it had cost £1,025 to build two years before.

In November 1814 orders were given for all coast stations in Norfolk, Suffolk, Essex, Kent, Sussex, Hants, Dorset and Devon to be 'broken up' as soon as possible. On 21 November Barrow was transmitting letters from Lieutenant Roche of Wakering Stairs station and Lieutenant Leckie of St. Peter's Chapel station, asking for disposal instructions for their apparatus.

By 30 May 1815 their Lordships were once more looking to their defences and directed that the posts between the North Foreland and Lands End should be re-established as semaphores. On 12 October 1816 Goddard submitted a report on experiments with semaphoring along the south and south-west coasts.

Conversion was slow. One report from Goddard covers progress up to 29 April 1820 in fitting out stations between Deal and Beachy Head. The entries include such items as 'Semaphor [sic] up and house ready'. About this time stations began to be converted from the three-arm system of 1811-12 to Popham's two-arm land semaphore already adopted for the Admiralty internal lines. By this time most stations were being required to communicate with Revenue vessels as well as with those of the Royal Navy.

Two stations between Deal and Beachy Head, at Fort Moncrieff (Hythe) and Dungeness beach, were square forts, as distinct from the round Martello towers, of which four became signal stations. A drawing by Goddard, dated 27 May 1820, shows the Dungeness station (No 2 Battery) with its Popham semaphore. When the building was used as a fort, the masonry reached only to the bottom of the upper row of loopholes. The first storey and roof were added later to accommodate Preventive men. Goddard added a small room on the roof for the semaphore operators and had windows made in the first storey.

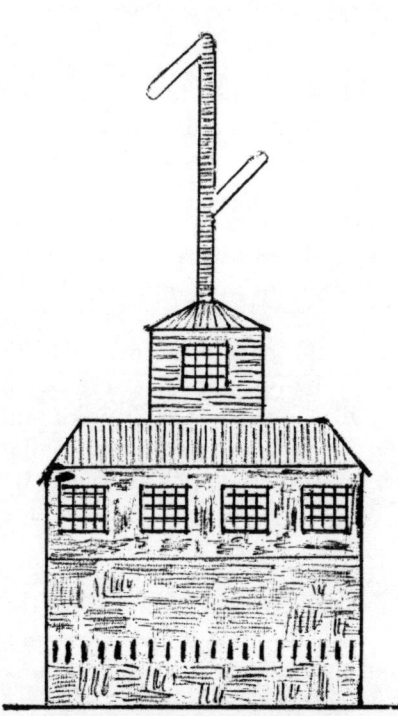

13. No. 2 Battery, Dungeness, with coast semaphore on top. (Sketch from a drawing by Thomas Goddard, 27 May 1820.)

At Deal the coast line would have connected with the projected semaphore line from London which was never completed. (See Chapter III.)

CHAPTER V

The Holyhead-Liverpool Telegraph

The Holyhead – Liverpool commercial telegraph functioned longer than any other visual telegraph in the British Isles. It probably handled much more traffic than the Admiralty line to Portsmouth, the only comparable system in Britain.

Its ancestor was a method of communicating shipping arrivals in the Mersey by flag poles on Bidston Hill in the Wirral. Each Liverpool shipowner had a pole there. If an owner saw his flag hoisted on his pole he would know that one of his vessels was coming into port.

Prints and other representations clearly show the curious scene. In 1929 the Mersey Docks and Harbour Board sent a handsome Christmas card containing a large folding plate of Bidston Hill summit with 76 poles and 150 flag signals all in colour. It might be more accurate to state that information was conveyed by means of balls and other shapes as well as by flags, and that the poles could also signal vessels in distress and needing assistance, and 'enemy in sight'.

The poles ran N. and S. and so presented a broadside view to observers in Liverpool. Liverpool Old Churchyard was a favourite place for seeing the poles and their flags. They were easily seen with the naked eye from such places as St. Domingo, Everton and St. James's Walk, and therefore quite discernible by the merchants for whom they were intended.

Seen from N. to S. there were first 27 poles, then a five-storey watchtower with one mast, and three more masts on an out-building; three more poles, then a very tall signal mast complete with yard and other contrivances; then 37 more poles as far as a windmill and a hut, and finally five more poles.

The signals displayed at Bidston were observed and reported to shipowners by a watchman at Liverpool, whose look-out post was first on a warehouse in Chapel Street, later on the tower of St. Nicholas's church, and finally on the top of Tower Buildings. The poles may have been used as early as 1763 but Liverpool Corporation did not lease the land at Bidston until 1771.

Although this system was superseded in October 1827 there were still eight poles left in 1856 and five in 1862. But long before 1862 all such methods of communication had been replaced by the electric telegraph.

In 1804 ball-and-flag Admiralty signal stations were erected in the Liverpool area and between Liverpool and Holyhead (see page 00).

Before 1827, messages could be received from vessels through Holyhead by a signal station on Holyhead Mountain, probably using Marryat's signal book of 1817. It had no connection with the Bidston arrangement. The information received through Holyhead was sent to Liverpool by ordinary post and arrived a day later.

By an Act of 27 June 1825 (6 Geo. IV cap. clxxxvii) 'for the further improvement of the Port, Harbour, and Town of Liverpool,' the Liverpool Dock Trustees were authorised and empowered to

> establish a speedy Mode of Communication to the Ship-owners and Merchants at Liverpool of the arrival of Ships and Vessels off the Port of Liverpool or the Coast of Wales, by building, erecting and maintaining Signal Houses, Telegraphs or such other Modes of Communication as to them shall seem expedient, between Liverpool and Hoylake, or between Liverpool and the Isle of Anglesea.

After much discussion in 1826, the survey for a line between Liverpool and Holyhead was entrusted to Lieutenant B.L. Watson, 'Royal Navy'. The estimated cost of establishing the stations was £1,700, a very low figure, but the machinery was not very elaborate or very technical, and the buildings were only small 'observatories' round the foot of the masts. The annual upkeep was to be £440 for seven men; £70 for one clerk; £50 for second clerk; £40 for stationery, oil, etc; £440 for one year's superintendence, and the cost of construction £1700 (£1300 for the stations, £100 for telescopes and £300 for plans and superintending the erection of stations). Watson received about £26 for his expenses, and another £100 for his plans.

On 1 August 1826 the sub-committee appointed by the Dock Committee reported in favour. The report was adopted and the sub-committee of five put in permanent charge of the telegraph. The mercantile associations favoured the scheme. The Underwriters Association was against it but when the telegraph was in service the underwriters were glad of it.

More will be said about the mysterious Barnard Lindsay Watson. Suffice it here to say that although at the beginning of his telegraphic ventures he is generally referred to as a Lieutenant, Royal Navy, his name cannot be traced in official Admiralty records.

The first exploit to usher in the line's working was the reporting of the American ship *Napoleon*, on 26 October 1827, from Holyhead to Bidston, which took 15 min. A message sent at 9 am on 5 November 1827 announced a change of wind at Holyhead from SW. to W. The change did not occur at Liverpool for an hour but the message took only five minutes to get there.

Unlike internal telegraph lines, this system, as well as communicating messages from end to end of the chain, had to be able to send and receive signals from intermediate stations, particularly those at Point Lynas (Anglesey), Great Orme and Hilbre Island (at the mouth of the Dee).

The apparatus consisted of a tall mast with a topmast above cross-trees. The mast was of Baltic timber, 40-50 ft. high and 22 in. in diameter (including oak fishes bound round it with iron hoops) and stayed by four strong chains. The topmast was 27 ft. high. The semaphore arms were of

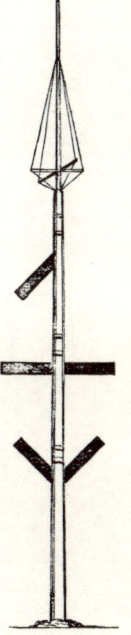

14. Watson's first telegraph.

African oak, painted black, 7½ ft. long by 16 in. wide. They 'played' into grooves in the lower mast. There were three pairs, working on three separate pivots, and displayed at the 45°, horizontal and 135° positions, by being hauled up by ropes passing through blocks up the mast. The same mast was used for hoisting the flags for signalling to passing ships when necessary.

Each pair of the three pairs of arms could take up 9 positions, to which the numbers 1 to 9 were assigned. The top mast pair represented *hundreds*, the middle pair *tens* and the lowest pair *units*. Ordinarily the arrangement would have allowed a total of 999 separate numbers to be indicated. But to increase the power of the machine, 9 different classes or groups were made, the operator first indicating the relevant group by making its number with the topmast arms and then following with the message number. All message numbers were referable to a vocabulary or code, in the following order: letters of the alphabet, compass points, portions of time, words and sentences, merchants' names, gazetteer, names of vessels using Liverpool. For example, the class I Sign followed by 1 (top pair), 9 (middle pair), 5 (lowest pair) signified No. 195 in the code — *Abandoned by the crew*.

Watson's *Code of Signals* was printed in several editions between 1827 and 1842, and his flags are said to have been among those in use at Holyhead even before he took over. This code used 12 flags, not well chosen optically, and vessels could use Marryat's flags instead if preferred. A ship in the offing would signal her news or requirements by flags relative to the signal code book, and the station operators would send on the message by semaphore. The masts could be trained round to follow a ship as she passed, when it was assumed that she could read the semaphore.

The distance from end to end of the line was frequently given as 72 miles. This may be so as the crow flies but as measured from station to station it appears to be nearer 75½ miles.

The original stations were:

Holyhead. Almost at the extreme W. end of Holy Island, on the shoulder of Holyhead Mountain at about 500 ft., between the North Stack and the South Stack. From here it was: 6.2 miles to *Church Bay*, Gareglwyd (Anglesey), also known as Mount Pleasant and close to the village of Llanfaethlu; 10.8 miles to *Llaneilian* (Anglesey), sometimes known as Paris Mountain, at about 300 ft., overlooking Point Lynas, the cruising ground of the Mersey pilots, called 'the Dungeness of the north'; 13 miles to the station at the NE. end of *Puffin Island* (Ynys Seiriol), also known as Priestholm; 7 miles to *Great Orme*, Llandudno, where the station was at the summit of the hill at about 670 ft.; 8.7 miles to *Llysfaen* (behind Old Colwyn), 670 ft. a.s.l.; 6.7 miles to *Foryd*, on the beach at the W. end of Rhyl, on the W. side of the estuary of the River Clywd; 6 miles to *Golden Grove*, a site somewhat inland from Prestatyn at about 770 ft.; 7.2 miles to *Hilbre Island*, at the mouth of the Dee; 6.4 miles to *Bidston Hill*; 3.5 miles to *Liverpool*. The terminal station was originally on the roof of Duncan's warehouse at the bottom of Chapel Street. In 1846 the warehouse was demolished and Tower Buildings erected in its place. The telegraph remained on the roof of these buildings until it was abolished in 1861. Tower Buildings were demolished in 1906.

Golden Grove is something of a misnomer and a mystery. The site, as marked on the 1840 one-inch O.S., is a good two-thirds of a mile WNW. of the mansion of that name and is nearer to Gwaunysgor, a mile inland and S. of Prestatyn. It seems to have been known alternatively, at least later, as Voel Nant, the name properly given to the station which superseded it in 1841 and was situated at a lower level and almost a mile distant on the slopes E. of Prestatyn. If stones which remain at the site are relics of the 1827 telegraph line, some at least of its stations may have been stone or brick structures from the outset or subsequently.

The Golden Grove station, the highest of any, communicated on both sides, at Foryd and Hilbre, with stations almost at sea level. Such a switchback was rare in telegraphy.

The Bidston telegraph was immediately adjacent to the lighthouse on the hill.

In 1830 the system was extended to the Tuskar Rock and Cape Clear, Ireland, and reports were sent by Wexford and Skibbereen respectively, presumably by post. On 5 May 1835, Watson wrote to the Dock Committee proposing a telegraph between Liverpool and Manchester, probably along the line of the new railway. He was given leave to make a survey and plans, but the decision as to whether he should superintend such a line was deferred until his plans were ready. In fact the line does not seem to have been built.

The new telegraph was described in the 13 January 1827 issue of *Billinge's Liverpool Advertiser* and the 24 November 1827 issue of the *Mechanics Magazine*. The articles listed its uses as: reporting the state of the wind at Holyhead daily; reporting vessels passing Holyhead inward and outward bound; reporting the state of vessels in distress or needing assistance, all along the coast; and reporting important political or commercial information from Ireland or America made known at Holyhead by home-bound ships.

The writer in the *Mechanics Magazine* asked if the telegraph was not well suited to the communications of impatient lovers. If Juliet had had a telegraph it would have made her happy and spared her nurse's bones, and might have saved Romeo's life.

> A sigh or a vow might be wafted from London to Bath or Cheltenham in a few seconds: a lover might thus conspicuously *signalize* his devotion to the fair one of his heart; and the pining mistress might learn from the expanded *arms* of the telegraph how soon she should be restored to the arms of her betrothed. . .

More prosaically, the writer believed that the telegraph was well adapted for conveying commercial intelligence and would be most useful in wartime.

Valuable details of the telegraph appear in the evidence taken by the Government Commissioners on the Corporation of Liverpool in November 1833. It was discussed before the Commissioners because of a complaint by a Lieutenant Richard James Morrison, RN, that Watson had appropriated his ideas. The commissioners declined to enter into the matter but heard evidence about the telegraph.

On 13 November 1833 Watson swore a statement about the working of the system and his duties. He said that communications were made with vessels directly they appeared off any part of the coast. Accidents, cases of distress and casualties were at once telegraphed and published in the Exchange and underwriters' rooms. No charge was made for messages sent. The sole expense to merchants was the cost of flags for signals. The state of wind and weather at Holyhead was frequently, during the day, reported in Liverpool, and the destination of vessels often changed to the Clyde, Belfast or other ports, at the request of their owners.

Watson disclosed that before the line was established about 300 vessels carried the Holyhead signals and communicated with the station there, which sent the information on to Liverpool by post. The number of vessels in the present list was upwards of 1,300 and their communications were sent to Liverpool within a few minutes. A whole code of flags, such as would enable the master of a vessel to send any communication, cost eight guineas, the vocabulary 12s. 6d; the price of flags for a number of a vessel only was £2. 5s.

A constant lookout was kept from daylight until dusk. At 1 pm a signal was sent along the line, to signify the exact time, so that the operators could regulate their timepieces. It also asked the question, 'Is anything to report?' An answer to the question was generally received within 1 min. — it had been received in as little as 23 sec., said Watson. On important occasions political news was telegraphed.

Watson declared he had no emolument from the Dock Trustees beyond his salary and travelling expenses of a guinea-and-a-half a day. He merely certified any bills. He appointed clerks and operators and considered himself responsible for the performance of their duties. The operators received between £40 and £55 a year.

Watson was appointed annually at first and paid £400 for his first year. He then received £250 until 1830 when he was permanently appointed. By a Dock Committee resolution of 7 October 1828 he continued at £250 and this rate was confirmed again in 1830.

On 23 November 1833 Morrison questioned Watson's chief clerk, W. B. Coleman, who said that money from flags went into Watson's own account, and averaged £500 a year gross. By no means all the 1,500 vessels on the list bought flags but many made their own.

Morrison alleged that Watson was paid for information supplied to certain associations. He had laid a memorial before the Dock Committee in January 1830, but the Committee, at its meeting of 21 September 1830, resolved to have nothing to do with a dispute between him and Watson over Watson's conduct in the erection of the telegraphs.

According to Morrison, Watson was carrying on a business at great charge to merchants and must have received much more than £500. Coleman replied that merchants were not obliged to take flags from them. He could not give the exact sum 'as there are no books'. Morrison then said that although the service was supposed to be free he understood that the Lycaeum paid Watson £21 a year and the West India Association, eight guineas. Coleman said he knew the Lycaeum received reports as well as other news rooms but did not know what Watson received.

On 25 November 1833 Watson addressed the commissioners:

'In consequence of seeing in the papers a report of what was said on Saturday, relative to the Telegraph, I have come from Wales, at considerable inconvenience to myself, for the purpose of affording some explanation on that subject'.

He said that masters often came to him on the eve of sailing, for a supply of flags. He had therefore to keep a stock in the office. He did not think the amount of money received was a matter for the present inquiry, but stated that in 1832 one of the steam packet companies had placed an order for flags worth £130, in consequence of information telegraphed which had saved one of its vessels from shipwreck. He admitted that the West India Association paid him, but said he had nothing from the East India Association. The American Chamber of Commerce used to receive information by post from the Harbour Master at Holyhead but as Watson was able to send them it a day earlier they transferred their allowance to him without solicitation.

Watson denied his telegraph was ever Morrison's property or invention. He devised it before he had heard of Morrison. He first built a model patterned on the Admiralty semaphore but found it too expensive and not extensive enough. He then turned to the design which Colonel McDonald [sic] had published in 1817 'and I received the thanks of Colonel McDonald for having his plan adopted'.

It seems likely also that Watson had taken a leaf or two out of the book of John Rowe Parker, a Bostonian who had set up a harbour system using similar apparatus at Boston, Mass., in 1822 (see Chapter XXVII). He may have been in touch with Parker, whose experience would have been of great value, or at least gained a full account of his system from the masters of the many vessels plying between Liverpool and Boston. Certainly Parker subsequently embodied the Liverpool – Holyhead code in his own three-part code book.

Lieutenant Henry Mangles Denham, RN, was a distinguished hydrographer who had carried out important surveys on the English, Welsh and Irish coasts, including the Bristol Channel and the approaches to Liverpool. In 1834 he received the Freedom of Liverpool for his local work and about the same time was appointed as Resident Marine Surveyor to the Liverpool Dock Committee. On 20 March 1835 he was promoted to Commander.

Denham's position at Liverpool must have been honorary for in 1842 he had to go back to the Service, and was advanced to Captain, RN, in 1846. At Liverpool he was in charge of the lifeboats and lighthouses belonging to the Dock Board. As the telegraphs at such places as Point Lynas, Great Orme and Bidston were sited with the lighthouses there, the personnel all came under the same management.

Denham was the author of *Denham's Mersey and Dee Navigation,* to give

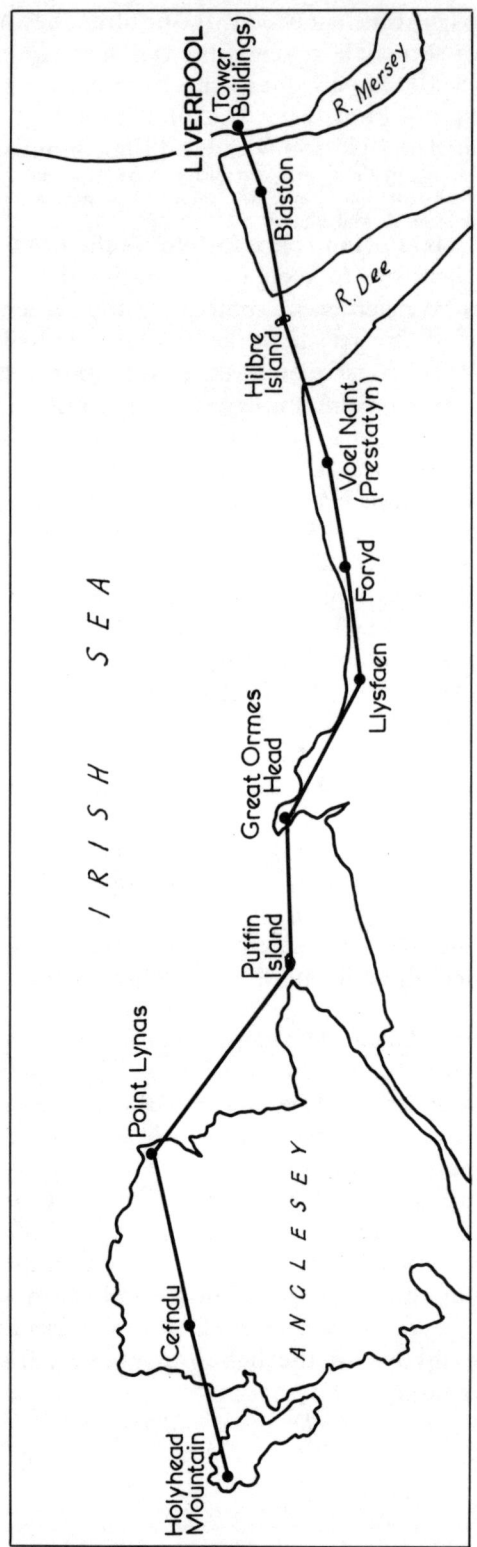

15. Holyhead-Liverpool semaphore telegraph.

it its short title, which was published in Liverpool in 1840. It was a 'pilot' or guide to sailing directions from Point Lynas to Liverpool and included pull-out plates with sketches of coast views, including several of telegraphs as seen from seaward and even showing the arms at work!

Point Lynas is stated to be the western cruising ground of the Liverpool pilot boats (sloops), of which there are 11 always on station. Point Lynas is distinguished by a lighthouse and 'adjacent telegraph'. Llenelion [sic] rises to 571 ft. and is rendered conspicuous by this telegraph. The mariner, approaching this headland, gives himself an opportunity of sending a telegraphic message to Liverpool, and a footnote emphasises that the local code (Watson's) should be superseded by Marryat's, otherwise none but Liverpool ships can make themselves understood. Hilbre and Voel Nant telegraphs are also referred to in the text.

Denham tells us that the NW. Lightship can take up the Liverpool code of signals and relay messages to Hilbre telegraph. Denham, incidentally, was in favour of an additional telegraph being erected at Crosby.

In 1839 Lieutenant William Lord, RN, was appointed Superintendent of the Telegraph in succession to Watson. Lord, at any rate, was a genuine naval officer! Watson had had a dispute with the Dock Committee over his remuneration and charges and the Liverpool Corporation paid a lump sum to be rid of him. By this time Watson was developing other ventures, as we shall see, and was probably quite happy to be freed.

The year 1839 was important in the history of the Holyhead telegraph and Lord's appointment may have had something to do with the improvements which then began. The first station across Holyhead Bay, at Church Bay, was augmented, if not superseded, by a new station called *Cefndu*, near the village of Llanrhyddlad. After Cefndu came into service, the original Church Bay station was used only as a subsidiary or when Cefndu was shrouded in mist or cloud.

Substantial buildings were erected at the stations, all to the same plan, which can easily be seen today in those that survive in excellent state and are occupied, at Cefndu, Llaneilian, Llysfaen, Voel Nant, and Hilbre. Little remains of the Holyhead station but that on Puffin Island is still recognisable, though ruined and entirely inhabited by puffins, razorbills and seagulls. Although most of the stations are reasonably accessible, it was always difficult to reach Puffin Island and Hilbre, and the transport of building materials and stores must have set a problem. Puffin Island is cut off from Penmon, Anglesey, by the turbulent waters leading into the Menai Strait. Even now, by motor boat, the passage is generally difficult and at times impossible. At Hilbre the sea goes right out and one can walk across between the tides, or be driven in a pony trap. Hilbre and its companion islands were bought by Hoylake Urban District Council from the Mersey Docks & Harbour Board in 1946. The keeper of the islands, a council employee, occupies the telegraph house.

Voel Nant, which replaced Golden Grove, stands at some 300 ft. about midway between Nant Mill and Gronant, and a mile E. of Prestatyn.

The new stations, of bungalow type, with a large, bow-windowed observation and signal room, were completed in 1841, as a slab in their gable ends testifies — for example, *Voel Nant Telegraph, Built in 1841 by the Trustees of the Liverpool Docks.*

The general layout comprised four main living rooms arranged in a square and sharing a common chimney, and the bay-windowed observation room. The houses were built of stone, with a slate roof. The observation room roof appears originally to have been flat but at some of the surviving stations it has been subsequently pitched.

The exception is at Hilbre, where the operator's dwelling and the operating room are separate buildings some distance apart. I am indebted to Mr J. Thomas of Hoylake, a former employee of the Mersey Docks and Harbour Company on Hilbre, for putting forward the likely explanation. As Hilbre is exposed and subject to heavy spray during westerly gales it was obviously thought wise to build the dwelling on the easterly side of the island, farther from the force of the wind. This position would have been difficult to see from Voel Nant, so the station itself was erected on the highest point of the island, close to the 56 ft. mark.

All the stations probably had wells. At Llysfaen the water supply is now taken by pump from a conveniently adjacent reservoir. (Incidentally, somewhat below this house, on the seaward side of the road through the village, stands the *Semaphore Inn*, whose sign unfortunately shows nothing more exciting than two crossed flags!)

All five surviving stations have been listed Grade II.

16. Llysfaen Station, Holyhead-Liverpool Telegraph.

In 1841 the tall, single, ship-fashion masts were replaced by twin lattice posts of wrought-iron work, 18 ft. high and 13 ft. apart, each mounting two pairs of arms. They stood on the roof of the operating room, which extended to seaward from the operator's dwelling. Watson had planned such changes in 1837 before he left Liverpool but the Dock Committee resolved not to entertain them 'at present'. He at once adopted a similar plan when he went into business on his own.

The eight arms, each capable of three positions, provided, one might have thought, a sufficient combination of signals. But both masts were also fitted with a board at the top which could be closed or exposed to raise the power of the machine, a variant of the class number system of Watson's original apparatus. If both boards were hidden, the signals were to be taken from Part I of Lord's *Code Book, The Liverpool and Holyhead Telegraphic Vocabulary*; if both were exposed, they were referable to Part II. If the board on the southern mast was exposed, real numbers were understood. Part I totalled some 9,620 numbers, Part II 5,695 numbers.

Each station, as well as lighthouses and lightships off the coast, had a flagstaff so that flag signals could be exchanged with ships in the offing, as the contemporary engraving of the Great Orme station shows. For this purpose Marryat's flag code was ingeniously adapted to Lord's Code. If the telegraphs were put out of action for any reason, messages could be flagged along the line by using the same hybrid system.

Ordinary messages, such as reports of vessels by name off Holyhead, usually took about 5 min. to relay. When a ship appeared off Holyhead she was telegraphed immediately and her progress notified as she passed the intermediate stations. An incoming ship might ask for instructions and the reply come back while she was still sailing past Holyhead.

A study of Lord's *Vocabulary* of 1845 reveals the completeness and ingenuity of such codebooks. It is in two parts, as follows:

Part I: Nos
  1-26   letters of the alphabet
  27-58   compass signals
  59-85   time signals
  86-249   signals in constant occurrence along Holyhead line
  250-9620 vocabulary

Part II:
  1-70   buoyage of Liverpool
  71-200   blank
  201-3064 vocabulary
  3065-5695 gazetteer

From Part I the following examples are quoted:
  1816   '*blow*'
  1824   if it continues to *blow*
  1831   shall I *blow* off my steam?

  4966   *fever* has not abated at. . . .; the daily average of death is. . . . (the numbers 4960-4971 related to *fever*)

A new note was struck by No. 9091 — I shall come by railroad. From the worldwide gazetteer section of Part II come, for example, 3065 — Aalborg and 5695 — Zuyder Zee.

Watson's compilation had included some arbitrary, easily recognised signs. On the single mast, the upper pair of semaphore arms in a straight diagonal position, 45° and 135°, signified 'Yes', those on the lowest pair in the same position, 'No'. The upper pair, horizontal and 45° represented 'Look-out'. The operators got to know these so well that when twin masts were introduced, 'Yes' and 'No' were signalled in the same way, by the lower pair of arms on the S. mast, and the lower pair on the N. mast respectively. 'Look-out' was indicated by the same old setting on the lower, S. arms. The operators called the arms 'stretchers'.

In 1830 occurred an incident in which one of the operators at Holyhead became impatient with another further along the line who seemed to be slow to respond. In exasperation the first man signalled in the current code, 'You are stupid!' (8-356 7-185). At once, or at any rate as fast as a person could look out a group in the book, came the retort, 'You are dismissed!' (8-356 3-227). Unfortunately, the overseer happened to be at the receiving end, and so discipline was kept. Telegraphing was so rapid that when a message of, say, nine groups was being transmitted, the first group would be in Liverpool while the ninth was leaving Holyhead.

There was a constant lookout by day. The telegraph provided reassurance to mariners who knew that lifeboats and steamers would be at hand as soon as 'ship in danger' was telegraphed. Its value to owners of outward-bound vessels was that it told them of the force and direction of winds in the open sea before they left the river anchorage.

With the new system, the procedure remained much the same, but the service, if anything, improved. Cornish's *Stranger's Guide through Liverpool*, 5th edn. (1847), referred to the office in Chapel Street, Liverpool, where the operating room was a small square room at the top of the building. In the centre was a winch on whose axle were wheels like pulley sheaves, of different sizes, to which were fixed a halyard attached to each signal arm. By one revolution of the winch all six arms could be set in motion, each arm arriving at its proper position at the same instant. The best time for visitors to see the telegraph at work was a few minutes before 1 pm 'as its extraordinary powers may then be witnessed in perfection'.

A pamphlet entitled *Who Were the Inventors of Telegraphs?* by Morris Griffith (of the South Stack, Holyhead) (1888), contains a good account of the Holyhead telegraph and has a pleasing woodcut of Llaneilian station featuring all the semaphore arms in action, and the mast displaying a flag to one of the four vessels in the offing. It also contains the names of the first lot of operators. Morris Jones, one of the Holyhead operators, was an extraordinary man who believed that one day ships would be controlled without crews.

A *Guide to the Beauties of Anglesey and Carnarvonshire*, published by J. Swinnerton of Macclesfield and undated (but probably about 1830), describes in glowing terms a visit to Puffin Island and Watson's 'newly invented telegraph' there. It is an excellent commentary on the original method of signalling on the line, but the author does not reveal what sort of an adventure he had to reach the island!

The guide tells us:

> Every accommodation his dominions can afford will be cheerfully accorded by 'the king' of the island, who is placed there to superintend the working of Lieutenant Watson's newly invented telegraph, and who will bestow every information relative to the capabilities of that surprising machine which the most inquisitive can desire. As many persons, however, might feel some degree of delicacy in addressing an entire stranger, and deprecate the questioning of one in his situation, whose whole thoughts, as well as hands and eyes, ought to be intent upon the due execution of the duties of his office, we shall take upon ourselves to furnish them with the theory of the science, not doubting that they will have an opportunity of observing it reduced to practice before they will have been many minutes on shore.

The writer then describes the construction of the original telegraph and its working:

> Each of the three pairs of arms is capable of assuming six positions, with only one arm extended at once, viz, an oblique inclination upwards at an angle of forty-five degrees, a horizontal position, and an oblique inclination downwards. Three other positions are assigned for each pair, which require both the arms to be extended at once. This telegraph is adapted to the numerical system; that is, all words, names, sentences, etc. of most frequent occurrence, are expressed by numbers or figures, previously arranged in a printed vocabulary. It is also capable of *spelling*, when a proper name or uncommon word occurs. The first twenty-six numbers signify the letters of the alphabet, the next thirty-two are appropriated to the points of the compass, the next one hundred and twenty-five to portions of time, from a second to two years, including the hours of the day, the day of the month, and the month of the year. The numbers beyond these to the extent of more than seven thousand are appropriated to a list of merchants' names, seaports, towns and countries, and the names of all the vessels frequenting the port of Liverpool in succession. These words are divided into nine classes, of nine hundred and ninety-nine each.
>
> Each pair of arms is capable of nine positions, and to these positions the figures from one to nine are determinately affixed; the uppermost pair of arms indicating hundreds, the middle pair tens, and lowest pair units. Thus suppose one of the upper arms to be placed so as to indicate number 2, the middle arm 5, and lower arm 3, the number is 253; or suppose the upper pair to indicate 7, the middle pair 8, and the lower pair 9, the number would be 789. But as the telegraph will only express three figures at a time the number indicated cannot be above nine hundred and ninety-nine, without the worker exhibiting the number of the class to which it belongs.

The guide explains that if the message requires more than one sign, the first is repeated before the second is made. As the operator is immediately below the arms of the telegraph, in the observatory, and cannot therefore see them move, he has no difficulty in operating with precision.

> At the end of each halliard is an iron weight, which just balances the arms; and in this weight is a bolt which fits into holes made in the mast at those exact points where the halliards should be pulled to, in order to raise the arms to the required positions.

The reader learns that a short message has travelled the length of the line, and an answer returned, in 53 sec. Nine different coloured flags are used to communicate with vessels, each of which represents a number or figure from one to nine. The flags are generally hoisted on the main-mast of the ship, the foremast being appropriated to the class flags.

A diverting account in *Hicklin's Illustrated Hand Book of Llandudno* (1858) tells of a visit to the Great Orme.

> ....a mile from the town brings you to the telegraph station, where the summit of the mountain is covered with mossy grass, and the velvet-like softness of nature's green carpet.... invites to a pleasant rest, and a leisurely enjoyment of the wonderfully grand views which this elevated position commands; the highest point of the promontory being 750 feet above the level of

the sea... [A detailed description of the telegraph follows.] The keeper of the station, Mr. Jones, is remarkably courteous and intelligent; not only affording ready information to every enquirer, but kindly permitting visitors to watch his telegraphic operations, which at convenient opportunities he is always willing to explain; and to take a peep through his powerful telescopes at the sublime scenery of the district....The solidity of the building does not more readily attract the stranger's notice than the domestic comfort and cleanliness of his dwelling, which are most creditable to the industrious management of his wife and family, every one of whom is also skilled in the use of the telegraph. (Mr. and Mrs. Jones have set apart a comfortably furnished room, where tea is provided at short notice, and a supply of good lemonade, soda-water and other beverages of the temperance class. Much in demand in summer by picnic parties).

By the time this idyllic description appeared, the days of the semaphore were numbered. A cable was laid at sea between Holyhead and the Great Orme and carried its first message on 9 August 1859. It was extended to Voel Nant on 5 October 1859 and thence to Hilbre in 1860. But the storms of 25-26 October 1859 so damaged parts of the cable that it was replaced by a wire overland and the semaphore was retained temporarily as a standby. It was in these storms that the iron ship *Royal Charter* bound from Melbourne to Liverpool was wrecked in Moelfre Bay with much loss of life.

A log book of Point Lynas station records its final message on 26 November 1860 but it may not have been the very last to be semaphored over the whole line.

A final word about this notable telegraph. Its varied and picturesque course well repays the researcher not only because of the number of surviving stations but also because, unlike the route of the London-Portsmouth line, there are few trees to obscure the fine views.

Plate 35: Liverpool waterfront, 1859, showing telegraph tower centre right.

Plate 36: Ruined telegraph station on Puffin Island off Anglesey.

Plate 39: Llysfaen station today.

Plate 40: Point Lynas station today.

Plate 37: Hilbre Island telegraph station house.

Plate 38: Hilbre Island telegraph station observation room.

Plate 41: Great Orme, Llandudno, telegraph station.

Plate 42: Operating room of Watson's telegraph, Southwark.

Plate 43: The fire at Topping's Wharf, Southwark, 19 August 1843. Watson's telegraph is ablaze at centre left.

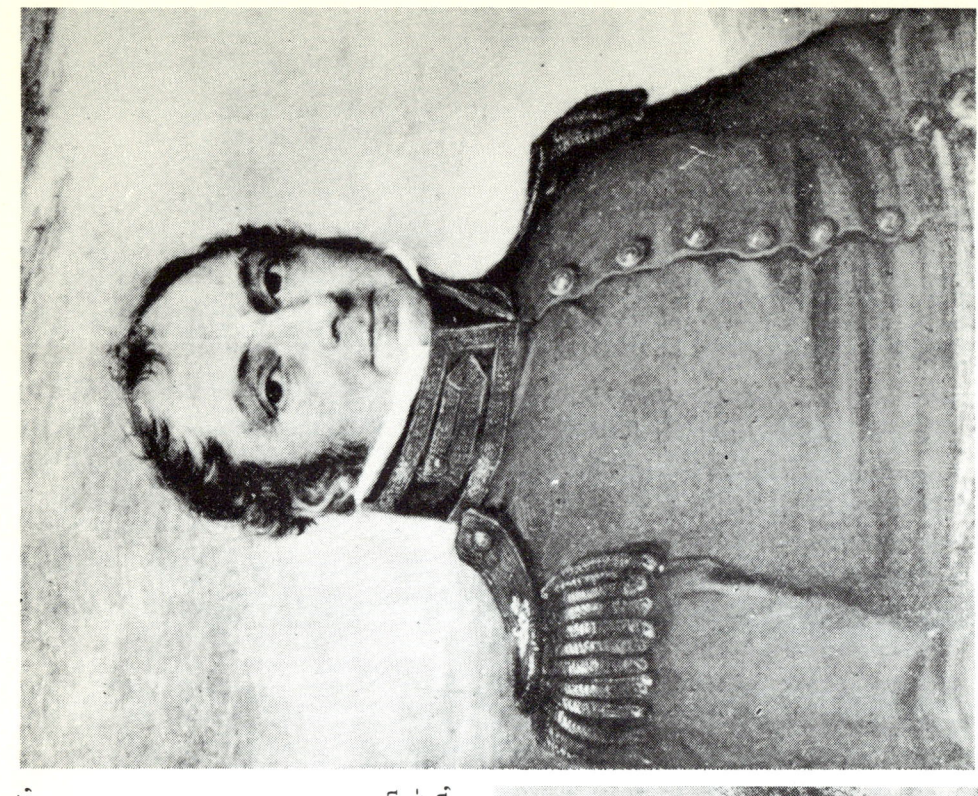

Plate 46: General Sir Charles Pasley, K.C.B.

Plate 44, left: Alderney, surviving telegraph tower.

Plate 45, below: Woolwich church tower, with Gamble's radiated telegraph, 1789.

CHAPTER VI

Watson's Other Ventures

As early as 1836 Watson was interested in other telegraphic ventures. In its November 1836 issue the *Railway Magazine* quoted from the *Liverpool Mercury* as follows:

> We understand that it is possible that a telegraph will be constructed along the Grand Junction and the London & Birmingham Railway, for the purpose of communicating commercial information and information connected with the management of the railway from London to Liverpool and Liverpool to London. The plan of such a telegraph has been formed by Lieutenant Watson, of the Liverpool and Holyhead telegraph; and it is, therefore, not unlikely that in three or four years we shall not only be able to travel to London in ten or twelve hours, but to receive information from London in as many minutes.

Such a scheme had been put forward, possibly by Watson, two years before. The *Mechanics Magazine* for 27 September 1834, quotes a suggestion in the *Liverpool Times* for a telegraph alongside the same railways, which had been incorporated only in 1833. The straightness and level character of a railway, it was believed, might be used 'with almost as much advantage as a chain of coasts and headlands'. Nothing came of the proposals, although, as has been said, Watson was concerned with a plan for a telegraph alongside the Liverpool & Manchester Railway. Incidentally, an early illustration of a London & Birmingham train shows a telegraph in the background, but of unmistakable French type.

It is tempting to link Watson also with a plan in 1836 for a telegraph to transmit stock exchange intelligence between London and Paris, but there is no firm evidence in support. (This line was to have had nine stations in England, including St. George's Fields and Folkestone, and 14 in France, the latter presumably independent of the Chappe telegraph. The estimated time of transmission was 1½ hours, from which it must be conjectured that one or two telegraph ships anchored in the Channel were proposed.)

At the meeting of the British Association for the Advancement of Science in Liverpool on 11 September, 1837, Watson began a paper on telegraphs, but, according to the *Railway Magazine*, 'was interrupted in consequence of his being unable to lay his plan before the section, because of his intention to patent it'.

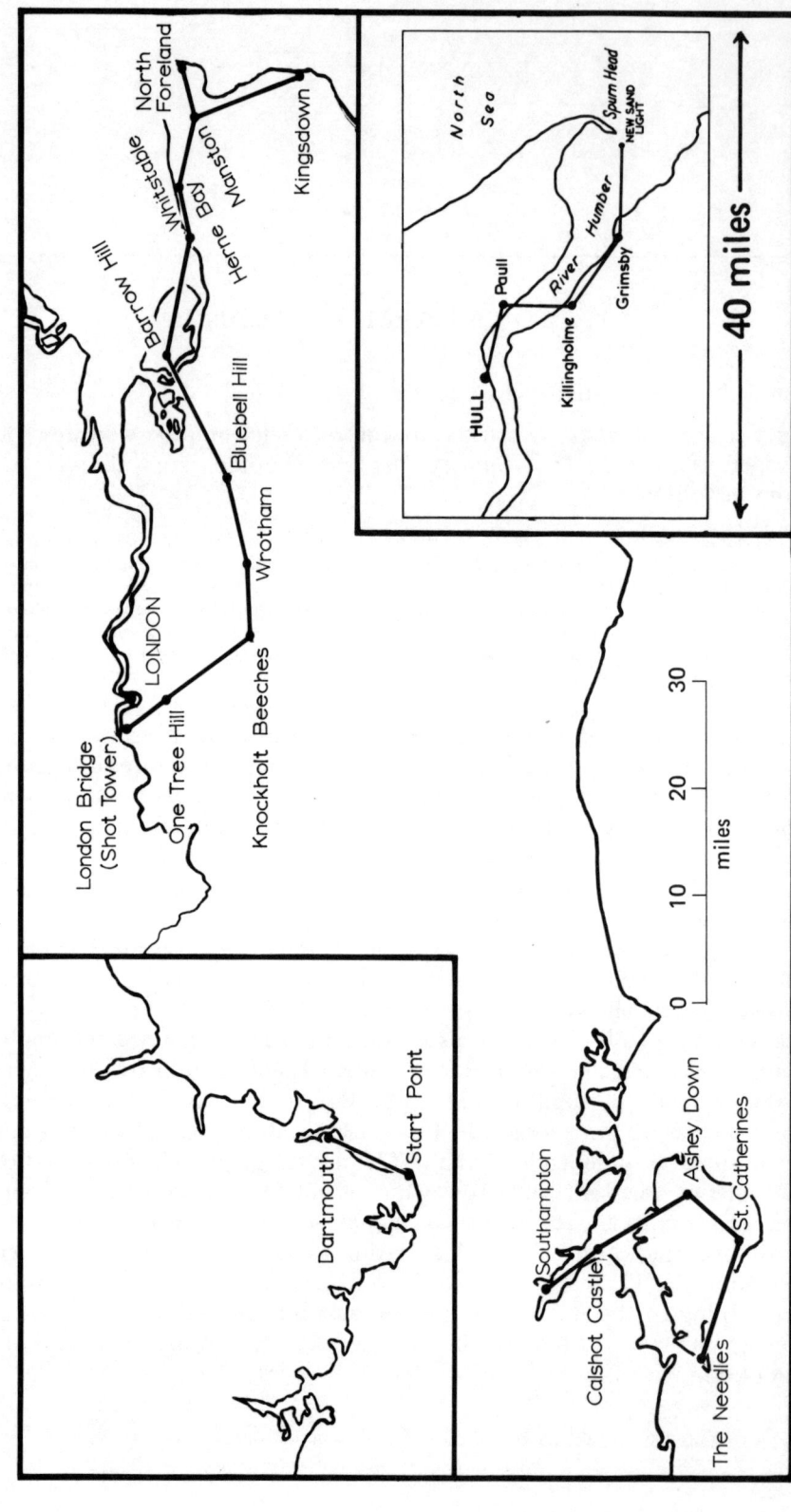

17. Watson's telegraphs, Southern England and Hull-Spurn.

> We have been at the Liverpool [telegraph] office, [continued the report] and were exceedingly pleased with the system and efficiency of his plans. A young gentleman, we believe a son of Mr. Watson, displayed to us and other gentlemen the whole process, with a clearness and urbanity that reflect great credit on him.

The 'process' was undoubtedly the two-mast telegraph subsequently adopted on the Holyhead line and also used by Watson for later lines.

The Hull Line

Impressed by the satisfactory working of the Holyhead line, the Hull Incorporated Chamber of Commerce & Shipping decided in April 1838 to pay Watson to survey and estimate the cost of a line between Hull and Spurn Head. He carried out his task in March 1839 and laid his report before the Chamber on 5 April. His plan was adopted and the Chamber sought the co-operation of the Corporation of Trinity House of Hull and the Commissioners of Pilots, in particular about using the New Sand and Spurn lights. At the monthly meeting of the Chamber on 4 June 1839, the Chairman announced that he did not intend to seek a Treasury grant until the line was ready, and then only for support in working it.

The *Hull Advertiser* reported in its 5 July, 1839 issue:

> A meeting of merchants and shipowners convened by the Chamber of Commerce, was held on Saturday last, at the George Inn for the purpose of devising means for bringing into immediate operation a line of telegraphs from Spurn to Hull, when a committee was appointed to carry that measure into effect. Lieut. Watson, of Liverpool, whose experience in telegraphs is well known, is expected here in a few days; under his superintendence the necessary work will be commenced forthwith, and it is anticipated that the telegraphs will be in use before the winter: vessels with the necessary signals, at a distance from ten to twenty miles from Spurn, will then be reported at Hull two minutes after arriving within sight. The Corporation of the Trinity-House have very handsomely offered a donation of £100, and the Corporation of the Dock Company £25 towards the necessary expenses, and other public bodies will, it is expected, follow their example. The plan will obviously be productive of great benefit, and might be extended along the coast. We anticipate that by means of the railroads now in formation a telegraphic communciation between Hull and Liverpool might be advantageously projected.

Work went ahead fairly rapidly. The same journal announced on 2 August that Watson had reported the results of 25 days' observations three times a day between the Hull terminus (the Pilot Office on the corner of Nelson and Queen Streets), Paull and Killingholme.

The line was opened towards the end of September 1939. From *Hull* it followed the Yorkshire shore of the Humber to *Paull* and then crossed to *Killingholme* and *Grimsby* on the Lincolnshire side, terminating at the *New Sand Light* off the Spurn. Watson was sole superintendent and his was the recognised code. The number 2,200, incidentally, represented the Russian national flag and 100 numbers were reserved for Russian ships, of which several were soon on the books.

It soon proved its value. The *Hull Advertiser* reported on 4 October, 1839:

> Yesterday afternoon the *Eclipse* steamer, which left on Wednesday evening with passengers for Newcastle, was reported to be off the Spurn, with loss of mainmast and chimney and with paddles damaged, having been run foul of by a vessel off Dimlington on Wednesday evening. This intelligence arriving just before the packets started for Grimsby, Lieut. Watson gave the information immediately to Captain Waterland, who promised to go to the assistance of the *Eclipse* after landing his passengers. As this vessel was not in the Hull list, she would not be known at the Floating Light until her name could be read, otherwise the intelligence might have been communicated perhaps two hours earlier; however, this circumstance shows how desirable it is that vessels with passengers should not only have signals to represent their number, but also to communicate to their agents what assistance they may stand in need of, as by prompt information both lives and property may be saved.

The report laid before the Chamber at its meeting on 28 February 1840 concluded:

> This great modern improvement, only possessed at present by one other port in the kingdom, is therefore now fairly before the public, and it only remains with the merchants and ship-owners to do their duty for insuring its successful continuance.

The next report waxed just as enthusiastic. The directors, it said, had cordially supported Watson's application to the Admiralty for a more extensive use of his Code, and the Hull enterprise had led, through Watson's perseverance to the formation of

> an important Company in London, which is erecting stations at the principal headlands round the kingdom, from all of which places, combined with cheap postage, intelligence respecting shipping will be communicated at trifling cost, of the deepest importance to those concerned, but of which they have hitherto been deprived.

Of the 'important Company' more anon. The reference to cheap postage is to the penny post which Rowland Hill introduced in 1840, a facility which, as we shall see, made possible some of Watson's subsequent ventures.

Apparently the Hull Chamber voted £10 a year to the telegraph, but the Report for 1845-46 implies that the merchants and shipowners were not warm enough in support. The directors had some success in raising extra subscriptions, but lamented the apathy.

Unfortunately no prints showing the Hull-Spurn telegraph have so far come to light. *The Handbook of Communication by Telegraph* (1842) (see page 222) confirms that it was of the imposing two-mast variety. The line was about 25 miles in length and it is on record that even before it was in regular service messages were sent from Hull to the light vessel and an answer returned in nine minutes, the *Victoria* being the first vessel reported while the operators were being practised.

The history of the line was apparently uneventful, but to judge from later Reports of the Chamber, it held its own, under the title of the Hull & East Coast Marine Telegraph Association.

The 1852-53 Report states that the new dock works at Grimsby had necessitated the erection 'at considerable expense' of a new station on the Grimsby Dock Company's land.

By 1857 the management of the Association had devolved on the Secretary, who soon arranged with the Electric Telegraph Company to transmit intelligence from Hull to Grimsby over its new wire, leaving only the Grimsby-Spurn section to the semaphore. The arrangement did not work well because the semaphore station and the electric telegraph office at Grimsby were far apart. The upshot was that the wire was extended to a new station at Cleethorpes, better placed than the Spurn semaphore to observe incoming vessels.

The Hull-Spurn telegraph, then, predeceased the Liverpool-Holyhead by two or three years but certainly gave much longer service than the other Watson lines which we are about to discuss.

Flamborough, North Foreland, Bristol and Norwich

While the Hull telegraph was in its early stages, Watson was maturing his other plans. On 19 March 1840, Trinity House wrote that it was willing to give him every assistance to erect signal stations on its land at Flamborough Head and the North Foreland, so long as he did not interfere with the lighthouses. It did not possess sufficient accommodation, possibly for the men required, to allow a similar request in respect of Lowestoft. A station at Orford Ness, further down the coast, appears on the map Watson subsequently published.

By the end of 1840 Watson was operating signal stations at both Flamborough and the North Foreland, reporting the passage of ships by the new penny post. He was erecting isolated stations on other prominent headlands. More ambitious were his continuous telegraph lines in progress to link Start Point with Dartmouth, the Isle of Wight with Southampton and The Downs with London.

Was he also interested in the Clyde and Bristol Channel? So reported the *Shipping & Mercantile Gazette* in its 2 October 1840 issue, but there is no evidence that a Clyde estuary signal station or telegraph materialised. Bristol remains an enigma. The *Railway Magazine* of 31 August 1839 quoted the *Bristol Mirror* as reporting that the City Council was considering 'the propriety of establishing a line of telegraphs in the [Bristol] Channel, in contemplation of its becoming the station for the South American and West Indian steamers'. It would have been strange if Bristol, with its traditions, merchant venturers and Brunel's broad gauge railway to London begun, had been content to lag behind Liverpool, Hull or Southampton in means of shipping intelligence.

In a letter to the *Nautical Magazine*, vol. I, (1842), 'Semaphore' wrote that 'the telegraph at Durdham Downs, near Bristol, which has been several years in practice, I believe, confines its intelligence solely to the steam vessels belonging to the companies of that city.' No other clues have yet come to hand. A station on Durdham Downs might well have communicated with Steep Holm or Flat Holm, and one can fancy other likely sites overlooking the Bristol Channel, but all is conjecture and the annals of this great port of the West, rich in other material, are apparently mute on the point.

It is said that Watson was concerned in a Norwich-Yarmouth project — the third since 1803 — but research has so far proved fruitless.

London to the South Foreland

Watson lived in Liverpool while he was connected with the Holyhead line. In 1834 his address was 3 Peel Street, Toxteth Park, and in 1837, 29 Hope Street. The telegraph office in the city was at 8 Chapel Street. By 1839 he had moved down Hope Street to No. 59 and was described as 'gentleman', his superintendence of the Holyhead line having by then ceased.

In late 1839 or early 1840 Watson moved to London, where his first private address seems to have been 58 Gordon Square. His correspondence with Lloyds in September 1840 was from 282 Regent Street but by 1841 he was operating from an office at 83 Cornhill. 'Watson's General Telegraph Association' had been formed with a managing committee composed of Wm. F. Black, M. Coventry, E. R. Foster, John Lindsay, Jr., Henry Wrench and T. C. Simmons. (Lindsay may have been the urbane young man at the Liverpool office.) Sometimes the title of the undertaking, of which Watson was superintendent, and Samuel Remmington secretary, is given as the Commercial Telegraph Association, at other times as the Shipping Telegraphic Association or the General Coast Telegraph.

The association's advertisement in the *Shipping & Mercantile Gazette* of 2 November 1841, reads that 'vessels of all nations entered in the Telegraph List will be reported off any of the stations, upon payment of 20s. each per annum' and that the undertaking 'has been established to afford the means of communication with vessels passing the principal headlands round the coast, being an extension of the system which has proved so much of value at Liverpool and Hull'.

In 1841 Watson was negotiating for sites for the stations between London and the Downs, a venture rivalling if not surpassing in importance the Liverpool – Holyhead line.

To avoid the fogs of the Lower Thames he chose an ingenious exit from London still further from the river than that of the Popham line to Chatham 25 years before. The following extract from the 4 January 1842 issue of the *Shipping & Mercantile Gazette* well summarises the progress:

> Fourteen years' trial of this admirable system on the line between Liverpool and Holyhead has fully established not merely the mechanical efficiency but the high commercial importance of this mode of intercourse. . .On every headland along this stretch of coast [from the Downs to Sheppey] a telegraph station has been established; and we are pleased to state that the liberal concession or co-operation of the Admiralty or the Trinity Board was never wanting to facilitate the completion of Mr. Watson's line. . .Keeping nearly parallel with the line of coast from the South Foreland to the Nore, the telegraph stations are then, very judiciously, diverted from the valley of the River Thames, in order to escape its frequent fogs, and are transferred to the higher range of ground to the south; the last of the intermediate stations, at Forest Hill, near Sydenham, communicating with the terminus on the Shot Tower at London Bridge. . .Even when the Thames fogs spread rather wide from the river, the railway communication between Sydenham and the Metropolis will prevent any serious interruption to the course of correspondence.

The communication referred to was that of the London & Croydon Railway, opened on 1 June 1839, whose station at Dartmouth Arms (Forest Hill) was reasonably near One Tree Hill, the 'Forest Hill' site.

The precise course of the telegraph was from the *Shot Tower* at Topping's Wharf, Southwark, to *One Tree Hill, Knockholt Beeches, Wrotham Hill, Bluebell Hill, Chatham* (so named), *Barrow Hill, Whitstable, Herne Bay, Manston* (with possible branch to the *North Foreland*) and *the Downs* (probably between Kingsdown and the South Foreland).

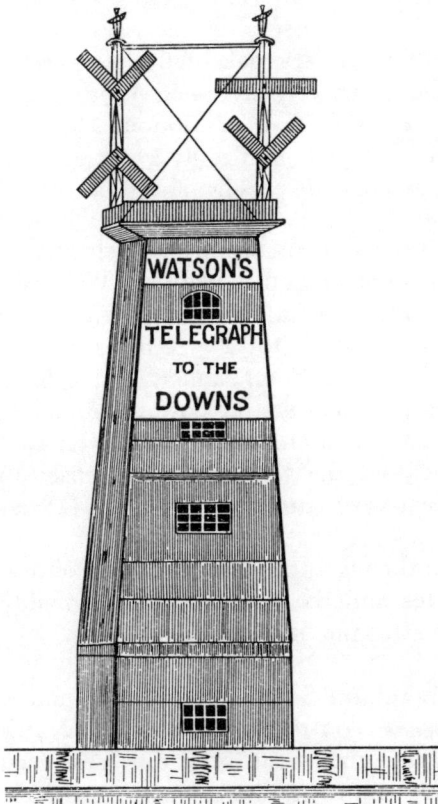

18. Sketch of Watson's telegraph on Shot Tower at Topping's Wharf, Southwark.

The wooden *Shot Tower* stood at the entrance to Duke Street, leading from London Bridge down to Tooley Street by the side of the approach to London Bridge station. According to Thornbury's *Old and New London*, it was completed in 1808, although a water colour in the Whitworth Institute, Manchester, shows it being built in 1796. It was no longer in use as a shot tower when Watson came on the scene.

Watson had the two masts of his new telegraph, each 20 ft. apart and 50 ft. high, erected on the top of the tower which was apparently about 100 ft. in height. Fortunately there is a contemporary engraving of the apparatus which shows it to have been similar to that of the remodelled Holyhead line. The masts, hollow columns of wrought iron, were cross-braced. Each mast carried two pairs of arms, one pair on each side of the mast, and each arm could incline upwards or downwards, point horizontally or drop down into the groove of the mast to be invisible (signifying out of operation, or indicating a cipher or ciphers in hundreds of thousands). A companion engraving of the operating room shows that each arm was worked by a wire rope, operated by a kind of windlass. Assuming the equipment was standard for all stations, Watson's telegraph must have been a most conspicuous landmark wherever it was erected and it is surprising that contemporary references to it are so meagre.

The Shot Tower and telegraph must have been one of the sights of London, exciting the interest and curiosity of both visitors and the army of City toilers crossing London Bridge. The toilers would have included the messengers who carried the reports from the telegraph tower to shipping offices in the City. Prominent at the top of the tower was the inscription: *Watson's Telegraph to the Downs: office 83 Cornhill.*

One Tree Hill is a 300 ft. rise forming part of the range at the southern end of which, at Sydenham, the Crystal Palace arose a decade later. It had long been a favourite resort. The ancient boundary oak of the Honour of Gloucester, in which the Manor of Peckham formerly lay, once stood on the summit. It gave rise to the alternative names of Oak of Honour Hill and Honor Oak Hill. The name One Tree Hill is not uncommon. It applied, as we have seen, to the hill at Hampstead before the shutter telegraph was erected there.

A committee was formed in 1897 to safeguard the hill from enclosure. It heard evidence from local inhabitants, some of them old. Walter S. Waghorn, born in 1842, recalled that when he was four his father went to live in the 'signalling cottage' on the hill. In this cottage, which had a garden round it, Waghorn senior lived rent free for 17 years. His son added that a local builder and brickmaker called Woodcock pulled down the cottage 'about 45 years ago.' Other inhabitants confirmed that there was a double-fronted cottage on the top, near the famous tree. One, aged 81, said that he used to help to work the signals. A Professor Skeat wrote in *Notes & Queries* (1909) that he remembered the telegraph at work.

No cottage, unfortunately, is visible in the views painted by T. A. Poulton about 1850. Watson only leased his sites and the stations were probably huts. Most would have been pulled down after the lines ceased working.

One Tree Hill is only four miles from London Bridge but 12 miles separate it from the next site, at *Knockholt Beeches,* on the S. slope of the North Downs. Little of the intervening ground had then been built over and even today there is a good view from One Tree Hill towards the North Downs. It is likely that Watson put his telegraph near the conspicuous clump of beeches at this prominent spot, at 753 ft. A field on the S. side of Brasted Lane, near the Waterworks, is sometimes called Telegraph Field. Bagshaw's *History, Gazetteer & Directory of Kent,* (1847), says that

> on the southern verge of the parish [Knockholt] are the remains of a telegraph formerly used for conveying intelligence between London and Dover; a little to the north of which is Knockholt Beeches.

Almost due E. of Knockholt Beeches, some eight miles away across the Darenth Gap, stood the next station, at *Wrotham.* Information is scanty, but the site was probably on the shoulder of Wrotham Hill, not far from the present television mast. From here there is a span of nine miles to *Bluebell*

Hill, midway between Maidstone and Chatham, looking over the Medway Valley and over 500 ft. up. Here also the exact site is not known, but it may have been near the Robin Hood Inn, at the end of Common Road.

Chatham, listed as the next site, has likewise defied identification. Possibly Chatham was the nearest post town. Dorland Hill, S. of the main road between Chatham and Rainham, or the rising ground behind Rainham are likely positions, maintaining the direction of the line towards Sheppey, where we are again on sure ground. The 100 ft. *Barrow Hill,* near Queenborough, one of the old shutter sites, was still owned by the Admiralty, with whom Watson made satisfactory arrangements in November 1841. He was required to fence the necessary ground and build at his own expense.

Barrow Hill and the Nore are too far apart for flag signalling to have been feasible. It is possible that Trinity House allowed Watson to erect a telegraph on the Nore light, or, more likely, he may have had a lookout post at Sheerness.

There seems to have been a second Watson station in Sheppey, at Harty, but the only evidence is cartographic. The Admiralty chart *Sheet II of the River Thames from Ramsgate to the Nore* (surveyed 1832-43 and published 1844) indicates telegraphs at both Barrow Hill and Harty by a two-mast symbol. Recent editions of the O.S. one-inch sheet No. 172 still show 'Telegraph' at Harty. An isolated house on the road to Harty Ferry marks the spot but it is not known whether it had a connection with the telegraph.

A station at Harty could have been short-circuited in clear weather, as with one or two of the Holyhead line stations. In any event the distance between Barrow Hill and Whitstable, the next point on the Kentish mainland, is not more that 12 miles, not exceptional on this line.

At *Whitstable,* the grounds of Tankerton Tower, or Castle, now the offices of the Whitstable Urban District Council, seem to have been the site, well placed for signalling to and from the offing and once occupied by a beacon.

We then come to *Herne Bay,* the best documented of all the stations on the line outside London. The Conyngham Papers in the Kent Archives Offices show that an agreement was sealed on 22 July, 1841, for a 21-year lease, from 25 March 1841, of 40 perches at Bishopstone Farm in the parish of Reculver. The farm was part of the estates of Elizabeth Dowager Marchioness of Conyngham. Named with Watson are James Hay and Erasmus Robert Foster, of London. Watson's solicitors are Messrs. Elderton and Phillott, of 25 Clements Lane. The annual rent is fixed at £5 and the land is required to be fenced.

The rent was a sore point with Watson. On 22 February 1841, he wrote to one Pilcher, the Marchioness's steward, with whom Thomas Skrymsher, landlord of the Pier Hotel, Herne Bay, had made contact on Watson's behalf:

> I beg to state that I have land for the same purpose from several noblemen and gentlemen who, considering the importance of the object to the Mercantile Marine and the public generally, have granted the accommodation at a mere nominal rent, and I have in no instance, on the line between the Downs and London even on most valuable land been asked more than £4 for ¼ of an acre.

The Bishopstone site has succumbed to coast erosion, the cliff edge now being further inland than a century ago. The station, generally called Reculver, though some way E. of the famous twin towers that were for long a noted seamark, was one of the stations in use before the whole line was opened; messages were then sent on by post.

Eight miles separate this station from the next, at *Manston*, where Watson selected Mount Pleasant, 178 ft. above sea level on the Thanet ridge and alongside the Canterbury — Ramsgate road. The site commands a fine view NW, and S. across the Stour estuary towards the South Foreland. The station may have been a junction, probably the only one on a Watson line, so as to communicate with both the North Foreland signal station and with the South Foreland, 12.5 miles to the south.

The exact site of the *South Foreland* terminal station is difficult to determine. Contemporary references call it 'The Downs', 'Deal' and even 'Dover.' But the fact that in one place it is named as the South Foreland, the discovery of a message from 'Kingsdown' and the existence on the 1871 six-inch O.S. of a site marked 'Semaphore (Disused)' on a 200 ft. hill overlooking the Channel one mile S. of Kingsdown, and now the site of a golf course, are useful clues. The O.S. position was also that of a former Admiralty coast signal station and was so shown on the 1819 O.S.

Watson's advertisements at the end of 1842 emphasised that he had the support of Government, the East India Company, Lloyds and Trinity House. He could report vessels in the Downs in London within 10 min. He announced that tugs of the Ship-Owners' Steam Towing Company were in communication with the stations 'between Deal and the Nore.'

The Admiralty did not disdain to make use of Watson's services. *The Handbook of Communication by Telegraph*, London, Henry Kent, Causton & Co., (1842), probably written or inspired by Watson, quotes a letter from Sidney Herbert, Secretary to the Admiralty, dated 23 May 1842:

> I am commanded by my Lords Commissioners of the Admiralty, to request you will send a message by your Telegraph to the Vice Admiral Commanding at Sheerness, namely — 'Stop *Rhadamanthus*.'

In the following month Queen Victoria thanked him for forwarding a report of the arrival of the King and Queen of the Belgians in *H.M.S. Ariel* off the North Foreland. On 20 August he reported that the Royal yacht *Royal George* had passed through the Downs that morning on her way from Portsmouth to Woolwich, and thereafter followed her progress.

The Admiralty paid Watson 10s. a message but could not be induced to make a grant of at least £1,200 a year which Watson hoped for.

The enterprise was brought to a spectacular and untimely end by the disastrous Tooley Street fire of Saturday 19 August 1843. Very early that morning fire broke out in Toppings Wharf and soon gained 'a most awful ascendancy.' For some time, the tower, although in the midst of danger, withstood the ravages, but at about 2.30 a.m. dense smoke began to issue from the windows and it caught fire shortly before three. As it was all wood covered with slates, it burned for some time, but about 3.30 it fell with a 'most tremendous crash.' Toppings Wharf and much other property perished and St. Olaves Church was badly damaged. Next Monday's newspapers discounted rumours that persons had been seen in the burning tower, as the building was untenanted except during the period of telegraphy.

Giving evidence before the Metropolitan Railway Commissioners, 1846, Alderman J. Humphery, M.P., who was advocating laying the electric telegraph alongside the South Western Railway between Southampton and London Bridge (to which it was proposed to extend the railway from Nine Elms), recalled that Watson's telegraph 'was burnt down about four years ago, and they have not had one since.' Humphery also stated that Tooley Street used to handle all the south coast trade between Lands End and Dover, adding: 'I have known vessels to remain that could not get round the Foreland a fortnight, three weeks or a month, on account of the north-east winds.'

Humphery's statement is proof enough that Watson must have given a useful service. It is therefore surprising that the loss of the London terminus, albeit costly, was apparently sufficient to terminate the venture. It held no records, and communications might have been kept going by using One Tree Hill as a temporary terminus and sending messages to and from there by railway until a new terminus could be found. Perhaps the calamity was the last straw added to a burden of inadequate support.

Southampton to the Isle of Wight

Southampton began to emerge as an ocean port in 1836, when the Southampton Dock Company was incorporated and bought 216 acres adjoining the Town Quay for docks. The London & Southampton Railway, authorised in 1834 and opened throughout on 11 May, 1840, stimulated the development of the port, which in 1843 replaced Falmouth as the chief foreign packet station. The newly formed Royal Mail Steam Packet Company secured the West Indies mail contract and was based on Southampton. In the summer and autumn of 1841, a year before the docks were completed, it concentrated a large fleet there.

Watson was alive to the possibilities. On 2 November 1841, he wrote to the Southampton Dock Company of his intention to construct a telegraph between the Isle of Wight and Southampton, and asked for suggestions. Later that month Watson and another member of the Association called on the Dock Company Secretary to negotiate for the erection of the telegraph on the company's premises. The roof of the Dock House or one of the sheds was considered suitable. It was arranged that Watson should pay for the installation and take a lease of seven, 14 or 21 years, or be admitted on sufferance. One man would be required. He would be paid direct or the Dock Company would receive gratis all telegraphic information about ships approaching the port.

Early in December 1841 the Chairman of the Dock Company recorded his view that the W. end of the top of the Dock House was the most suitable place for the telegraph, which he thought would bring considerable advantages not only to the company but also to commerce in general. Overlooking both Liverpool and Hull, he concluded that it would give Southampton a feature not possessed by any other dock establishment in the kingdom.

By 16 December he and Watson had met and Watson pronounced the Dock House roof to be the best site in Southampton. In January 1841 he advertised that the line was in progress and that intelligence would be sent on from Southampton to London by train if required. The Association's map shows that he engaged a cutter to cruise off the S. of the Isle of Wight, as also in the Downs.

The line began at the *Needles* and ran by way of *St. Catherine's, Ashey Down* and *Calshot Castle,* to *Southampton.* The three Isle of Wight sites — though their precise locations cannot be established — are superb vantage points. The downs near the Needles rise to 463 ft., St. Catherine's Hill rises to 780 ft. and commands a magnificent view of the Channel and over all the island to the north. Ashey Down, conspicuous by its sea mark, and 429 ft. high, was a coast signal site in Napoleonic war days. The only likely position at Calshot was on or alongside the Tudor castle. Unfortunately no views showing the stations have come to light. The distances between the stations are approximately: the Needles — St. Catherine's 13.5 miles; St. Catherine's — Ashey Down, 8.5 miles; Ashey Down — Calshot, 10.5 miles; Calshot — Southampton, 7 miles.

Dartmouth to Start Point

If it is not known how long the Southampton venture lasted, the Dartmouth — Start Point line is still more of a disappointment. It is not even certain whether it was completed, let alone went into service. Mr. Percy Russell, a local historian of repute, has failed to find any record of it. It may have linked Compass Hill with Start Point; the intervening eight miles are flat and no intermediate station would have been needed.

Watson also erected signal stations at Scrabster, the port of Thurso; at Skirsa Head, N. of Wick; and at Peterhead, but there was no continuous telegraph at those places.

The London directories know neither Watson nor his association after 1843. Did he retire and take no further direct interest in telegraphy? All we do know is that he lived for 20 more years, dying at the age of 64 on 27 February 1865, at the village of Monks Eleigh, in Suffolk, where he is buried. No member of his family was present when he died, although he had a daughter Frances, to whom in 1849 he inscribed the copy of his Code, Watson's *Code of Signals,* 6th edn. (1842), now in the British Museum, and, as stated, he may also have had a son.

Duckham's Falmouth Project

Prompted perhaps by news of Watson's enterprise, Alfred B. Duckham of Falmouth proposed in 1842 to install a telegraph between the port and the Lizard. Lloyds, refusing him a grant, asked their Falmouth agent to contradict part of an announcement by Duckham in *The Times,* 27 December 1842, stating that

> in addition to the patronage so liberally bestowed by the Admiralty, Ordnance and Lloyds on Mr. Duckham's proposed line of telegraphs to the Lizard, for communicating the vessels of all nations, the Postmaster General has been pleased to cause to be transmitted to him a handsome set of flags, to be used when in communication with merchant vessels having mails on board from foreign countries, or with Her Majesty's packets.

There is no record that Duckham's project was carried out. The West Indian and South American packets were transferred from Falmouth to Southampton in 1843, an event which would have weakened Duckham's claim for official backing, even though Falmouth long remained a first class port. No doubt the change in Falmouth's fortunes, and possibly also in Watson's, were grounds enough for Watson's lack of interest there.

CHAPTER VII

Channel Islands

In view of the position and strategic importance of the Channel Islands it is not surprising that in Britain's wars with Revolutionary and Napoleonic France there was need for early warning of French naval movements in St. Malo Roads. The first step seems to have been taken by the Jersey Chamber of Commerce, whose suggestion was approved by the Admiralty in London and the Island Defence Committee. Rear-Admiral d'Auvergne, Prince of Bouillon and C.-in-C. of the Jersey naval squadron, proposed to establish flag-and-ball signal stations around Jersey. An Act of the States of Jersey for their erection was passed on 18 January, 1798. Stations were built at *St. Helier, St. John's, Noirmont, La Moye* and *Tât de Geon (Trinity)*. The system was discontinued in May 1802 with the Peace of Amiens but re-established on 22 May 1803, when war was resumed with France.

G. R. Balleine records the next developments in his *A Bibliographical Dictionary of Jersey*.

> In 1806 Lt. General Sir George Don was appointed Lt.-Governor of Jersey. Sworn in on 21 June, he at once showed extraordinary vigour. He established a signalling system by which, if the French fleet left St. Malo, news could be flashed from a look-out ship to Mont Orgueil Castle, and thence, via Grosnez [a headland on the NW coast of Jersey] and Sark, to Guernsey, where a British fleet was stationed. It was found that a message could be passed from Mont Orgueil to St. Peter Port in a quarter of an hour, using flags by day and blazing tar barrels by night.

It seems likely that General Don merely reorganised the existing stations and extended the service to Sark and Guernsey. This is borne out by a MS. report in the library of the Société Jersiaise in which Don states that at a trial at Grosnez on 14 August 1806, he distinctly saw the Union flag hoisted on Fort George, Guernsey, and also flags hoisted on Sark, some 14 miles distant.

Further proof that Don's plan was an extension of the old system is adduced by a letter to d'Auvergne dated 8 August 1809. He writes that soon after he took command in Jersey the Admiralty authorised him to build 10 signal stations round the coast of Jersey, to be commanded by naval Lieutenants. To avoid much expense, he had applied the balls in use for common signals to a telegraphic code, which answered the purpose round the island, but without a telegraphic communication to Sark, Guernsey and

Alderney it was 'almost impossible to apprize the Squadron of the Enemy's sailing' from Granville, Cancale, St. Malo, Havre de Grâce [Le Havre] and Cherbourg. This desirable object encouraged him to try many experiments for communication between Grosnez and Sark, 'which tho' not entirely successful, convinced me of its possibility.'

Eight of the stations were newly built under the supervision of Nicholas Vass in the style of the coast signal stations in England. An existing house on *La Montagne de la Ville* (Town Hill), St. Helier, was adapted and suitable rooms were fitted out at *Mont Orgueil*. In November 1806 Vass was instructed by Sir James Saumarez to build a signal house in Guernsey to link with Jersey – presumably that at *Jerbourg*. The Admiralty directed Vass to proceed as Saumarez wished. On 16 January 1807 the Navy Board approved the erection of a house at *Varde Noirmont*, [Les Vardes, Port Grat] Guernsey, and on 7 September that of houses at *Mont Hérault*, *Prévôté* and *Icart*. Vass also built a station on *Sark*, to Saumarez's direction. No houses were built to serve the signal posts at *Castle Cornet, Fort Saumarez* and *Fort George*, where the military had charge of signalling.

In 1806, the Printer to the States of Jersey produced a booklet entitled *Signals for the Better Communication between the Islands and the Squadron – Signals to be made by the Signal Posts to the Squadron without any regard to the common Island Signals*. He lists the ten signal stations, in clockwise order round the coast from St. Helier, as: *La Montagne de la Ville* (Fort Regent, St. Helier), *La Pointe de Noirmont, La Moye, Grosnez, Le Mont Mado, Le Bouley, St. Martin, Mont Orgueil, Verclut* and *Herket*. The signal flags are shown in colour and their significance is described. An explanation of the signals, with a drawing of the signal post, is given in a signal book published by the printer under Don's authority.

The limitations of the system, in particular for inter-island communication, must soon have been felt, but a remedy was forthcoming. It is perhaps best set down in a report by d'Auvergne on *Experiments made to establish a Telegraph Communication between the Islands of Jersey, Guernsey, Sark and Alderney* dated 1 September 1809.

> The distance for the Telegraphic Communications with the ordinary Machinery appeared almost too great to make our recurrence to that expedient much better than doubtful. But about a year ago a commercial gentleman of ingenious Mechanical Ability who had turned his mind that way, exhibited to me a contrivance he had imagined, for applying the principle of the French Telegraph, to communicate in a distinct and simple manner at great distances.

The ingenious gentleman was Peter Archer Mulgrave (1778-1847), whose origins are obscure. As he later went to Van Diemen's Land (see Chapter XXV) and applied the telegraph there, he seems to have had a special flair for insular telegraphy.

In 1809, d'Auvergne, by then C.-in-C. of the Squadron at the Guernsey station instructed Mulgrave to apply his telegraph at Grosnez, Sark and Jerbourg Point, Guernsey, with the approval of General Don and the equally energetic Lt.-General Sir John Doyle, Lieutenant-Governor of Guernsey. According to d'Auvergne, its application was 'attended with the most satisfactory success, with a distinct celerity almost inconceivable.' On 8 August 1809 Don wrote to d'Auvergne that he believed

> that the many official communications which have already been made between this Island and Guernsey, prove that this Telegraph will answer every military purpose, and should you be of the same opinion, I hope will have the goodness to recommend to the Lords of the Admiralty, the establishment of this Telegraph on the Chain of Posts abovementioned.

D'Auvergne lost no time in asking for Admiralty sanction to establish 'this expedient Machinery' in place of 'those more expensive ones' already used and sought 'such decision upon Sir John Doyle's application for the establishment of one in Alderney.'

Mulgrave was not overlooked. Remuneration was sought for his zeal and in 1809 he was appointed an Inspector of Telegraphs, with the same status as Roebuck, the Inspector of Admiralty telegraphs in England, and granted a salary of 15s. a day.

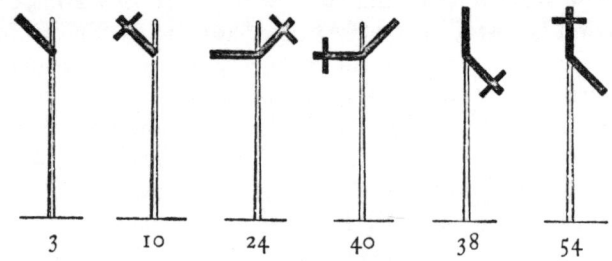

19. Mulgrave's Channel Islands Telegraph.

Mulgrave's telegraph had a mast with a pair of arms of deal board. The result was a species of semaphore very like Pasley's, except that one arm carried a cross piece at the end, doubling the power and giving the apparatus 56 positions. The Admiralty Library in London contains logs kept at the Sark station in 1809-14. Some have rough drawings of the masts and arms, and give the values of the arm positions. The masts originally had topmasts and yards, but these were removed. The Sark station was a house near Beauregard, inland from Havre Gosselin, and still standing; it has recently been extended. It is called appropriately Le Mât. The cross-trees remained, and the arms were worked by blocks attached to them, on the same principle as Macdonald's 'British Semaphore'. As the mast was a thick piece of timber, it would obscure the arm in the upright position, so that position was omitted. This reduced the number of available settings to 42. According to the Sark logs, there does not seem to have been any method in the allocation of the arm positions.

D'Auvergne's report gives a good example of the efficiency of the system. Transports had arrived with orders to embark three battalions of infantry from the islands for Portugal. As was the custom, they arrived for one division at Guernsey and for the other at Jersey. D'Auvergne was in Jersey at the time and used the new telegraph to send orders to Guernsey for a division to embark and its transports to sail for three hours in a given direction. The two battalions from Jersey were embarked separately and the

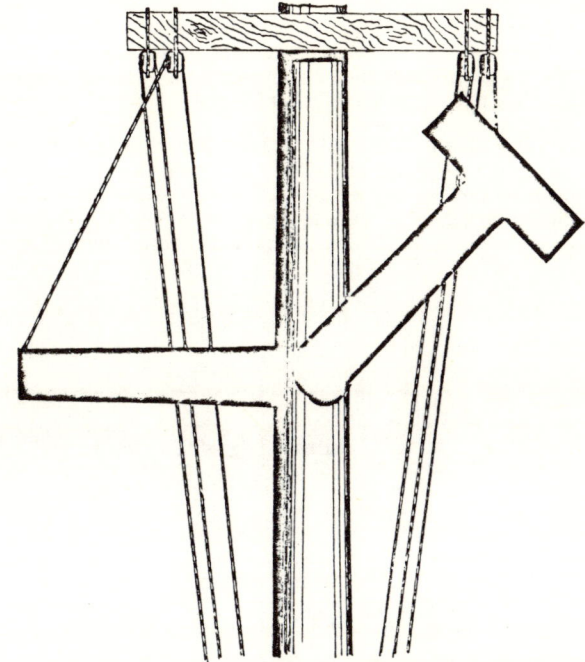

20. Sark Semaphore, 1809-14.

vessels met and proceeded in convoy. Without the telegraph, the operation would have taken days, not hours, and a favourable wind would have been lost.

John Stead's *A Picture of Jersey; or, Stranger's Companion Through that Island,* (1809) gives the complement of the Jersey stations as a Lieutenant, a Midshipman and two seamen, and has the following details of some of the stations:

Montagne de la Ville (Fort Regent): On this Hill is a Signal Station, which commands a most extensive View of the Coast of France, the Bay of St. Aubin's, etc.

Herket: Upon the Hill above [Pontac] is the Signal Station, called Arcot [sic], a neat convenient Dwelling, with a good Garden; a small, but excellent retreat, for a Weather-beaten Seaman; his Mast Yards and Rigging remain, and he is still able to show his Colours; his Birth [sic] is amply furnished, his supply of 'Fresh Provisions and Grog' is certain; he may enjoy all the Conveniences of the Shore and all the Amusements of the Sea, without the Dangers attending upon the Profession.

Verclut (St. Clement's): Passing by the Houses of Mr. Le Maistre and Mr. T. Falle, the signal Station of Verclute [sic] is discovered; the View from it of the Vallies beneath is enchanting.

Mont Orgueil: It is now occupied by the Lieutenant who commands the Signal Station erected on its Tower, and his attendants. It is also garrisoned by Troops and expected to be furnished with fourteen Days Provisions.

Rozel: Near the Manor House is a Signal-station, that communicates from Boulay Bay to Mont Orgueil. [The reference is to St Martin's.]

At Grosnez Signal-station a distinct Prospect may be obtained, in a clear Atmosphere, of the Islands of Guernsey and Sark, and the Ships of War may be discovered riding at their anchors in Guernsey Roads. From thence Intelligence is conveyed from one Island to another by Means of telegraphic Signals that are made from the different Stations.

There are ten Signal-posts erected on the most elevated and conspicuous Parts of the Coast of the Island; at each of which is a Lieutenant of the Navy, a Midshipman, and two Seamen. They form a regular Line of Communication around the Island, and are intended to give timely Notice of the Approach of an Enemy, British Ships of War, merchant Vessels, and Packets; and to alarm the Country in case of a menaced attack, for which Purpose they have Cannon, Fire Beacons, etc.

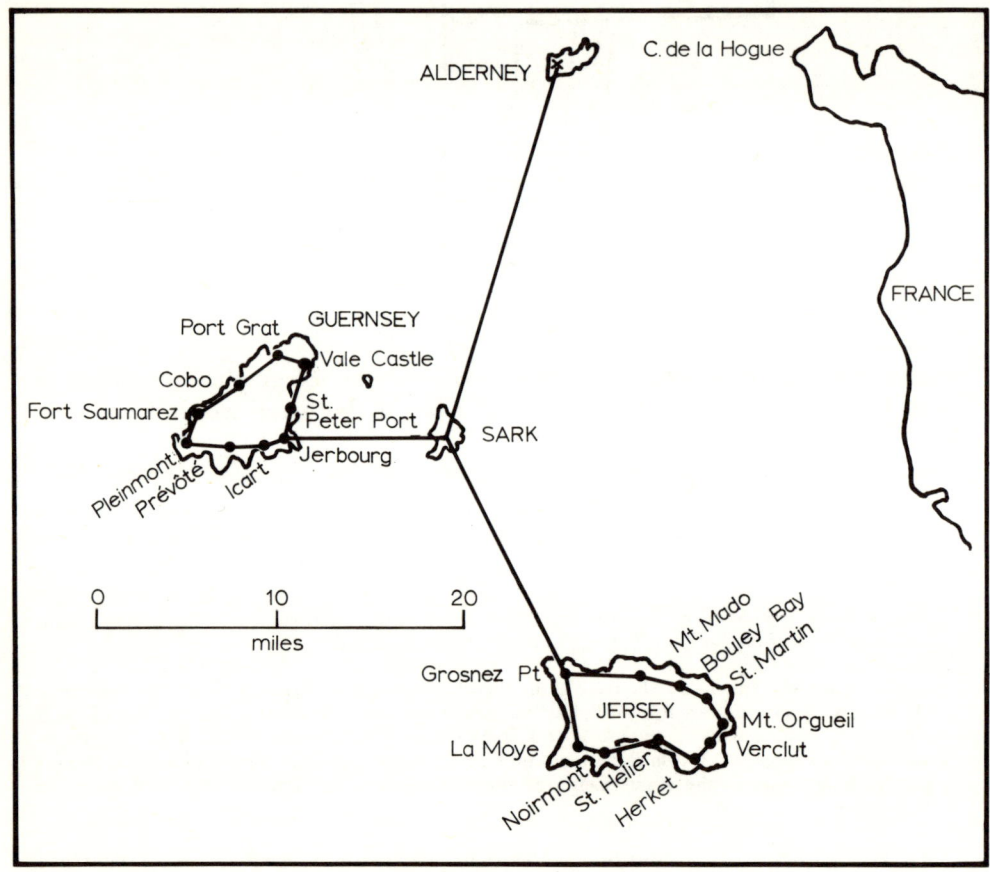

21. The Channel Islands telegraph routes.

A coastwise system was completed in Guernsey, as a counterpart to that of Jersey. A map in William Berry's *The History of the Island of Guernsey* (1815), shows stations as follows, in clockwise order from St. Peter Port: *Fort George* (half-mile S. of St. Peter Port), *Jerbourg, Icart, Prévôté,*

Pleinmont, Fort Saumarez, Cobo, Port Grat (Les Vardes) and *Vale Castle*. In an article entitled 'Jerbourg and Its Fortifications' in *Transactions of the Guernsey Society of Natural Science and Local Research for 1903*, G. T. Derrick stated that in 1801 a Captain White had charge of a signal station at Jerbourg, with quarters nearby. Becoming dilapidated, the building was replaced by a watch tower like others along the S. coast of Guernsey. But a map of 1832 shows a 'signal house' SW. of Doyle's Column at Jerbourg — the Germans removed the column in the 1940s.

Fort George may have been regarded as the hub of the Channel Islands semaphore system, as the following extract from a letter dated 25 August 1810, from General Don to an unnamed correspondent (presumably the C.-in-C., Guernsey) may indicate:

> ...the Duke of Bouillon and myself have written a joint letter to the Secretary of State soliciting him in the strongest manner to appoint Mr. Mulgrave to the Signal Station at Fort George agreeably to your wishes. He is certainly well qualified for the situation and as the project is in a great degree his own he will naturally exert himself to the utmost to secure success.

It is not clear whether the appointment was synonymous with, or extra to, that of Inspector of Telegraphs for all the islands.

The origin of the Alderney station is referred to in a letter from Doyle to d'Auvergne of 28 August 1809. Doyle writes that he requested Mulgrave to examine on the spot the feasibility of building a telegraph in Alderney. Mulgrave 'minutely inspected the place, accompanied by some of the most intelligent inhabitants and Pilots' and finally recommended a site on the Beacon heights, where a tower might answer the purpose of a residence for the officer and men employed. The building was apparently finished in 1811, for records of the Court of Alderney show that proceedings were taken that year against a farmer who had blocked the road to the site of the tower by digging a trench to deny the contractor access.

The Alderney station is in the SW. of the island, beyond the airfield, on the 280 ft. contour. It is still in good condition and inhabited. The bay below it is called Telegraph Bay. The station is a round tower, with three floors, of elongated Martello type. It is probable that it was the only Channel Islands station to be built new for the Mulgrave system, and that all others except le Mât were the flag stations modified or existing fortresses.

It is 21 miles between Sark and Alderney. That telegraphy could be regularly practised over such a distance speaks well for the simplicity and clarity of Mulgrave's apparatus and code.

A history of Alderney written in 1810 says:

> A Telegraph on a new construction has lately been erected here by the ingenious Mr. Mulgrave, which reflects great credit on his abilities; its object is to communicate with Guernsey and Jersey, through the medium of the island of Sark, where a repeating Telegraph is stationed. It has succeeded far beyond the most sanguine expectations and has already been found of great utility and benefit.

On 11 May 1814 the Channel Islands stations were ordered to be given up. On 25 May Mulgrave reported on their condition. His services were discontinued in a letter of 23 June from the Admiralty to the Navy Board.

The stations were ordered to be resumed by an Order dated 12 April 1815 and Mulgrave was reappointed as inspector. It was stated that a 60 ft. mast was to replace the former 50 ft. mast at Prévôté and that new large telescopes would be needed at Alderney, Sark, Grosnez and La Moye stations. The following stations were readopted: *Alderney, Sark, Les Vardes, Jerbourg, Mont Hérault, Icart, Prévôté, Fort George, Mont Mado, Bouley, Verclut, St. Martin's, Mont Orgueil, Herket, St. Helier, Noirmont, La Moye* and *Grosnez*.

On 5 May 1815 Mulgrave transmitted a letter from Doyle's secretary stating the terms on which the houses at Mont Hérault and Prévôté were again granted by the States and particulars as to terms for Sark and other stations.

The resumption was short-lived. On 13 July Barrow ordered the Navy Board to suspend the stations again. On 20 September Mulgrave again reported on them, stating that he had prevented repairs from being carried out to some of them on learning that the supply of stores had been temporarily suspended. His services were finally dispensed with on 4 August 1815.

Fort Regent had been built on Town Hill in 1804. There seems to be some doubt about the position of the Herket station. The name Herket looks like a corruption of (La) Hougette, behind Pontac, on St. Clements Bay. But M. B. Kavanagh, Flag Signalman at Fort Regent, in his monograph *Signal Stations in Jersey*, extracted from the *Bulletin of the Société Jersiaise*, 1970, places the station further inland on Mont Ubé, a 180-ft. hill two miles from St. Helier, just N. of the road to St. Clements, and a map on silk presented to General Don, and marking the signal stations, certainly indicates such an inland site.

It would appear that one or two stations continued in use for some time after. A report issued in February 1828 states that the La Moye station was lent to the States after 1815 for use as a signal station. That at Verclut was still occupied — the officer had added a room to the existing five — although the ground had been made over to the public. The Alderney station, also on Government land, had been let — correspondence in 1821 reveals that the inventory then included a mast with two arms and a yard. The report also refers to the stations at Bouley, Herket, Les Vardes, Icart and Jerbourg and adds that all other station houses had been returned to the Lords of the Manor.

Until 1888-9, when shipping intelligence began to be relayed by electric telegraph from Corbière Lighthouse to St. Helier, communication between La Moye, Noirmont and St. Helier, presumably a signal station on Fort Regent, was maintained by flags, pennants and balls. Kavanagh gives the signal code and private signals of merchants and of steamship lines regularly serving Jersey. He says that the Fort Regent station, a landmark 300 ft. above sea level, still uses private signals and must be one of the oldest extant in Britain. Its signals are reviewed each year. A special committee of 1967 recommended that it be retained for its historic interest.

J. L. V. Cachemaille, *The History of Sark* (1928) records that the officer in charge of the Sark station killed himself after the station closed down.

CHAPTER VIII

Ireland

Richard Lovell Edgeworth (1744-1817) is remembered today, if at all, as the father and collaborator of Maria Edgeworth, the novelist, second of the 21 children of his four marriages. Yet he was a man of remarkable versatility as well as philoprogenitiveness — enlightened landowner, roadbuilder, inventor, educationalist, politician and writer, friend of many of the leading men and women of his time. He occupies a secure place in the history of telegraphy.

He brought his first wife to live at a house he had taken at Hare Hatch, between Maidenhead and Reading, near enough to Town to allow him to mix in fashionable company. At the salon of the Misses Blake in Great Russell Street he met Sir Francis Delaval, rake and adventurer but one who also possessed an inquiring mind. One evening in 1767 at Ranelagh Lord March regretted to Delaval that he could not leave London to attend a race meeting at Newmarket. Delaval said that he would station fast horses along the Newmarket road to bring news of the winner by about 9 pm and would bet accordingly. Edgeworth then offered to lay £500 that he would name the winner in London at 5 pm. Delaval knew his friend would not bet lightly and wagered an equal sum.

Later, Edgeworth explained to Delaval that he proposed to employ a sufficient number of persons with 'machines' on high ground and visible by telescope up to 16 miles. Delaval doubled his bet and March and his friends took him up.

When the parties met next day Edgeworth explained to March that he intended to rely not on swift horses but on a mechanical system. By this time March evidently knew his man and he and his friends withdrew their bets.

So much for the 'telegraph line' between London and Newmarket, which some authorities have asserted was built and a few have even ascribed to the Admiralty, though Newmarket seems an unlikely source of official naval intelligence! The siting of a shutter telegraph station near Newmarket in 1808, on the Admiralty's London — Yarmouth line, may have compounded the error.

Delaval, who alone shared Edgeworth's secret, urged him to try out the system. Edgeworth erected apparatus to enable Delaval to communicate between his house in Downing Street and Piccadilly. He also experimented with a night signalling system between a house of Delaval's in Hampstead — some accounts say Hampstead churchyard — and that of a Mr. Elers in Great Russell Street to which Edgeworth had access. These forms of communication were apparently large wooden letters displayed in succession and illuminated by lamps. They are said to have served well but to have been too expensive for general use.

Edgeworth's *Memoirs* tell us that he had early been acquainted with Wilkins's *Secret and Swift Messenger* and had read Hooke's works, although in a contribution to *Nicholson's Journal* (1799) he says that in 1767 he tried showing wooden letters in succession before he had read Hooke.

Back at Hare Hatch he continued his experiments. He thought of using windmills which, he believed, might be erected for common economical uses and yet afford easy communication on special occasions. A neighbour called Perrot went with him to Assy Hill, between Maidenhead and Henley, to observe by telescope a windmill at Nettlebed which Edgeworth believed in recollection years later to have been 16 miles away, although in fact the distance is little more than eight miles.

Nothing lasting came of these essays for Edgeworth was soon involved in the management of the family's Irish estate at Edgeworthstown (Meathas Truim) which, by unremitting industry, he transformed into a model property. Not until 1794, when Ireland seemed to be under threat of invasion by revolutionary France, did he resume his experiments, in the belief that Ireland needed swifter intelligence to allay rumours. News of the success of Claude Chappe's *télégraphe* in France seems to have rekindled his old interest, just as it also prompted others to try their hand for the first time.

Edgeworth perfected an apparatus with four separate pointers in a row, each of which pointed like the hands of a clock to different situations in the circles they described. Of eight possible positions, seven were used to denote

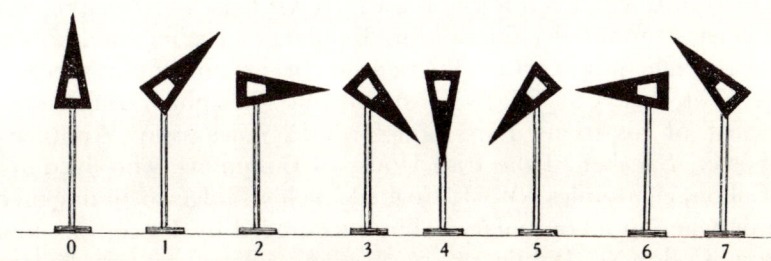
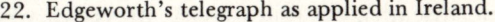

22. Edgeworth's telegraph as applied in Ireland.

figures. The upright position of the pointer represented 0 and the directly downwards position, 4. The figures were referred to a numbered vocabulary. The pointers represented, in turn from left to right, as seen by an observer, thousands, hundreds, tens and units. No number above 7,777 nor a group with 8 or 9 could be signalled. For permanent stations 'which may be seen clearly with tolerable glasses at twenty miles distance', stone, brick or wooden pillars 15 – 20 ft. high were needed. On top a movable platform turned horizontally. On the platform an axis moved vertically carrying the pointer with it.

There were also portable machines with pointers 10-12 ft. high, on a light triangular stand, fastened with tent pegs to the ground.

When communication was to begin, the pointers showing the thousands and units were whirled round until the corresponding station repeated the action. When this figure had been answered, the first operator began to send. As soon as he began, the hundreds pointer at the opposite station was turned to 2, and kept so until the word was made out from the vocabulary. It was then turned to 0. The initiator, on seeing that he was understood, turned all the machines to 0, this procedure being always adopted at the conclusion of every word.

When all his machines were in that position, the corresponding operator turned his hundreds pointer to 2 where it remained until he received another word. To show that communication was finished the thousands and units machines were vibrated to and fro with points downwards.

A large permanent station needed one man per machine, plus one at the telescope and one at the vocabulary. A single pointer could be worked with one man sending and another at the telescope. A smaller vocabulary would suffice and the transmission would be at the rate of a word a minute.

The vocabulary comprised 49 double pages, its seven parts each having seven pages. The numbers '8' and '0' were omitted. There were also eight classes, as follows: common words; less common words; technical terms; persons; officers; places; ships; phrases and sentences.

By this time Edgeworth's apparatus, which he called a *telelograph* (later contracted to *tellograph* and signifying a machine describing *words* at a distance, as distinct from a telegraph, which wrote at a distance) consisted of large triangles mounted either on brick pillars or timber frames. Some sets were 25 ft. high. Edgeworth first tried out his new invention successfully in August 1794, between Edgeworthstown and Pakenham Hall (now Tullynally), seat of his friend Lord Longford, 12 miles away. Another friend, John Foster, Speaker of the Irish House of Commons, who lived at Mount Oriel, Collon, eight miles NW of Drogheda, helped Edgeworth in experiments and in drawing up a vocabulary. Thus encouraged by Foster, he prepared a line from Collon to Dublin, more than 40 miles away, via Bellewstown, Ratoath and Mulhuddar. In November 1794, a trial was made between Collon and Skreene (Meath), 15 miles apart; messages were sent and a reply received in 5 mins. Foster advised him to bring his invention to the attention of government.

In the same month, Edgeworth relates, his sons 'talked ... across the channel between Ireland and Scotland', using machines 30 ft. high. The experiment was evidently repeated a year later for on 24 August 1795 young

Lovell Edgeworth sent four messages from Donaghadee to Portpatrick, 21 miles, and back, 'before a vast concourse of people'. The machines were again 30 ft. high and were 15 ft. across at the base. In clear weather a 12 ft. high pointer could be plainly distinguished across the channel.

On 11 December 1794 he wrote to Dr. Erasmus Darwin,

> I have been employed for two months in experiments upon a telegraph of my own invention ... I tried it partially twenty-six years ago. It differs from the French in distinctness and expedition as the intelligence is not conveyed alphabetically. I propose to government, to raise a corps of vedettes, and to station them in fifty or sixty posts, from which a constant correspondence may be kept up with the capital. I also propose to carry my visual communication across the channel to London ... Suffice it that, by day, at eighteen or twenty miles distance, I shew, by four pointers, isosceles triangles, twenty feet high, on four imaginary circles, eight imaginary points, which correspond with the figures: 0, 1, 2, 3, 4, 5, 6, 7. So that seven thousand different combinations are formed, of four figures each, which refers to a dictionary of words; by an additional contrivance, seven different vocabularies are referred to — of lists of the Navy, Army, Militia, Lords, Commons, geographical and technical terms, &c, besides an alphabet. So that every thing one wishes may be transmitted with expedition. By night, white lights are used.

Darwin replied on 15 March 1795,

> ... The telegraph you described I dare say would answer the purpose. It would be like a giant wielding his long arms, and talking with his fingers; and those long arms might be covered with lamps in the night. You should place four or six such gigantic figures in a line, so that they should spell a whole word at once, and other such figures within sight of each other, all round the coast of Ireland; and thus fortify yourselves, instead of Friar Bacon's wall of brass round England — with the brazen head, which spoke — 'Time is! Time was! Time is past!'

In February 1795 the Bishop of Ossory presented a memorial on Edgeworth's invention to Lord Fitzwilliam, the Lord Lieutenant. Had Fitzwilliam remained in Ireland, something might have come of it.

On 30 May 1795 Edgeworth presented a memorial to Lord Camden, the new Lord Lieutenant, explaining the military utility of the invention and the aid to commerce from the rapid communication of prices in the chief markets. He suggested 30 permanent stations, tenable against musketry, and costing about £300. They should be manned by men subject to military discipline.

Camden did not consider such a provision necessary but said he would consult Speaker Foster, who afterwards told Edgeworth that the plan would not be pursued.

On 8 September 1796 Edgeworth wrote to Lord Carhampton, Camden's private secretary and a wit and conversationalist who had seen the tellograph at work at Edgeworthstown. He pointed out the danger of French invasion and need for rapid communication, and offered to pay for a line from Dublin to Wexford, the area thought most vulnerable.

Carhampton replied that Camden would be glad to talk on the subject and asked Edgeworth to come to Dublin — 'I trust you will be employed'.

Edgeworth had an audience of Camden on 12 September 1796. Camden asked the cost of a line between Dublin and Cork and how soon it could be ready. Edgeworth thought a temporary establishment might be formed in a month and kept up for a year, at a cost of £700, compared with the £3000 a year which the London-Portsmouth shutter telegraph cost, according to Edgeworth. Camden then directed Edgeworth to prepare for an experiment before him, told him that Pelham (later Lord Chichester), the Chief

Secretary, had already engaged somebody from the Admiralty to establish the Admiralty telegraph in Ireland, and regretted that Edgeworth's proposal of 1795 had not been adopted.

In seven days Edgeworth built seven machines of improved design. They were portable, contrived to close like umbrellas and had a portable stand. They were of varying sizes, from 6 to 20 ft. high.

On 19 September 1796 Edgeworth wrote to Pelham that he was ready to demonstrate before Camden. He had collected some earlier machines for immediate use between Dublin and Cork. He said that his tellograph could be erected round the whole coast of Ireland — 'besides inland correspondence' — for £4,000 – £5,000, with a corps of 400 men. Part of the corps could keep up communication by one-man portable tellographs with the troops of a cordon. He wrote:

> I suppose that with a corps of four hundred men, and with an expense of four or five thousand pounds, a communication might be established between Dublin and the following places, Bray Head, Wicklow, Wexford, Waterford, Dungarvan, Youghal, Cork, Limerick, Galway, Sligo, Lough Foyle, Belfast and Carlingford, besides an extensive inland correspondence; part of the corps might be employed in keeping up a speedy intercourse between the troops of a cordon with my portable Tellograph, which a man may carry and can set up in five minutes, and which is legible in weather when other Telegraphs are ineffectual.

On 26 September 1796 Pelham replied that Camden was willing to see the tellograph. On 2 October 1796 Edgeworth waited on Pelham at his house in Phoenix Park, Dublin, where he was breakfasting with the Secretary at War and the Secretary for the Civil Department. Pelham asked for a trial there and then but Edgeworth had brought only his servant with him. The two secretaries then obligingly offered to assist with a experiment with a portable machine Edgeworth had brought. They went to the butts in the park 'where they were arrested by a *centinel* [*sic*] who considered them as persons that must have some evil design in view from the strange apparatus they had with them'. They had some difficulty in establishing their identity but took the affair in good humour. Pelham, on his lawn, received a 'tellograph' message from them.

All were much satisfied and Edgeworth was asked to go the next day to Collon to meet the Lord Lieutenant. Edgeworth's machines were on the road to Dublin and he therefore sent his servant to take them across country by night to Collon. Two of his sons carried machines to Bellewstown, 11 Irish (16 English) miles from Collon, four miles south of Drogheda.

The day of the trial, 3 October 1796, was cold and stormy but Camden condescended to stay for two hours on a bleak hill to watch the experiment. Satisfied, he confirmed that no other telegraph would be preferred to Edgeworth's. At Pelham's wish, Edgeworth submitted a written proposal, presented on 6 October, for tellographic communication. He undertook to convey intelligence between Dublin and Cork by means of 14 or 15 stations, at a rate of £100 a year per station, and from Dublin to anywhere else at the same rate in proportion to the distance, and asked for a contract to cover his costs.

At Collon Pelham had told him that the Duke of York, whose interest in telegraphs has been noticed, had asked for a reconnoitring telegraph. It was thought that Edgeworth's machine might exactly suit him. By 10 October Edgeworth had made a new portable tellograph 'in the handsomest manner' which he presented to Pelham. He made a second machine, which Lovell Edgeworth took to England.

The new tellograph contained a telescope in its axletree 'so as to be manageable by a single person'.

Pelham had become quite enthusiastic and mastered all details of the tellograph, including the vocabulary for common use and 'one which might be constructed for an universal language'. He told Edgeworth he had written to Lord Grenville recommending the use of his tellograph between Ireland and England, adding that 'if the English ministry did not concur in such an establishment, it was still in the power of the government here, to do what they thought proper'.

On 29 October 1796 the portable tellograph was shown to the Duke in Kensington Gardens. He approved of it and his Secretary, Colonel Brownrigg, engaged Lovell Edgeworth to meet him next day at the Admiralty so that he and the Duke's A.D.C. might learn how to manage the machine. General Sir William Fawcett thought the tellograph might be used in the West Indies. The Duke gave Lovell a 'handsome retracting telescope.'

Edgeworth expected that Pelham would be making arrangements to establish the tellograph in Ireland and 'extending it to Scotland to meet the English telegraph'. Alas for such hopes! When Parliament adjourned, Edgeworth went again to Dublin, saw Pelham and at once perceived a change of heart.

On 14 November 1796 he had to write to his wife from Dublin: 'The Admiralty in England do not chuse to extend their telegraphic fame: the government here does not chuse to extend mine'. Three days later Pelham wrote to Edgeworth that Camden had received word from Lord Spencer and the Board of Admiralty that they did not think they could encourage His Excellency in making the experiment.

> His Excellency thinking the invention a very ingenious one, and wishing to show every degree of attention to you in the business, consulted the Commander-in-Chief . . . and not receiving more encouragement from him, than he had done from the Admiralty in England . . . [does] not see any purpose in this country, for which he could be warranted in incurring the expense.

In his reply dated 21 November 1796 Edgeworth said: 'My invention however had been adopted at the Admiralty by the Duke of York's chaplain [Gamble] with such slight alteration as cannot blind the public'. He was at a loss to conjecture why his services were no longer acceptable. But he learned later that the machine then set up at the Admiralty had been constructed some weeks before his own had been taken to England. He said that Pelham had brought over someone from the Admiralty on purpose to establish a telegraph between Dublin and Cork even though he knew of the cost of the Admiralty system.

Edgeworth judged that expense could not have been behind the rejection of his plan for whereas the shutter telegraph between London and Portsmouth cost £3,000 a year — Edgeworth's estimate — his own estimate of a Dublin-Cork tellograph covering twice the distance was only £1,400. He generously considered that Pelham had been compelled to act against his better feelings.

Nevertheless Edgeworth rightly felt he had been sacrificed to patronage. The Irish administration was filled with nepotism and graft. He said as much in a long letter dated 13 December 1796 to the Earl of Charlemont, the 'volunteer Earl,' patron of the arts and first president of the Royal Irish Academy,

> ... had I introduced my telegraph in the form of a lucrative job in which there might be good pickings for others, I might have increased the number of my friends, and have gratified those who are in power, by an opportunity of increasing patronage.

This letter was published in 1797 as a *Letter to Lord Charlemont on the Telegraph*, with a postscript dated Edgeworthstown 17 January 1797. In the postscript Edgeworth says that on 30 December 1796 he received authentic news that the French were on the Irish coasts. He sent Pelham an offer to erect tellographs on a line that government should direct, bringing his own men with him, or to join the army with his portable tellographs to reconnoitre and convey intelligence. Pelham promised a quick answer but none came.

Edgeworth wondered how long it took for news of the first appearance of the French to be known in Dublin Castle. By his tellograph it would have been known in minutes.

In *Transactions of the Royal Irish Academy* vol. 6, (1797), appeared Edgeworth's 'An Essay on the Art of Conveying Swift and Secret Intelligence'. He had read it before the Academy on 27 June 1795 and read a supplement on 3 December 1796.

The essay concluded:
> ... I will venture to predict, that it will at some future period be generally practised, not only in these islands, but that it will, in time, become a means of communication between the most distant parts of the world, wherever arts and science have civilised mankind.

In the supplement he refers to the 'four' new tellographs he had built in September 1796 for the experiment before the Lord Lieutenant. He had found that the 30 ft. high machines with which his sons had communicated between Ireland and Scotland in September 1794 were liable to accident in storm and he had therefore contrived a means of furling their canvas when not in use and of using cordage instead of wood braces. He wrote that he had used windmill sails in 1767 but now found that a triangle whose base equalled half a side was the most distinct of all figures.

Nothing came of Edgeworth's importunings and by the time the new century dawned he must have given up hope of seeing his telegraph applied.

During the Peace of Amiens, Edgeworth and Maria went to Paris, staying there from October 1802 to January 1803. There they met Abraham Niclas Edelcrantz, a Swedish courtier who had pioneered telegraphy in his country.

Edelcrantz, 'of superior understanding and mild manners', and Maria formed an attachment, but Maria, rather plain and by then 35, decided her place was with her family and that she could not live in Sweden.

During the visit Maria wrote of the telegraph on the Louvre, which she could see in operation from her hotel window in the Place de la Concorde, that it was 'as nearly as possible my father's'. Only filial loyalty could have prompted such a comment in a normally observant woman, for the two machines were about as dissimilar as they could be.

Meanwhile, events in Ireland had at last been working in Edgeworth's favour. The Irish Parliament was no more, as Great Britain and Ireland had been united on 1 January 1801.

The new administration at once took a more vigorous line on defence. In 1803 Ireland was again threatened with a French invasion. In October the British government began to erect Martello towers on parts of the Irish coast and subsequently signal stations. Edgeworth renewed his free offer of a tellograph, this time to Lord Hardwicke, the new Lord Lieutenant.

On 13 October he submitted a plan for an all-Ireland system of 727 miles, radiating from Dublin to Malin Head, via Dundalk and Londonderry; to Erris Head via Sligo, Ballyshannon and Killala; to Slyne Head via Athlone and Galway; to Dingle via Limerick and Tarbert; and to Bantry Bay via Wexford, Waterford and Cork. There were to be two men to each of the 71 stations.

Edgeworth stressed the advantage that his apparatus could be seen at a great distance yet could be speedily removed in case of danger. He considered that large, permanent machines would be needed at a few principal points. Elsewhere, portable machines would suffice. In a letter dated 18 October he offered his services 'without a view to any honour or emolument'. He said that he feared delay, as he had earlier knowledge of Bonaparte's designs than any other Englishman who had been in Paris and had sent intelligence to England 'on Shrove Tuesday'. Bonaparte, he affirmed, aimed to make Ireland the theatre of war between France and England.

At last Edgeworth received an encouraging reply. He was asked to form a line between Dublin and Galway, 120 miles, and Captain Francis Beaufort, RN, his brother-in-law, was engaged to assist him. (Beaufort, then aged about 30, was later Hydrographer to the Navy, became a Rear-Admiral and was knighted), Beaufort declined remuneration for his services and applied himself to the task with vigour.

The State Paper Office in Dublin preserves interesting correspondence to and from Edgeworth in 1803-4. On 15 January 1804 Edgeworth wrote that he had drawn up the course of the Dublin-Galway line and that the apparatus and men to work it were ready. On the 27th he announced that he would arrange for a temporary communication between Dublin and Galway to be ready within 48 hours. On 12 March he wrote to say he could find no part of Dublin Castle from which the Royal Hospital at Kilmainham, named as the Dublin terminus of the line, was visible. In the same month Lt.-Colonel Bagwell was appointed to go with Edgeworth to take land for the stations. Edgeworth was named Superintendent at £300 a year.

The Lord Lieutenant was eager for the telegraph to be placed on the footing of a military establishment, with responsibility clearly defined. He wanted to see a superior officer 'of confidence' appointed at Dublin, Athlone and Galway, and also at Limerick when the telegraph was extended there, 'as also a person who can be relied on at every other station'. The question of manning the line by invalids from Kilmainham arose in December 1804 when the stations were at last ready for service.

Temporary guard houses were built. A telegraph corps was formed from yeomanry raised by Edgeworth and from tenants he judged suitable. The men received extra pay and the government met their living-out expenses.

Messages were said to have taken only eight minutes to pass from Dublin to Galway at a trial before Lord Hardwicke, who expressed his gratification.

The machines could be erected or dismantled in a few minutes. In this way the course of a line could be varied to suit demands, at the will of the commander-in-chief. Two men could easily carry on their shoulders the whole paraphernalia of one station.

The 'portable' nature of the installation is no doubt the reason why, unfortunately, no record has apparently survived of the exact course of the line and the station sites. We know that the site of the third station out of Dublin was 'Cappa', where, as the events of 1798 had shown the need for a military post, Edgeworth fortified a ruined stone windmill on the hill. (This site was possibly near Cappagh House, NW. of Kilcock, County Kildare).

By this time fears of invasion had receded. Although Edgeworth and Beaufort were 'diplomatically thanked for their exertions', it does not appear that the service lasted long, if indeed it ever regularly functioned.

On 27 June 1808 the Military Secretary at Dublin Castle wrote to Lord Hawkesbury, Home Secretary, about the difficulty of keeping the signal posts of the W. coast of Ireland in repair. He suggested instead building

> signal towers upon the lines from Dublin by Athlone to Galway; and from Dublin, by Athlone and Limerick, to Cork, and to establish communication between Limerick and the mouth of the Shannon and between Cork and Cape Clear, by the Old Head of Kinsale, and to keep up the communication between Cape Clear and the mouth of the Shannon, and to establish communication between the Isles of Arran and Galway, and to put down all the other signal establishments. By the adoption of these means, intelligence will be received in Dublin of the arrival of a fleet off any part of the South of Ireland, or in Galway Bay; and an useless and large expense will be saved by the discontinuance of the establishments along the North-West and West Coasts.

It is surprising that there is no reference to Edgeworth's recent venture and that there was still some fear of a possible enemy descent on the Irish coasts.

A subsequent letter, dated 10 August 1808, enclosed an estimate of the probable expense of forming such a system and its annual operation — upwards of £33,000. This sum covered the purchase of land for stations between Dublin, Athlone, Galway Bay, Cork, Cove (Cobh) Harbour, and the mouth of the Shannon, etc., and for building 'towers' and supplying apparatus.

It was proposed to acquire one acre at each of the 48 stations suggested by Colonel Benjamin (later Lt.-General Sir Benjamin) d'Urban, AQMG, Dublin, at a cost of two guineas per acre, plus the same at five extra stations suggested by a Major Taylor. The purchase of land and the cost of erecting stations was put at £20,961 10s. There were to be seven Superintendents, at Dublin, Galway Bay, Loop Head, Spike Island, Athlone, Limerick and Cork, receiving 4s. 8d. a day each. Each was to have an assistant, paid 2s. ¾d. a day. The operators at other stations would receive 3s. 8d. a day and their assistants also 2s. ¾d. A General Superintendent and a Paymaster were also proposed. There is no mention of the system to be used.

It appears that one of d'Urban's staff appointments was to establish such a system but there is no record that it was ever begun. Possibly his early departure for the Peninsular War unfortunately put an end to the project, which would have given Ireland a remarkable network.

CHAPTER IX

Other British Inventors

This chapter is concerned with various British telegraph inventors whose ingenuity did not result in regular systems within the British Isles. For this reason it excludes Richard Lovell Edgeworth whose pertinacity was at last rewarded by the application of his system, if ephemerally, in Ireland, in the chapter on which his whole work is therefore recounted.

Macdonald

Colonel John Macdonald, FRS, fourth son of Flora Macdonald, the Jacobite heroine, was 'by far the most persevering and voluminous writer on telegraphy', to quote Barrow's contribution to the *Encyclopaedia Britannica,* Supplement to the 6th edn. (1824). Unfortunately his enthusiasm led him to such lengths that he developed into an eccentric bore. He repeatedly bothered the Admiralty, the War Office and others to adopt his plans and finally angered them, to his disadvantage.

His military career included service in the 84th Royal Highland Emigrants, Bombay Infantry, and Engineers, Lord Macdonald's Regiment of the Isles and the Cinque Ports Volunteers. He was Colonel of the Royal Clan Alpine Fencible Infantry, Commandant of the Royal Edinburgh Volunteer Artillery and Captain, Royal Engineers.

His telegraphic industry shows his obsession with complication, the alleged superiority of the coding to spelling method and – for long – the alleged superiority of the shutter method to any other.

In 1797 he advocated shutters as being the mode best suited to the English climate and a nation-wide telegraphy system, saying that 'a less simple plan, the establishment of mail coaches, was at the time deemed impracticable and visionary'.

The Admiralty Report of 19 August 1806 included a section:

> The Lords Commissioners of the Admiralty gave a decided preference to the alphabetical mode, and the experience of many years has abundantly justified that preference...there is seldom reason to complain of its tediousness; and on the other hand, as the alphabetical key is soon retained by heart, the person who writes and reads the messages can perform the operation with nearly as much celerity as in common writing.

All this was a red rag to Macdonald who thought it 'little more than a laboured panegyric on the spelling telegraph.'

Macdonald thought he could improve on Murray's six shutters by devising a machine with 12 shutters, a top board, a vane and a ball. He called it a 'Terrestrial Telegraph' and quoted long messages which would require only two movements. But the system needed a complicated and extensive code book which the operator had to consult as each combination of signals was made.

It took much to discourage Macdonald, as extracts of letters received from the Admiralty from time to time show.

'Lord Melville [First Lord of the Admiralty, 1804-5] presents his compliments to Colonel Macdonald and has to thank him for the trouble he has taken in addressing Lord Melville.'

'Having laid before my Lords Commissioners of the Admiralty your letter relative to Telegraphic Communication, I have their Lordships' commands that they will not give you any farther trouble on this subject.'

'...you suggest that you should be authorized to examine the plans received at the Admiralty Office respecting Telegraphic Communication. Having no reason to suppose that the public service would be benefited, I do not find myself at liberty to recommend to the Board a compliance with your suggestion.'

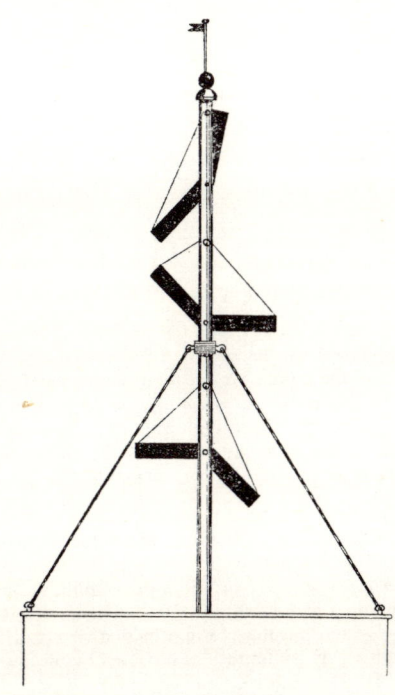

23. Macdonald's British Semaphore.

Macdonald remained obstinately convinced that a shutter was better than a semaphore and numbered sentences better than spelling. He said so in the 1817 edition of his book in which he foolishly called the French coast semaphore 'that wretched contrivance.'

This book contained a description of his new British Semaphoric Telegraph, with a 50 ft. mast, tapering from 15 to 10 in. in width and carrying six arms each 6 ft. in length and 18 in. wide, or wider than the mast. Instead of the neat, all-enclosed gearing of Popham, the movement was by ropes attached to the outer ends of the arms. The mast was topped by an iron rod up and down which a black leather ball could move, so doubling the range of the machine. No mention is made of a shelter for the operators. Otherwise the whole thing is a close copy of Pasley's Second Polygrammatic Telegraph.

Macdonald said that by placing his new telegraph alongside one of his Terrestrial Telegraphs, one machine could express any word or phrase, and the other machine the class! (In the accompanying sketch, the machine shows 6, 11, 13).

In his 1817 book Macdonald also described his 'Minor Semaphoric Telegraph', which had one post with six arms on a common pivot. One arm was plain, the others bearing such devices as a disc, cross, or a disc and cross, as shown in the accompanying sketch.

Macdonald's *Dictionary* (1817) ran to hundreds of pages. Unlike an ordinary lexicon it contained thousands of phrases of which some might occur once or twice in a lifetime. For instance, there are 1,500 expressions under 'great', from 'great absurdity' to 'great wrong' and including 'great goose', 'great rascals' and 'great rhapsody of nonsense'! Each would need at least five variations of aspect. If only one in five were corrupt, the receiver might well take in 'great nourishment' instead of 'great expectation'!

But his tedious labours were not wholly in vain. As we have seen, the principle of his British Semaphoric Telegraph was taken up and applied by Barnard Watson, whose own variant was adapted by the German Pistor for the Prussian Government.

Gamble

Gamble, passed over by the Admiralty in favour of Murray, next tried to interest the War Office in a 'Radiated Telegraph' he had devised and sent one to the West Indies. It originally had four arms on the lazy tongs principle but he modified it so that there were five simple arms with discs at their ends. There were five arm positions and 31 variations.

In 1797, Gamble produced his *Essay on the Different Modes of Communication by Signals.* Some of his observations merit quotation in full:

> It is impossible to give directions on the distance from each other at which telegraphs may be most advantageously placed, as this will necessarily be governed by the surface of the country over which the communication is intended; in many cases also the dimensions of the machine itself may be commodiously varied. That their number should be diminished as much as possible is obvious, not only on account of the original cost of erecting, and the subsequent expense of attending them, but also from the increased probability that all the assistants may not be equally on the alert when the intelligence is to be transmitted; and negligence at any one post interrupts the whole series.
>
> The general remark of situation, is, that such, if possible, should be chosen, where the telegraph may be seen in both directions clear of a background, and yet not so elevated as to be frequently obscured by the clouds. It requires much trouble and caution to fix on the most eligible spot; as after any point is chosen, and appears perfectly adapted to send intelligence in one line, yet when it is to be returned in a contrary direction, a more elevated hill may form a back ground to the telegraph, and render it obscure; an inconvenience which may sometimes be obviated by placing it more to the right or left.
>
> The apparent altitude of an object depends not only upon the absolute height to which it is raised, but also upon the relative situation of the person who views it, this double advantage may be obtained in the last described construction; for the machine being considerably elevated, and the operators remaining at the bottom part, they will have the best opportunity of seeing the next telegraph clear of a back ground. In any line of communication, the smoke of cities, and the thick vapours frequently hanging over the course of rivers and low marshy grounds, are, as much as possible, to be avoided: for which reason, should it ever be found expedient to establish a permanent correspondence by means of telegraphs from London to the extremities of the kingdom, some elevated spot near the metropolis, as Shooter's Hill, or Hampstead Hill, should be chosen as the grand depot of communication; at which place might be erected a telegraph of large dimensions, and, turning upon a centre, to give or receive intelligence in every direction. Several, at

small distances, may be placed in a line from this, to convey it further to any of the offices in town. By this mode the unnecessary multiplication of machines is avoided, which must occur when the intelligence is brought from several points by a distinct series of telegraphs.

As no science, further than being able to count five numbers, is required in the persons absolutely working the telegraphs, soldiers or sailors, who have suffered in the service, are certainly the men who ought to be employed at the different stations. Greenwich and Chelsea would undoubtedly furnish deserving objects, who, by a small addition to their pension, might be induced to attend much better than men unused to discipline. From the ease with which the radiated telegraph is worked, a man and his wife would be sufficient at each machine; and even children may be taught to keep a look-out, or pull the ropes.

To convey a message with the utmost degree of velocity, three persons are certainly required at every intermediate station; one to receive the signal, a second to repeat it, and a third to observe when it is passed forward. But as it can rarely be a matter of any importance, whether the communication is made in five or in ten minutes, one person might easily be so placed between two telescopes, and the ropes of his own telegraph hanging before him, as to be fully equal to manage the whole operation, and probably with greater accuracy, than where many eyes and hands are employed. Certainly, by this means, the most material expense of attendance would be greatly reduced.

Early in 1798 a committee of four Major-Generals, four Colonels and four Lieutenant-Colonels of Royal Engineers was convened to report on the operation of radiated telegraphs which Gamble had erected on the tower of St. Mary's church, Woolwich, and elsewhere in the vicinity. On March 15 1798 they reported to the Marquess Cornwallis, Master-General of the Ordnance, in favour of the telegraph, which they found to be communicating perfectly between Shooters Hill, Blackheath and Woolwich.

Gamble wrote on 18 April 1798 to the Duke of York, as Commander-in-Chief, with an estimate for 12 portable telegraphs at £50 each, to communicate between London and the coast of Essex, Kent or Sussex. Cornwallis had already sent the committee's findings to the Duke, who had referred them to William Windham, Secretary at War. The Duke's secretary also referred the question of the portable telegraphs to Windham, saying that the £600 was to be paid and a telegraph formed between London and the East Coast. It is not clear whether such a line was begun, let alone completed, but Gamble was sufficiently morally recompensed for his rebuff at the Admiralty. On 30 July 1798 the Duke instructed him to install a temporary telegraph between London and the army camp near Windsor, at a cost of £200.

In 1803 Gamble tried out a variant of his telegraph, with four arms on a common centre, installing it on one of the towers of Westminster Abbey.

Pasley

Lt.-Colonel Sir Charles William Pasley, RE, KCB, was born in September 1780. He became a Royal Engineer in 1798 when he entered the 'Shop' at Woolwich. After being badly wounded at the Siege of Flushing in August 1809 he was debarred from further active service. He became a specialist in sapping and mining. One of his exploits, the clearance of wreckage in the Thames, gained him the Freedom of the City of London in 1838. Between 1839 and 1844 he removed the wreck of the *Royal George* in Spithead.

In his pamphlet *Description of the Universal Telegraph for Day and Night Signals* (1823), he dated his interest in telegraphy from 1804, when he first considered both Murray's and Gamble's systems to be inadequate and Chappe's to be not clear at a glance.

His influential uncle Sir Thomas Pasley introduced him to Lord Melville. In 1807 he submitted to the Navy Board his 'Polygrammatic Telegraph', a massive affair with four uprights each bearing two arms. With the outermost arms extended it spanned 40 ft.! The diagram shows that he proposed to bracket the ends on each side of the operating hut.

He proposed to have two operators, one to two machines. The arms would be worked on a wheel and ratchet system. When a sign was completed, the ratchet could be disengaged and the arm would close of its own gravity. Each arm bore a counterpoise rod.

As the contrivance was for daytime use only, Pasley also proposed a night telegraph in which six fixed lights were hidden or revealed by blinds and could give 41 combinations. He wrote

'I was first induced to offer the above project, by having lately, on enquiring, partly for amusement and partly for information in the Progress of the Telegraphic Art, found to my great Astonishment, that all correspondence of this kind is cut off with daylight, and entirely suspended during a time when often it may be most necessary, and that whilst Diurnal Telegraphs have been brought to a great Degree of Perfection and are in constant use, a Nocturnal Telegraph has no where been established, and the schemes hitherto proposed to that purpose at least as many as come to my knowledge seemed so imperfect, that it was natural to attempt an improvement.'

Pasley received the usual answer from the Navy Board, which professed itself well satisfied with the shutter system, and 'a still more repulsive answer' from the Society of Arts. Tilloch's *Philosophical Magazine* gave details of the invention in its January 1808 issue.

While on the ill-fated Walcheren Expedition he first saw the French coast semaphore. It taught him a valuable lesson in simplicity. Back home he first reduced the overall width of his machine to 22 ft. by making the first and third masts tall ones so that the arms of adjacent posts could overlap. He described both his modification and the French semaphore in Tilloch's magazine in May 1810. He next progressed to the idea of having one tall post carrying, first, four pairs, and then three pairs of arms. He called his new apparatus the 'Second Polygrammatic Telegraph'.

When the Admiralty adopted Popham's telegraph in 1816 Pasley began to show his resentment towards Popham. He asserted that although Popham's machine was avowedly an imitation of the French semaphore it even more resembled his own. In 1821 he complained when the Society of Arts gave a silver medal to Sir Nicholas Nicolas (see page 118) for a telegraph identical, as he alleged, to his Second Polygrammatic. He conceded that 'of course the same idea may have occurred to him, without having heard of my plan, but the priority of invention rests with me'.

The *Penny Cyclopaedia* supported Pasley and added that Macdonald had also adopted Pasley's Second Polygrammatic with only slight variation. Pasley decided to concentrate on a single-post telegraph with two arms on a common pivot. He regretfully abandoned his Polygrammatic, which although it had been 'a pleasing speculation' was rendered less useful on account of its expense and need of extra operators.

Pasley's new telegraph was admirably simple and he was wise to have the description lithographed for the benefit of his fellow sapper officers. Already matters were moving in his favour. He was in touch with Vice-Admiral Sir Benjamin Hallowell, who on 26 October 1822 ordered all his ships to carry the new Pasley telegraph instead of Popham's Sea Telegraph which had proved not altogether satisfactory.

Pasley describes his 'Universal Telegraph for Day and Night Signals' as follows:

> For the day signals, the Telegraph consists of an upright post of moderate height, of two moveable arms fixed on the same pivot near the top of it, and of a mark, called the indicator, on one side of it. Each arm can exhibit the seven positions 1,2,3,4,5,6, and 7, exclusive of its quiescent position, called "the stop", in which it points vertically downwards, and is obscured by the post... The indicator merely serves to distinguish the low numbers 1, 2 and 3 from the high numbers 7, 6 and 5... The arms and the indicator for the day signals are made of wood, framed and pannelled, for the sake of lightness. (In a very hot climate, plates of light copper may be used for the pannels). The indicator plays in a mortise, cut in the upper part of the post, and is let down into its horizontal, and raised into its vertical, position, by means of a small rope, and a small pulley. The arms must be fixed externally, one on each side of the post, and must be exactly counterpoised, by means of light frames of open iron work, which become invisible by day, at a little distance, and which, even when viewed closely, do not impair the clearness of the Telegraphic signs.

In 1826 an Admiralty committee proposed that Pasley's telegraph, with two modifications, should be generally substituted for Popham's on ships. It came about in this way that an Army officer became responsible for the general form of telegraph at sea which the Navy adopted and proliferated in Victorian days. The Army does not seem to have made great use of his telegraph on an extended scale. The only important application which has come to notice was in South Africa (see Chapter XXII).

Pasley became a K.C.B. in 1846. Promoted Major-General in 1841 he became a Lt.-General in 1851 and died on April 1861 as a General.

24. Conolly's telegraph for stations in India, 1825.

Conolly

Joseph Conolly, Master-at-Arms, RN, who had proposed a field telegraph to the War Office in 1807, published the *Philanthropic Vocabulary and Code of Signals* in 1821. It proposed a semaphore telegraph in imitation of Popham's, even to the 48 arm settings. The mast was 10 ft. high in the signal room and 30 ft. clear above, and 18 in. in diameter. Each of the two arms was 9 ft. 6 in. in length and 18 in. wide. He considered that to avoid confusion in speaking of 'right' and 'left' of the sender it would be better to refer to N., S., E. and W. He recommended using four extra positions for each arm which he called the '8 degrees' positions. Instead of the arm being folded out of sight against the post, it could appear at 8 degrees to the vertical. Each arm could therefore assume ten positions and the apparatus give two distinct numeral tables.

Conolly proposed to have stations about nine miles apart but said they could be up to 16 miles apart in hilly country with a clear atmosphere. He may have had India in mind, as he seems to have been accepted as something of an authority in Calcutta, where his venture will be noted in Chapter XXIV.

Nicolas

Also worthy of notice is Lieutenant Sir Nicholas Harris Nicolas, GCMG, RN, editor of *The Despatches and Letters of Lord Nelson* and an early campaigner for the preservation of British Natural Archives. In 1821 when attending the Inner Temple he gained the silver medal of the Society of Arts for 'an improvement on the Vertical Semaphore, and for his method of adapting a shifting key to Telegraphic Communications, for the purpose of insuring their Secrecy'.

He proposed a mast with four pairs of arms placed equidistantly. The lowest pair indicated units, the next above tens, the next above that hundreds and the topmost pair thousands. The apparatus gave a range of numbers between 1 and 16999.

The Society noted the similarity to Macdonald's British Semaphoric Telegraph of 1817 but observed that 'the invention is no doubt original with Lieutenant Nicolas'. Also, as we have seen, Pasley complained of the resemblance to his own Second Polygrammatic Telegraph. Nicolas seems to have been more original with his secret code, apparently suggested by Bonaparte's secret cipher.

Boaz

In February 1802 the *Philosophical Magazine* published details of a 'Patent Telegraph' devised by James Boaz of Glasgow. His plan was to have letters of the alphabet or other characters arranged in a frame and made visible or invisible by means of shutters or blinds, and lamps. In 1804 he proposed a commercial telegraph network with lines linking London with Liverpool and Hull and a line between Holyhead and Liverpool. He put the cost at £150,000 with initial outlay of £6000 for 20 stations and profits of 300-400 per cent!

Law

It is pleasant to record that at least one serving member of the British telegraphic and signal service was rewarded for zeal in seeking improvements. Alexander Law, a Midshipman discharged from *HMS Monmouth* because of wounds, was in charge of the signal duty at the Port Admiral's office at Deal. In December 1814 the Society of Arts awarded him 20 guineas for his 'Improved Telegraph, or Method of Conveying Intelligence to and from Sea and Land'. Thomas Foley, Vice-Admiral at Deal, and Lt. David Cree, former Superintendent of the telegraph at Deal, were among those who produced certificates in support of Law.

CHAPTER X

France

Guillaume Amontons, a member of the Academy of Sciences in the reign of Louis XIV, may be considered the first Frenchman to attempt seriously to transmit signals overland by means other than smoke or fire. Fontenelle describes experiments which Amontons conducted in the Luxembourg Gardens, Paris, in 1690 when the Dauphin was present.

> The secret was to station men in several consecutive positions who, with long-range telescopes, having perceived certain signs from the preceding post, transmitted them to the following post. The signals were really letters of the alphabet. The maximum range of the telescopes was made the distance between each consecutive pair of stations.

Amontons seems to have based his methods closely on those of the Englishman Dr. Hooke, as described in the *Philosophical Transactions* (1684) and almost certainly the first serious attempt at telegraphy in modern times. He also laid down precise conditions for mechanical telegraphy a century before its time. His ideas, like those of Hooke, were published but not developed. (See Chapter I).

During the next century other Frenchmen tried their hand at mechanical signalling. A naval commissioner called Marcel is said to have invented a 'telegraph' at Arles in 1702 which could operate by day or night. He petitioned the King to pay for its transport to Paris but without result, so he destroyed his work.

Other 'telegraphs' were allegedly devised by the philosopher Dupuis and the publicist Linguet. About 1788 Dupuis is reported to have corresponded across Paris from Belleville to a friend at Bagneux, and Condorcet and Franklin are credited with having recommended the device to the government. Linguet is said to have been freed from the Bastille on condition he destroyed a 'telegraph' he had invented.

Chappe's Télégraphe

The French Revolution brought to the fore two remarkable mechanical inventions which the nation's new masters adopted in the interests of efficiency. One was destructive, the notorious 'philanthropic apparatus for decapitation' named after Dr. Guillotin. The other was constructive, the *télégraphe aérien,* or T-type telegraph devised by Claude Chappe which became the most important and best-known of any means of visual communication.

There were five brothers Chappe, sons of prosperous parents: Ignace Urbain Jean (Chappe l'Aîné), Claude, Pierre-François, René and Abraham. All owed some distinction to their uncle, a celebrated astronomer named Jean Chappe d'Auteroche.

Claude, born at Brûlon (Sarthe) on 25 December 1763, was trained for the church but was attracted instead by science. Leaving college with no religious obligations but two lucrative benefices he was able to devote himself to scientific investigation, including the study of what we now call telecommunication. He first attempted to use electricity, but, at that primitive stage of electrical development, insulation was poor and wires had little mechanical resistance. He therefore turned to perfecting a reliable visual instrument.

In 1790 he first tried an opto-acoustic system, using two large clocks synchronised. Their dials showed agreed signs. When the hand of one dial reached the signal to be sent, two copper pans emitted a sound which could be heard 400 m. off. He then made an instrument wholly optical. A pivoted wooden plate, 4 m. high, painted black on one side and white on the other, was displayed when the hand passed the transmission mark. On 2 March 1791 local officials watched an exchange of messages over the 15 km. between Brûlon and Parcé.

The Revolution deprived Chappe of his church benefices and at the end of 1791 he went to Paris. There, thanks to the influence of Ignace, a member of the Legislative Assembly, he gained authority to erect his machine at the Étoile barrier. A mob fancied it was a device for communicating with enemies of the State and destroyed it during the night. One account says that masked men broke into one of the little pavilions at the barrier which Chappe was using and carried off the apparatus. Chappe was not dismayed. He fixed upon an apparently safer site, the estate of the assassinated deputy Le Peletier de Saint-Fargeau, at Ménilmontant.

Chappe had improved his apparatus again, by dispensing with clocks and using a rectangular board which could show six colours. He introduced his invention to the Legislative Assembly on 22 March 1792 and his petition was passed for examination to the Committee for Public Instruction.

Once again the apparatus was maliciously destroyed, seemingly because the Chappes were thought to be communicating with the Royal prisoners in the Temple. Chappe promptly asked the Assembly for protection and an indemnity to cover the cost of repairs.

By this time Chappe had become convinced that moving arms were more easily seen. With the help of the noted clockmaker and watchmaker Abraham-Louis Breguet he perfected his T-type moving-arm machine. Breguet made three iron arms worked by cords and pulleys, with a small repeater exactly reproducing the position of the main instrument.

At first Chappe called his invention a *tachygraphe* (rapid writer), which certainly conveyed the idea of its chief recommendation, speed. The name *télégraphe* is, as stated earlier, said to have been suggested in April 1793 by

Miot de Mélito, then departmental chief in the Ministry of War and a classical scholar.

Chappe submitted details of his latest invention but they were pigeon-holed while political turmoil grew and the republic's frontiers were menaced. On 15 October, 1792 (24 Vendémiaire, An I) the Convention, successor to the Legislative Assembly, referred the matter to committee.

Romme, president of the Committee of Public Instruction, unearthed Chappe's plans and brought them to the Committee's notice. On 1 April 1793 (12 Germinal, An I) he reported most favourably on the telegraph to the Convention and asked for an experiment to prove its worth.

Three members of the Committee, the scientist Lakanal, the legislator Daunou and the mathematician Arbogast, were appointed to observe the trial. Chappe was voted 6,000 francs towards the cost. The intelligent Lakanal foresaw a great future for the invention. He warmly supported Chappe, winning over both his fellow members — at first hostile to Chappe — and Cambon, the Finance Commissioner, whose advice was sought. Safeguarded by a decree of protection published by the Convention, Chappe went ahead with a 35km. line between the park of Saint-Fargeau and Saint-Martin-du-Tertre, with an intermediate post at Écouen.

The fateful trial took place on 12 July 1793 (24 Messidor, An I). Two operators (*stationnaires*), one at the handle of the machine, the other with a telescope, stood ready at each station. Chappe and his brothers, vocabulary in hand, were at the terminal stations. Lakanal and Arbogast watched from Saint-Martin-du-Tertre, Daunou from Ménilmontant. Savants and artists also came to watch the proceedings.

At 4.26 pm Saint-Martin station signalled: 'Prepare'. Ménilmontant at once began to send:

> Daunou has arrived here; he announces that the National Convention has just authorised its committee of general security to affix the seals to the papers of the representatives of the people.

This took eleven minutes to send. Saint-Martin came back in nine with:

> The inhabitants of this beautiful country are worthy of liberty because of their love for it and their respect for the National Convention and its laws.

The Commissioners then exchanged conventional messages.

The experiment was a complete success. Lakanal's eulogistic report to the Convention on 26 July 1793 (8 Thermidor, An I), when Danton presided, ended:

> What brilliant destiny do science and the arts not reserve for a republic which, by its immense population and the genius of its inhabitants, is called to become the nation to instruct Europe.

The Convention approved the adoption of the telegraph as a national utility and instructed the Committee of Public Safety to map suitable routes. Chappe was styled *ingénieur-télégraphe* and granted an Engineer-Lieutenant's pay.

Chappe was grateful to Lakanal and a lively correspondence, in extravagant terms on Chappe's side, grew up between them. Lakanal wisely observed that the establishment of the telegraph was the best answer to publicists who thought France too big for a republic. Perhaps he foresaw that the telegraph would forge close links between the central authority and its representatives in the departments that had replaced the provinces.

Inspired by the energetic Carnot, the 'organiser of victory', the Committee of Public Safety decided on lines from *Paris* to *Lille,* chief city of the north, where the Republic's Army of the North was fighting, and from *Paris* to *Landau,* a strategic fortress town in the Rhenish Palatinate then held by France. From the start the government viewed the telegraph as primarily, if not wholly, a military instrument.

On 4 August 1793 (17 Thermidor, An I) the Minister of War was instructed to give orders to acquire sites that Chappe had selected for the Lille line and arrange for the manufacture of equipment.

On paper at least the Committee of Public Safety facilitated Chappe's work, giving him powers to erect telegraphs on existing buildings where necessary and cut down trees blocking the view. Owners were to be compensated for land taken and trees felled. Local authorities were ordered to contribute men and materials.

In fact Abraham's letters from Lille to Claude in Paris in late 1793 and early 1794 show that they were beset by difficulties of raising money to build the lines – money was dropping fast in value – and of procuring materials and stimulating local bodies. The budget for the Lille line was fr. 166,240 in assignats, which soon lost 40% of their value. Delivery of materials was complicated by the fact that the army had requisitioned many horses. Some inhabitants showed such ill-will that agents and workmen had sometimes to carry arms.

On 16 July 1794 (28 Messidor, An II) the Committee instructed Chappe to complete the Lille line without delay, to enable the operators to become proficient. The first station sites chosen out of Paris (Belleville) were those of the trial operation: Écouen, and Saint-Martin-du-Tertre. Thence the line ran via Clermont, Belloy, Boulogne-la-Grasse (east of Montdidier), Lihons and Thélus (east of Arras). To Lille (St. Catherines).

By an order of the Committee dated 2 May 1794 (13 Floréal, An II) Chappe was enabled to bring the line from Ecouen into the heart of Paris, with an intermediate station on the old church of Montmartre and a terminal station, with the apparatus painted in the three national colours, on the Pavillon de Flore of the Louvre. Claude Chappe's apartment at 23 Quai Voltaire commanded a good view of the terminal headquarters.

The line, 210 km. in length, appears to have operated regularly from the end of Thermidor, An II (mid-August 1794). Although correspondence between Claude and Abraham Chappe deplored hitches in transmission the new medium was soon working well enough to announce news which dispelled all doubts of its value.

It has been said that the news of the liberation of Landrecies, announced to the Convention by Barère (Bertrand Barère de Vieuzac), of the Committee of Public Safety, on 1 Thermidor, An II (19 July 1794), reached Paris by the new telegraph but this seems unlikely. The distinction almost certainly goes to the news of the capture of Le Quesnoy on 28 Thermidor, An II (15 August 1794). Barère announced it to the Convention on 30 Thermidor, An II (17 August 1794) '... a machine by means of which the news of the recapture of Le Quesnoy has been brought to Paris, two days ago, an hour after the garrison had entered [Le Quesnoy]'. The speaker then described and praised the invention. He would hardly have done so on that occasion had the telegraph already been in full service for a month.

On 30 Thermidor, Abraham wrote to Claude regretting difficulties but saying he was able to telegraph the news of Le Quesnoy: 'We have announced its surrender ten hours before the courier could have arrived at the Convention'.

A final triumph came on 15 Fructidor, An II (1 September 1794) when Lazare Carnot told the Convention: 'Here is the telegraph report which has just reached us. Condé is restored to the Republic. Surrender took place this morning at six o'clock'. The Convention decided to telegraph Lille to authorise Condé (Condé-sur-l'Escaut) to change its name to Nord-Libre, the message also assuring 'the brave army of the North that it continues to deserve well of the fatherland.'

The place of the telegraph was assured. Many travellers visited the Louvre to see the new invention at work and descriptions in various languages were circulated. The fame of the invention soon spread.

Carlyle, writing his *French Revolution* more than 40 years later, recalls the trials and successful debut of the Chappe telegraph:

> What, for example, is this that Engineer Chappe is doing, in the Park of Vincennes? In the Park of Vincennes; and onwards, they say, in the Park of Lepelletier Saint-Fargeau the assassinated Deputy; and still onwards to the Heights of Ecouen and further, he has scaffolding set up, has posts driven in; wooden arms with elbow joints are jerking and fugling in the air, in the most rapid mysterious manner! Citoyens ran up, suspicious. Yes, O Citoyens, we are signaling: it is a device this, worthy of the Republic; a thing for what we will call *Far-writing* without the aid of postbags; in Greek it shall be named Telegraph – *Télégraphe sacré*! answers Citoyenism: For writing to Traitors, to Austria? – and tears it down. Chappe had to escape, and get a new Legislative Decree. Nevertheless he has accomplished it, the indefatigable Chappe; this his *Far-writer*, with its wooden arms and elbow-joints, can intelligibly signal; and lines of them are set up, to the North Frontiers and elsewhither. On an Autumn evening of the Year Two, Far-writer having just written that Condé Town has surrendered to us, we send from Tuileries Conventional Hall this response in the shape of Decree; 'The name of Condé is changed to 'Nord-Libre, North Free. The Army of the North ceases not to merit well of the country' – To the admiration of men! For lo, in some half hour, while the Convention yet debates, there arrives this new answer: 'I inform thee, *je t'annonce*, Citizen President, that the Decree of Convention, ordering change of the name Condé into *North-Free;* and the other declaring that the Army of the North ceases not to merit well of the country; are transmitted and acknowledged by Telegraph. I have instructed my officer at Lille to forward them to North-Free by express. Signed Chappe.'

As perfected, Chappe's apparatus consisted of a mast, some 5m. high, at the top of which was a wooden beam, about 4.62m. long and 0.35m. wide, called the *regulator* and forming the top of the 'T'. The regulator could be turned at 45° on each side of the mast, as well as horizontally and vertically. At each end of it, and forming, in one position, serifs to the 'T', was an *indicator,* a short, mobile arm about 2m. long and 0.33m. wide. The arms were made like a window shutter, with alternate slat and aperture, half the slats being set to the right and half to the left, to lessen wind resistance and produce light and shade.

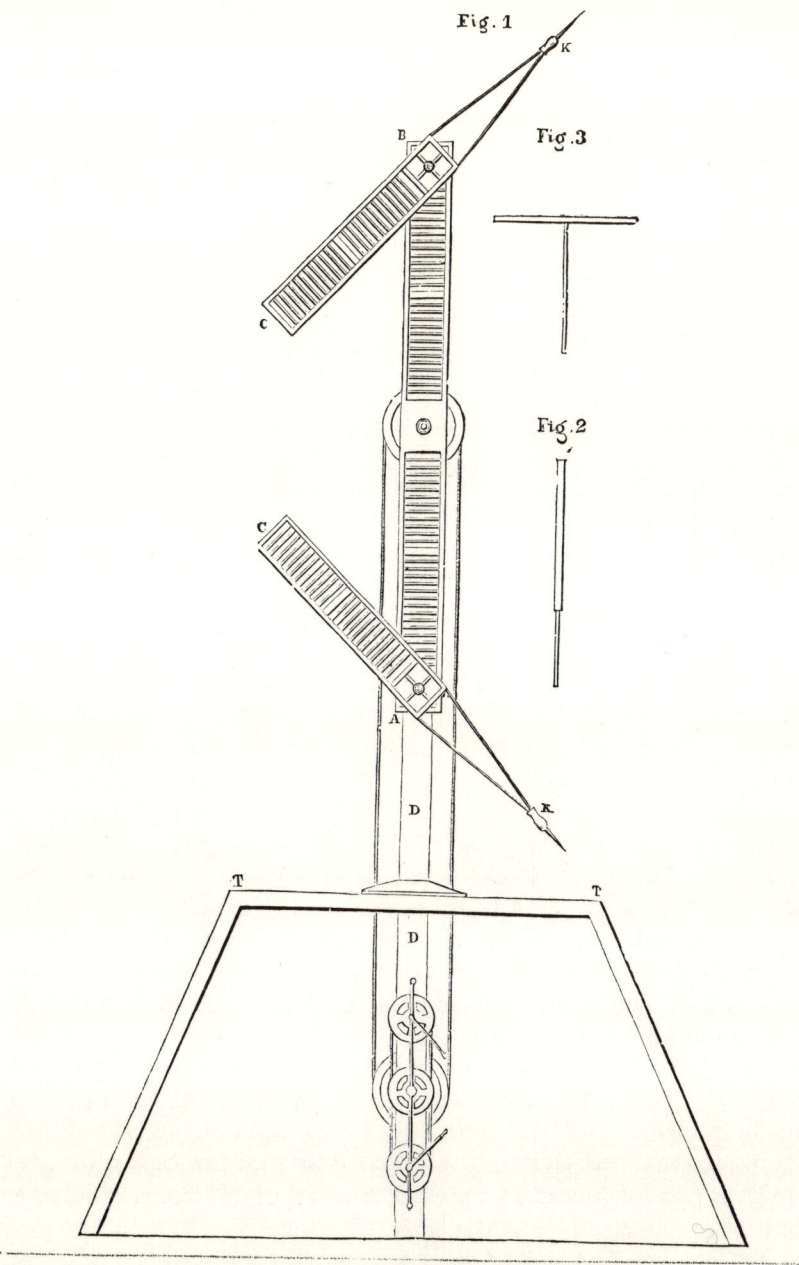

25. Arrangement of Chappe telegraph in France.

The indicators were compensated by thin iron counter-weights, invisible at a distance. Generally, regulator and indicator were painted black. They were worked by rope or copper wire, the movements being reproduced by a miniature apparatus inside the station.

The indicators could be turned right round to take up positions radially from the beam ends at the usual settings of 45°, 90° and 135°, on each side. The 180° position was discarded as it appeared only as a prolongation of the regulator and was, therefore, indistinguishable at a distance. Thus, each indicator could assume seven different positions. Using the two indicators, the apparatus could form for each position of the regulator (horizontal, vertical, right inclined, left inclined) 7 x 7 = 49 signals, thus 4 x 49 = 196 in all.

British readers can study the operation of the Chappe telegraph from the excellent working model in the Science Museum, South Kensington. The model was evidently designed from an illustration in Alexis Belloc's *La Télégraphie Historique* (1888).

Léon Delaunay, related to the Chappes and a former French consul at Lisbon, drew up the first vocabulary for the telegraph. It contained 9,999 words, phrases and expressions, each one rendered by a number. It soon proved to be too slow and inconvenient as from one to four signals were needed to transmit a group of one to four ciphers, and in 1795 Claude Chappe substituted a new code, to speed transmission.

The horizontal and vertical positions of the regulator were reserved for 'assuring' the signals, which were first executed with the regulator oblique, then reported and thus confirmed on the horizontal and vertical. The effective number of 'working' signal positions was thereby reduced to 96.

In the new code Chappe reserved 92 of the 96 for sending information, by means of a vocabulary with three categories. Each category had 92 pages and 92 expressions, or a total of more than 25,000 different significations. The first vocabulary contained 8,464 words, the second one had 8,464 expressions or parts of phrases, such as degrees of urgency, incidence of fog and destination of despatch; one sign was needed to show the category (*phrasique*), another for the page and a third for the number of the expression or phrase. The third category was geographical, with names of places and phrases used in correspondence.

The code was again revised in 1830. A new codebook in two parts was published, with 704 pages of vocabulary. Part I ran from 'A' to 'hier' and Part II from 'hippodrome' to 'zélé pour'. If, for example, it was desired to telegraph 'Canton Berne', the sign for 19 followed by that for 28 was sent, as 'Canton Berne' was to be found on page 19, line 28 of the code. Similarly, signals 53 and 13 in succession signified page 53 line 13 – 'the end of this message has not yet reached me'. Again, to send the message: 'I reply to your last message', the despatching station would show the signal indication ⌐ followed by ⌐, these signifying 53 and 21, corresponding to page 53, line 21, where the phrase was found.

First of all, the preparatory signal would be shown by placing the indicator at oblique. When the position had been seen and repeated by the next station, the message would follow. The system allowed for spelling out words, but this method was used sparingly.

The two operators who did duty turn and turn about at each intermediate station did not know, or need to know, the meaning of the signs they relayed, except for a number of basic 'service' signs — originally 92 and later cut to 88 — which they were expected to memorise. These signs indicated such things as 'end of transmission' 'error in transmission', 'one of the operators is ill', 'rain (or fog) prevents transmission', 'fire has destroyed the station', and so on. Otherwise they merely followed the lead of their neighbour and displayed and wrote in a register the signal positions according to a formula which included the words 'ciel' or 'terre' to show that an arm had been inclined upwards or downwards. In this way an error in transmission could be traced to the offending operator.

Normally, messages emanating from Paris had priority but if a message from, say, Lille was preceded by the signal denoting 'urgency' it was allowed to pass first.

Short periods of break were allowed, when the operators could set their machines at the inert position, one of the service signs.

At the terminal stations and also, as the system expanded, at important intermediate places designated, a Director and an Inspector were stationed. When a message was handed in, the sender was given a receipt stating his name, the date and the time of the deposit and the destination of the message. The Director encoded out-going and decoded incoming messages and handled administrative correspondence, staffing, equipment and accounting. The Inspector supervised the operation on a given section of line and saw that the machinery at each station was in good order.

The Committee of Public Safety finally decided on 3 October 1794 (12 Vendémiaire, An III) to build the *Paris-Landau* line, via *Metz* and *Strasbourg*. A particularly detailed general instruction from Claude Chappe regarding this project has been preserved.

Chappe advises on the manner of choosing sites, saying that a prospective location must be visited on different occasions and at different times of day to establish all circumstances. Forests, marshes and water are to be avoided if possible, because of their vapour. Local inhabitants are to be asked about the prevalence of mists between adjacent prospective sites.

The 'managers' of the divisions of the line are responsible for the siting of the stations, of which there are to be up to 12 per division, each to be placed as near to a main road as possible. The managers are also responsible for the woodwork of the telegraphs — dry wood is essential for the construction of the machines and for the stations.

It was intended at first that the Landau line should use a new type of machine, with seven indicators, five of them like the small arms of the conventional Chappe instrument and arranged in a row. After some prototypes were made, the idea was given up because of cost and orthodox machines were reverted to.

Even so, news of the 'Landau' type telegraph was conveyed across the Channel. A description of the new apparatus erected on the Pavilion of

Unity at the Tuileries appeared in the *Star* on 10 October 1796. It is reported as comprising a large beam, painted black and fixed horizontally on four large posts. Attached to the beam are five distinct arms, similar to the 'secondary' arms of the Chappe telegraph. Two other arms are stated to be attached to the two central posts. The five arms can each assume two vertical, four inclined and two horizontal positions.

When the Ministries were abolished, Claude Chappe was attached to the Committee of Public Works with the title of *ingénieur-en-chef*. Ignace and Pierre-François became his technical assistants. The telegraphic administration was removed from Chappe's apartment to the Hôtel Villeroy, Rue de l'Université, where Ignace and Pierre-François lived. In 1800 the telegraph service was transferred from Public Works to the Ministry of the Interior.

On 27 April 1795 (8 Floréal, An III), in view of the army's advances in the NE., the Committee of Public Safety ordered the Lille line to be extended via Roubaix and Oudenarde to *Brussels*, where René Chappe was appointed director, and to St. Omer, *Dunkirk* and, possibly, *Ostend* at the expense of the Ministry of Marine. It is not conclusive that the coast extension was carried out. It does not figure on an official map of 1805, though Ignace, writing in 1824, gives 1798 as the date of extension to Dunkirk.

Although the line to Landau had been sanctioned at the same time as that to Lille, it had run up against even greater difficulties of shortage of money and materials. On 24 August 1797 (7 Fructidor, An V) the Directory, which had replaced the Convention in October 1795, ordered all work on it to cease. Rabaut (Pomier) had proposed on 22 July 1795 (4 Thermidor, An III) that the Paris terminus of the line should be at the Palais-National but the convention decided on the Tuileries. St. Sulpice church was the final choice.

The Directory soon changed its mind. In view of the impending opening of the congress at Rastatt, work was resumed in November 1797 on the Paris-Metz-Strasbourg section. At the same time the construction of a great new trunk line, of 870 km., from Paris to *Brest*, via Dreux, Avranches, St. Malo and St. Brieuc was begun. One of the most interesting sites on the Brest line, built at the Navy's expense, was at *Mont-St. Michel*, where the telegraph was erected on the tower of the abbey church, at that time without a spire. It is said that the station on the summit of Mont-Dol was formed of stones from an old chapel which had been built as a Mithraic temple in late Roman times. The tower is now the Chapel of Notre-Dame de l'Espérance.

At the Paris end of the Brest line a telegraph was erected on the roof of the Ministry of Marine in the Place de la Concorde.

The opening of the Congress of Rastatt on 9 December 1797 (19 Frimaire, An VI) spurred the rapid construction of the Strasbourg line. It made remarkable progress. In six months after the Directory's order of 17 November 1797 (27 Brumaire, An VII) the 46 stations were erected and the line, 480 km. in length, opened on 31 May 1798 (12 Prairial, An VI). It cost fr. 176,000.

Plate 47, above: Claude Chappe.

Plate 48: France: St. Pierre de Montmartre church in 1832, surmounted by telegraph of Paris-Lille line.

Plate 49: Telegraph station at Dilbeek, near Brussels, in 1804 (from a painting by Paul Fitzthumb). The station was on the Lille-Brussels extension operated during the First Empire.

Plate 50: Rue de Rivoli, Paris, about 1830, showing (centre background) Ministry of Marine with its telegraph (Paris-Brest line).

Plate 51: St. Malo, showing telegraph of Paris-Brest line on cupola of the cathedral.

Plate 52: Present-day view of Notre-Dame de l'Espérance tower, Mont-Dol, which formerly carried a telegraph of the Paris-Brest line.

Plate 53: A fanciful English engraving, 1798, entitled 'The French Raft', and intended to represent a raft in Brest Harbour for an invasion of England. The artist has mistakenly included an English shutter telegraph.

Plate 55: Old engraving of the tomb of Claude Chappe in Père Lachaise, Paris.

Plate 54: The Strasbourg optical telegraph, 1825.

Plate 56: Telegraph on lighthouse at Calais.

Plate 57: Telegraph on church tower at Ardres (Pas-de-Calais).

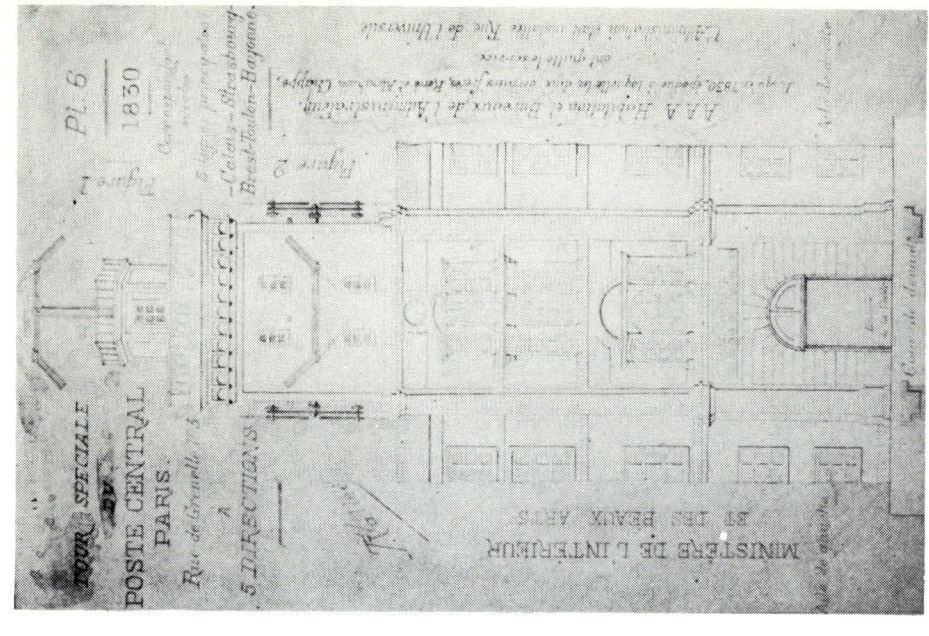

Plate 59: The later central headquarters of the telegraph in the Rue de Grenelle, Paris. The drawing shows the apparatus for four of the five lines of which this was the terminus.

Plate 58: Rennes Cathedral with telegraph of Avranches-Nantes line on north-west tower.

Plate 60: St. Sulpice church, Paris, in the 1830s with Lyons line telegraph on north-west tower and that for Bordeaux on south-west tower.

Plate 61: Early illustration from *Illustrated London News* showing abstract of the Indian Mail being telegraphed from Marseilles to Paris.

Plate 62: Former statue of Chappe in Paris.

Plate 63: Telegraph tower at Trou d'Enfer, Bailly, near Versailles, on Paris-Brest line, from a water-colour of 1842.

Plate 65: Haut-Barr telegraph, Alsace, as reconstructed to form a monument to the Chappe telegraph.

Plate 64: Modern view of Trou d'Enfer station, Bailly.

Professor Henri Gachot of Strasbourg has undertaken much research into the history of this line and its associates, and much of the information on it is drawn from his book *Le Télégraphe Optique*.

Various forms of construction were adopted for the stations, according to the site and availability of materials. Some stations were stone-built towers, either round or square, while others were of wood: square, squat buildings like cabins, some raised on timber piers. The Strasbourg terminus was on the tower of the transept of the cathedral and the main station at Metz was on the Palais de Justice.

Like other provincials, the people of Strasbourg soon came to know that important news was passing when they saw the telegraph arms on their cathedral constantly agitating. Parisians learned of important telegraphed news through news bills posted up in public places. After 1800 the Strasbourgeois enjoyed a similar service with regard to messages relayed from Paris which the Prefect of Bas-Rhin deemed proper to publicise.

For better or worse the telegraph had become a predominantly military instrument, though Chappe had hoped it would also benefit commerce and industry.

Indignation reigned in Paris when the telegraph brought the news, sped by courier from Rastatt to the Strasbourg telegraph, that Austrian hussars had murdered two of the three French delegates as they were returning with their families to France after the congress had broken up, on 28-29 April 1799 (9-10 Floréal, An VII).

The Strasbourg line produced two offshoots, both short-lived. The first was a continuation from *Strasbourg* to *Huningen* (Huningue), near the Swiss border N. of Basle. It had 14 stations, some, if not all, on church towers. After taking only three or four months to complete it was opened on 16 August 1799 (30 Thermidor, An VII). The last-known message it carried announced Moreau's victory at Hohenlinden, sent on 3 December 1800 (11 Frimaire, An IV).

During the campaign against Austria in 1800 the Strasbourg line gave valuable service. When Austria again sued for peace and a congress was convened at Lunéville, Bonaparte, as First Consul, ordered a branch to be built from Vic, six stations S. of Metz. So rapidly was it built that he was able to give orders by telegraph to General Clarke at Lunéville about accommodation for the Austrian plenipotentiaries only 13 days later. The branch had three intermediate stations. Messages began to pass regularly from early October 1800. The line probably ceased working soon after the treaty was signed on 8 February 1801 (19 Pluviôse, An IX).

The *Brest* line, most ambitious so far, with 55 stations, was completed in the remarkably short time of seven months and opened on 7 August 1798 (20 Thermidor, An VI). Yet another important trunk line was soon ordered, to *Lyons*, passing close to *Auxerre* and serving *Dijon*.

By order of 9 December 1798 (19 Frimaire, An VII), news of the French successes against the Neapolitans was relayed over all telegraph lines. Two days after the fateful coup d'état of 9 November 1799 (18 Brumaire, An VIII), when General Bonaparte took command of the armed forces of Paris, Chappe asked permission to telegraph the news that a Consulate composed of Sieyès, Roger Ducos and Bonaparte had replaced the Directory, but bad weather apparently stopped it from being sent — appropriately as 'Brumaire' signifies 'foggy.' A message was eventually telegraphed on 12 November (21 Brumaire) announcing the transfer of the legislative corps to St. Cloud and Bonaparte's appointment.

The new century opened with three trunk lines in service: *Paris-Lille, Paris-Strasbourg* and *Paris-Brest*. The extension of the Lille line to *Brussels* was completed in 1803. It was further prolonged in 1809 to *Antwerp* and thence via *St. Nicolas, Sas* and *Breskens* to *Flushing*. Its subsequent extension from *Antwerp* to *Amsterdam* is described in Chapter XII. The Lyons line was still under construction.

In October 1801 Villaret-Joyeuse, admiral at Brest, sent a letter under a flag of truce to Admiral Cornwallis, whose squadron was blockading the port, with a copy of a telegraphed message from Paris. Signed 'A. Chappe' and dated 5 October 1801 (13 Vendémiaire, An X), it read, 'Peace has been signed in London on 9 Vendémiaire.' The reference was to the signing of the Preliminaries of London by France and the United Kingdom — the Peace of Amiens was not concluded until the following March.

Depillon's Sémaphore

At this point we must digress from Chappe and his rapidly spreading invention to notice a new form of mechanical communication that was to have a profound effect.

In 1801, *Les Annales des Arts* contained a contribution from a former artillery officer named Depillon on his three- or four-arm machine, with details of its construction and working. Depillon recommended the three-arm variety as being more than sufficient as it furnished 301 signals, 105 more than the Chappe telegraph. He saw three important uses for it. The first was for coast work, to communicate between shore and ships in the offing and to follow ships proceeding 'dans leur marche'; the second for cross-country telegraphy, Depillon giving as an advantage over the Chappe telegraph the fact that the mast could be erected on any house or tower, thus saving cost. The third purpose was as a portable machine for the use of armies. The second point was hardly fair, as at many places Chappe machines had been placed on church and other towers.

The Minister of Marine almost at once adopted Depillon's invention for coast signalling, installing it along the whole of the French coastline from about 1803. Mindful of the special application of the invention, the precise French devised the term *sémaphore* to distinguish it from the true, continuous *télégraphe*. Later modified and improved, it remained in use until early in the present century. A book entitled *Signaux de la Ligne Sémaphorique établie sur les Côtes de l'Océan, depuis Flessingue jusqu'à Bayonne,* was officially published in 1807 by order of Vice-Admiral Decrès, Minister of Marine. The cover is labelled *Océan,* emphasising that there was a separate organisation for the Mediterranean coast (later also for French North Africa).

The coast was divided into 22 *arrondissements* between Flushing and Bayonne, including Dunkirk, Boulogne, Le Havre, La Hogue, Cherbourg, Granville, Saint-Servan, Cap Fréhel, Ile de Bréhat, Ile de Bas, Brest, Audierne, Penmark, Lorient, Le Croisic, Les Sables d'Olonne, Rochefort, Royan and La Tête de Buch. Each *arrondissement* had a headquarters lookout, not necessarily in the arrondissement town: for Dieppe it was at Pointe d'Ailly and for Rochefort at Saint-Pierre d'Oléron.

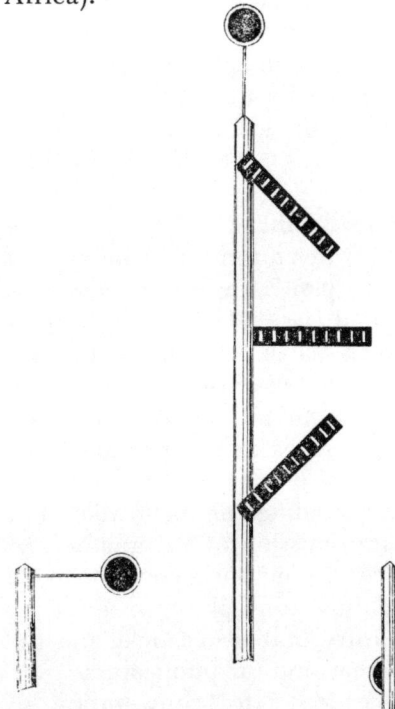

26. French coast semaphore.

The signal communication round the coast was not continuous, nor was it intended to be Even some neighbouring *arrondissements* were not in touch, because of a break in the coastline or for some other reason. At first the main object of the *sémaphores* was to report English fleet movements along the coast and pass details to the nearest naval base. Not until a later period were messages sent inland. The *sémaphore* was the forerunner of such telegraphs as Popham's, Pasley's, Parker's, Watson's, and Pistor's.

The *sémaphores* featured largely in Napoleonic War naval events, particularly after Trafalgar, when the British Admiral Cochrane was harrying and harassing French forces. Cochrane would descend suddenly on a *sémaphore* post, burn the building and machinery, remove the signal books and throw the guns of any protective battery into the sea.

Cochrane's first recorded anti-*sémaphore* exploit was when he commanded the *Pallas.* In May 1805 he wrote,

> Having found by experience that the French had organised a system of signal-houses, by means of which they were able to indicate the exact position of the enemy, so as to warn their coasts of impending danger, I resolved on destroying one on the Isle Rhe [Ré], at the town of St. Martin.

It appears he was not satisfied with one but treated others, including Port Vendres, in a similar fashion. At the end of January 1808, he was given a roving commisssion in the Western Mediterranean. He blew up numerous batteries, lighthouses, signal posts and towers. Sir Walter Scott wrote:

> Lord Cochrane, during the month of September 1808, with his single ship the *Impérieuse,* kept the whole coast of Languedoc in alarm, destroying the numerous semaphoric telegraphs [sic] ...

After the Napoleonic Wars the French *sémaphores* continued their work, in the service of merchant shipping. Much later they became postes *electro-sémaphoriques,* being connected with the inland electric telegraphs.

The original arrangement still applied, *Océan* relating to stations Dunkirk-Bayonne (numbering 97 in 1882), as distinct from 32 stations along the Mediterranean and six along the Algerian coast. But though the working was the same, machines and masts had been modernised, as the signal books of the *International Code* and early picture postcards and other illustrations show.

Later Schemes of the Chappe Brothers

Claude Chappe, who had always hoped that his invention might be applied to commerce, proposed to Bonaparte a pan-European commercial system stretching from Amsterdam to Cadiz and even taking in London, as he claimed to be able to correspond between Calais and Dover. He also proposed to relay stock exchange news daily. Yet another of his ideas was an official journal to be sent from Paris by post to all *départements* and supplemented by a telegraphed summary of the news of the day, approved by the First Consul and also relayed daily.

Unfortunately all these schemes were rejected as impracticable, but Bonaparte at least consented to the weekly transmission of the numbers of the winners in the national lottery. In this way the numbers could be made known in the provinces the same day as the draw took place, so upsetting the machinations of secret offices in the country that issued tickets in the intervals between the closure of the official bureau and the publication.

The new function brought in a much-needed extra contribution to operating revenue, for Bonaparte would agree to a vote of only fr. 150,000 a year — by 1799 the network was costing fr. 434,000 to run.

In 1804 Napoleon, as Emperor, prepared for his descent upon England with the Grande Armée encamped at *Boulogne*, to which the telegraph had been extended via Saint-Omer in 1803. He ordered Abraham Chappe to devise means of telegraphing across the Channel by day and night. For some three years the ends of the arms of the machine on the Louvre had experimentally carried lanterns for night use. Abraham adopted a simplified form. To overcome distance and fog, he dispensed with the end arms, made the indicator 5.5 m. long and 76 cm. wide and divided it into two separately moving parts, and erected a mast 9m. in height.

Each moving part was extended by a balancer which acted as a counterweight and bore at its ends a lantern with a parabolic reflector, which could move with the arms. Hydrogen, oxygen and carbonate of lime provided the fuel. Its light was said to be visible eight leagues away. The abandonment of the invasion plan ended the experiment. (In 1801 Abraham had devised a mobile field telegraph for the army and a dozen were made for the Army of the North. The machine was a small version of the standard design with hand-worked arms and mounted on a vehicle.)

On 23 January 1805 (3 Pluviôse, An XIII) Claude Chappe, then aged 42, took his life by throwing himself down a well at the telegraph headquarters. He had long suffered from a painful ear complaint that may have deranged him, although contemporary reports quoted a farewell message reading, 'I kill myself because I am weary of a life that burthens me. I have nothing to reproach myself with.'

Some authorities have alleged that Chappe was discouraged by a belief that others had anticipated his invention, but the supposition is unlikely. It is more probable that he was worried about the high costs of the telegraph as well as his health.

He was first buried in Vaugirard Cemetery but later re-interred in Père Lachaise with his brother Ignace, who died in 1829. On the reverse of his tombstone were engraved the signals of the telegraphic alphabet. The stone was later placed in the entrance of the Hôtel des Télégraphes (telegraph headquarters, Rue de Grenelle) at the base of the tower. The tomb is surmounted by a lead reproduction, now in bad shape, of the Chappe apparatus.

Ignace and Pierre-François succeeded Claude as joint administrators. Abraham was attached to the État Major General of the Grande Armée by Imperial Decree of 30 August 1805 (12 Fructidor, An XIII), with the rank of Colonel and a special uniform. His task was to translate messages for or from the Emperor and his immediate staff and keep the Emperor informed of troop movements reported by the telegraph. He is said to have taken precedence over even the Marshals!

The *Lyons* line had got only as far as the vicinity of Dijon when Cretet, Director-General of Highways and Bridges, ordered work to stop. Works already completed were to be safeguarded.

It had been proposed to erect two stations in Dijon, one on the Observatory tower, the other at Sainte-Chapelle. But when work was resumed on Napoleon's order of 19 June 1805 (30 Prairial, An XIII), the line was deviated well west of the town to give a more direct route and further slight modifications were subsequently made in the area.

When war was renewed with Austria, the Lyons line, completed in 1807, was further extended over the *Mont Cenis* and by way of *Turin* to *Milan*. Napoleon attached the greatest importance to this ambitious undertaking. In a letter of 18 March 1809 to Prince Eugène, Viceroy of the Kingdom of Italy, he said that he had ordered the line to be ready by 1 April. On 16 March he had instructed the Minister of the Interior, 'I desire that you will have the telegraph line to Milan completed without delay and that in a fortnight it will be possible to communicate with that capital.'

Eugène was told to hasten the works in Italy and extend the line as far as *Mantua,* from which it would be continued to *Venice*. This was to be an achievement indeed. Milan was some 1100 km. from Paris, via Lyons, Chambéry and Turin, and extension to Venice would add another 320 km.

In spite of Imperial insistence on speed, the line was not opened to Milan until the end of 1809, hardly surprising in view of the difficulty in taking it over the Mont Cenis Pass, by far the highest point ever reached by a regular optical telegraph line. The Col du Mont Cenis is 2082 m. and the Hospice, a telegraph site, 1929 m. above sea level. In winter, heavy snow in the pass delayed the regular Imperial couriers and must have upset telegraphy. Farther east, the mists of the Po valley must also have been a hindrance at times.

Between *St. Jean-de-Maurienne* and *Turin* there were stations at least at *Valloire, Modane, Bramont, Lanslebourg, L'Hôpital* (Mont Cenis hospice), *Susa* and *Rivoli*. The present Fort du Télégraphe, one of the cliffs rising 900 m. above the valley of the Arc, recalls the Valloire telegraph.

The final extension from Milan to Venice was put in hand without delay. The order for the execution of Andreas Hofer, leader of the Tyrolean revolt against Napoleon, was telegraphed from Paris to Mantua in February 1810. Incidentally, Mantua is sometimes said to have been served by a branch, but the list of stations and a contemporary map show it as almost certainly on the main line. Between 1810 and 1812 the Venice line was further extended along the shores of the Adriatic, as recounted in Chapter XI.

The *Lille* line again came into the limelight in 1806 and 1807. Over it first passed the news of Napoleon's entry into Berlin (4 November 1806), the capture of Stettin five days later, the capitulation of Magdeburg (19 November), the armistice with Russia on 5 July 1807 and the Peace of Tilsit on 19 July.

The events of 1809 once again drew particular attention to the *Strasbourg* line. On 4 April 1809 Napoleon sent an order over it that Oudinot's 12 battalions were to set out as soon as they had assembled. On 10 April he telegraphed his chief of staff Berthier (Alexandre Berthier, Prince of Neuchâtel, later also of Wagram, and Marshal of the Empire):

> I think the Emperor of Austria is going to attack soon. Proceed to Augsburg to act in accordance with my instructions and if the enemy has attacked before the 15th you must concentrate the troops at Augsburg and Donauwörth and everything must be ready to march. Send my guard and horses to Stuttgart.

Bad weather delayed receipt of the second message at Strasbourg by three days. Throughout the month the telegraph was sending Paris details of troop moves and subsequently the news of the victories of Abensberg (20 April) and Eckmühl (21 April).

In an instruction of 4 September 1809 Berthier entrusted General Guilleminot with the construction of a rudimentary telegraph to link *Strasbourg* with *Vienna*, which Napoleon had entered on 11 May. Professor Gachot has devoted much research to this hitherto little-known venture, which may have been inspired by a system that had enabled the defenders of Danzig to hold out long against Napoleon. The 'telegraph' consisted of a long pole on which were displayed either a white, red or black flag 12 m. in length. The stations were only some 3 km. apart and on church towers. There were more than 200 on a route via *Durlach, Pforzheim, Göppingen,*

Landshut and *Schärding*. They required more than 700 soldiers, allocated normally three to a station but sometimes four.

The significance of the signals was known only to Napoleon, Berthier and Clarke (Henri Jacques Guillaume Clarke, Minister of War 1807-14). The colour displayed or to be displayed was signalled by the Chappe telegraph between Paris and Strasbourg. In fog a soldier would carry a small piece of material of the appropriate colour from one station to the next. It is not known how long the service lasted.

The momentous news of the birth of the King of Rome, heir to the Imperial crown, on 20 March 1811 was telegraphed from Paris over all lines — the 20-word despatch took only 22 min. to reach Strasbourg.

A line between the Strasbourg line and Mainz had been projected about 1799 under the stimulus of Lakanal, then government commissioner for the four French Rhenish *départements*, but lapsed after some work had been done. Stations were apparently completed at the *Scherholm* near Wissembourg, *Eschkopf, Blokülb* and *Mont Tonnerre*.

On 13 March 1813, when the French armies were withdrawing from Germany, Napoleon ordered a branch to be built from *Metz* to *Mainz*, where Marshal Kellerman, Duc de Valmy, then commanded the fortress. The work was carried out with remarkable speed under Abraham Chappe's direction, for on 29 May the first message was sent. Directors and inspectors not only contributed to the cost of the line — fr. 105,000 — but also helped to build it. Some 225 km. in length, the line diverged from the Strasbourg one on the roof of the Palais de Justice in *Metz* and ran via *Tromborn, Siersberg* (Rehlingen), *Duppenweiler, Humes, Leitersweiler* and *Kreuznach*. There were probably 18 intermediate stations. The Mainz terminal was at first on the citadel on the Windmühlenberg but was soon moved to the tower of the Stephanskirche.

Although its life was fated to be only seven months the line gave great service to the Emperor and the army. For a short while the function of the Mainz station was akin to that of Strasbourg in 1799. News from Saxony and Silesia went by courier to Mainz, whence it was telegraphed to Paris. On 27 August 1813 the line carried the news of the Battle of Dresden to the Empress Marie-Louise.

On New Year's Day 1814 Blücher crossed the Rhine at Kaub and entered Kreuznach. By mid January the speed of the Allies' advance had put all the line in their hands. But the stations were not taken without a bitter fight. The operators destroyed the apparatus and, weapon in hand, were either taken or killed. A woodcut in Alexis Belloc's *La Télégraphie Historique* shows Allied troops raining missiles on a station and attempting to break down the door with a tree trunk while four telegraphists resist with pistol and flintlock. Gachot makes the point that as they knew only the significance of service signals the operators had no means of alerting Metz of their peril.

The fall of the Empire made no difference to the utility of the telegraph service, now shorn of its eastern extensions across the frontiers.

On 18 April 1814 the Ministry of Marine telegraphed to the naval chief at Boulogne orders to alter the name of the vessel *Le Polonais* to *Le Lys*, in readiness to embark Louis XVIII.

The First Restoration did not affect the status of the Chappes even though they had stood high in Imperial favour. They were first and foremost servants of France and were respected by all. Only they could translate and transmit political despatches and they were told State secrets before even the chiefs of Staff.

Hardly had Louis XVIII returned when on 6 August Baron de Vitrolles, his Secretary of State, asked for a statement on the extent of the telegraphs. The situation was then that lines radiated from Paris to Lille and Boulogne; Metz and Strasbourg; Brest; and Lyons.

The Chappes said that a branch from *St. Malo* to *Cherbourg* would be useful and cost fr. 40,000, and that an extension from *Lyons* to *Toulon* would cost less than fr. 200,000 for more than 100 leagues. The Channel and Mediterranean would then be linked.

Paris heard the news of Napoleon's escape from Elba by telegraphed despatch from Lyons at 11 am on 5 March 1815, four days after he had landed at Golfe-Juan. Three hours later the astounded Marshal Soult, created Duke of Dalmatia by Napoleon and Minister of War in the First Restoration, telegraphed General Bayer, commanding the 19th Division at Lyons, for more details. Thereafter, the Lyons line was working constantly at pressure, with the King ceaselessly demanding information from his nephew the Duke of Angoulême, in Lyons.

During the Hundred Days the telegraphs once more served the Emperor under whose rule the waving arms had been carried far beyond the confines of France itself. On 18 June 1815 the Lille line relayed a message that Napoleon had defeated Wellington and Blücher but on the 20th it carried the news of Waterloo. On 23 June all lines transmitted the intelligence that Napoleon had abdicated and on 4 July that the French army before Paris had capitulated.

On 31 July the lines were relaying the numbers of the winners in the Royal lottery. But there were more sinister tasks as they were pressed into the search for former Imperial high officers and partisans who had not made their peace with the royalists. News of the verdict on and the execution order for General Labédoyère was sent by telegraph on 19 August 1815, followed the next day by that of the arrest of Marshal Ney.

On 7 December the news of the verdict on Ney and his execution were relayed.

The Comte de Lavalette was more fortunate. On 9 January 1816 a message was telegraphed ordering his arrest at the frontier. But night fell before the despatch could get through and Lavalette, helped by English officers, made good his escape.

The telegraphs did not escape physically from the turmoils of the times. For instance *Concoeur* station near Dijon was attacked in June 1815 by four disguised men with rifles and another station was destroyed. In the same area, the operator of *Millery* station was accused by the local mayor in 1817 of passing on information relayed by telegraph. The charge may have been false or trumped up for, of course, operators were supposed to know the significance of only a few standard service signals.

The Second Restoration ushered in a new era for the telegraph. Lines were modified to suit the changed conditions. The Boulogne branch, built mainly with the invasion of England in mind, was replaced by a line from *Lille* via *St. Omer* to *Calais,* which was increasing its importance with the resumption of relations and travel between France and England. On 5 July 1821 the Camp Marshal at Calais sent the Minister of War a message from the British consul at Calais reporting the death of Napoleon on St Helena on 5 May 'des suites d'un abcès dans l'estomac.'

Abraham Chappe once more became Inspector General of Telegraphs, Ignace and Pierre-François keeping their functions as administrators. Louis XVIII made all three Chevaliers de la Légion d'Honneur.

The foundations of a truly national telegraph service were being laid, with Paris being linked with the capitals of many *départements,* and serving administrative, as well as military and naval, purposes.

In September 1821 the Council of Ministers voted for an extension from *Lyons to Marseilles* and *Toulon*, 760 km. from Paris, which was brought into service as early as 14 December. It ran by way of *Valence, Orange* and *Avignon.* A trunk line linking *Paris* with *Bordeaux* and *Bayonne* was first considered in 1820, when war with Spain seemed likely. It was finally decided on at the demand of the Ministry of War. Begun in 1822 it was completed on 3 April 1823. The route lay through *Orleans, Tours, Poitiers* and *Angoulême* to *Bordeaux*. The Ministry of Marine had preferred a route via Nantes and thence coastwise through Rochefort to Bordeaux. Nantes was later to be brought into the network, but Rochefort was never served.

M. J-F Massie has devoted much research to the Bordeaux-Bayonne section and its subsequent extension to Béhobie, close to the Spanish frontier. The present author is indebted for information to M. Massie's monograph, reprinted from the *Bulletin* of the Société de Borda.

At Bordeaux the telegraph used the old tower of St. Michel, an isolated tower, 54.7 m. high, west of the church of that name. Built in 1472-92 the tower originally bore a spire, which fell in a storm in 1768. The present spire was erected after the telegraph was abandoned.

The two machines necessary for operating in the Paris and Bayonne directions respectively were later joined by two more, one for the *Blaye* branch (1832), the other for the long cross-country line to *Toulouse* and *Narbonne* (1834). Until the Blaye line was given up in 1845 the summit of the tower must have presented a remarkable sight with its four telegraphs and the operators must have found conditions very congested, particularly as the station was a divisional point where despatches had to be encoded and decoded.

In a crypt of the tower lie 40 mummified bodies. Victor Hugo wrote in 1843 of the contrast between them — 'a council of spectres' — representing the eternal, and the telegraph, representing the ephemeral.

The municipality had agreed to the use of the tower in an access of patriotism, expecting the telegraph to be taken down after the Spanish war. When the telegraph remained, it sought legal redress but not until 1845 was bad feeling removed by the government agreeing to pay a modest rent. In 1830 some Bordelais took matters into their own hands. In an uprising on 30 July rioters dismantled all the telegraphic apparatus and cast it, with furniture from the Préfecture, into the Garonne, As Massie observes, had they kept calm, like the Lyonnais, they would have heard on 1 August the telegraphed report of the proclamation of the Duke of Orleans (the future King Louis-Philippe) as Lieutenant-General of the Realm.

There were nine stations in the *département* of the Gironde, 20 in the Landes and four in Basses-Pyrénées. One only survives, at Gradignan. Most of the towers were square, as at *Bayonne* (St. Etienne) and *Biarritz*. At some places, church towers were used, as elsewhere. At *Labouheyre* a new site was chosen in 1845 — a new church tower. The sceptical parson said of the proposed transfer of the apparatus that it would be done 'when geese have crests.'

Some accounts suggest that *Béhobie*, the final terminus, to which the line was extended in 1847, and *La Croix des Bouquets* nearby were two distinct stations but Massie decides that they were one and the same. The director at Béhobie was appropriately named l'Espagnol.

Much of the Landes was then a great desert, very well suited to optical telegraphy, save that in very hot weather the haze was troublesome. The *département* did not begin to be afforested until after the end of the Chappe telegraph.

In October 1833 a line was proposed from *Bayonne* to *Perpignan* via Tarbes and Pau, but was not built.

Massie records that in 1827 one signal took 12 min. to pass from Bayonne to Paris. A message containing 150 signals, equivalent to half a page of writing, passed in 1½ hours. The most rapid message he discovered was relayed on 16 June 1847, from the French Minister in Lisbon to the Minister of Foreign Affairs in Paris, which took only 1hr. 23min. to go from Béhobie to Paris.

Because of religious unrest in the S. the Chappes proposed a line from *Avignon*, on the Toulon line, to *Nîmes*, *Narbonne* and *Perpignan*. It was accepted in principle, but deferred.

The Later History of the French Telegraphs

Ignace and Pierre-François Chappe were retired on pension in 1823 and Count Kerespertz was named administrator under the authority of the Director-General of Highways and Bridges. René and Abraham, by then better known as Chappe-Chaumont and Chappe des Arcis respectively, became second and third administrators. In 1824 Ignace brought out his *Histoire de la Télégraphie* in which he set out some of the difficulties with which the family had to contend.

After Charles X had declared war on the Dey of Algiers in May 1827 government despatches were telegraphed from Paris to the maritime *prefect* at Toulon, for onward transmission by ship to the French squadron off Algiers. News of the capture of Algiers on 5 July 1830 was brought by ship to Marseilles and telegraphed to Paris and thence to provincial cities — it reached Strasbourg on 9 July, a telegraphic journey of 1,308 km.

Towards the end of Charles X's reign Alexandre Ferrier de Tourettes, a historian born at Draguignan in 1810, is said to have devised an acceptable means of making the Chappe telegraph visible at night. The Revolution of 1830 ended his experiment but in 1833 A. Ferrier et Cie., of 14 Boulevard Montmartre, issued a prospectus for a commercial telegraph network of eight lines, some apparently duplicating Chappe routes but others to serve Le Havre and Nantes. There were to be 350 stations. The capital was put at fr. 1,000,000 and annual working costs were estimated at fr. 900,000.

Information is conflicting about the method Ferrier planned to use. One account refers to two posts 3 metres apart, each bearing one arm, another to the use of a modification of the Chappe system. Again one refers to the adoption of two fixed lamps with two others moving round them like satellites, for night working, while another says that Ferrier proposed to use one fixed and one stationary lamp to each arm, with a fifth lamp able to move horizontally between them.

Ferrier tried his telegraph out before a government commission and sent 20 messages each of 12 — 15 words in an hour. But capitalists proved coy and the government soon put obstacles in Ferrier's way, *Le Moniteur* pursuing him with venom. It is said that Ferrier managed to complete a line between Paris and Rouen and sent stock market news over it before Norman peasants destroyed one or two stations.

A Law of 2 May 1837 confirmed that telegraphs were a government monopoly. Meanwhile Ferrier had removed his equipment and men to Belgium, where a rich senator is said to have commissioned a private telegraph for the pleasure of receiving news of the Spanish civil war before anyone else. Ferrier seems to have given sufficient satisfaction for a line to be built between Antwerp and Brussels, although it cannot have lasted long in view of the introduction of the electric telegraph between those towns in 1845.

An indication of the efficiency which the Chappe system had reached is given in an article in the 11 April 1829 issue of the *Journal des Débats*, which reported that the news of the elevation of Pope Pius VII left Rome by courier by 8 pm on 31 March, arrived at Toulon at 4 am on 4 April and reached Paris by telegraph four hours later. On 13 July 1830 news of a *Te Deum* to celebrate the capture of Algiers reached Strasbourg from Paris in 40 min.

In January 1829 the Director General of Posts proposed to take over the telegraph service but the Ministry of the Interior turned down the plan.

When the Revolution of 1830 broke out, Count Kerespertz resigned. He was replaced on 19 October by a *député* named Marchal, styled Commissaire du Gouvernement près les Télégraphes. Meanwhile both Abraham and René Chappe had received their *congé* at the hands of the new government. René had refused to hand over their records to the future King Louis-Philippe during the three days of the Revolution (*Les trois glorieuses*). Abraham had been absent during the fateful July, but Louis-Philippe decided that no Chappe should henceforth be in the telegraph administration. The indignation of the brothers comes out in a new edition of Ignace Chappe's book which Abraham published in 1840. At least the king was gracious enough to provide Abraham with an adequate pension for his 35 years of unstinted devotion to the nation's service.

Marchal soon gave place to the competent Alphonse Foy, nephew of a famous general. Under him the system reached its zenith. The centre of administration was moved to a special square tower at No. 103 rue de Grenelle, Paris, which still stands. One machine was mounted on top of the tower and others were installed one on each of the four sides of the tower, to serve each of the five lines radiating from Paris.

Foy drew up an ambitious scheme for a new radial line from *Paris* to *Le Havre* — not carried out — and concentric lines linking the trunk routes to enable messages to travel by an alternative route if the normal route were congested or out of action. The concentrics he proposed were:

> Avignon — Montpellier — Narbonne — Toulouse — Bordeaux
> Dijon — Besançon — Strasbourg
> Metz — Valenciennes — Lille — Boulogne — Caen — Avranches

In spite of the importance that the Government attached to the Foy plan, it was not carried out in full because of the many political complications at home and abroad, the Chamber of Deputies voting the credits sparingly and then only successively.

The *Avignon — Bordeaux* line received priority; it reached Montpellier by March 1832 and was completed throughout by August 1834. The *Dijon — Strasbourg* line got as far as *Besançon* in 1840 and no further. The great northern arc from *Metz* to *Avranches* was built only as an extension from Calais via Boulogne to *Eu*, where the king had a chateau, in 1842. Branches from Avranches to *Rennes* and *Nantes* and to *Cherbourg* were opened in 1832 and 1834 respectively, and from *Narbonne* to *Perpignan* in 1840.

It may be helpful here to recapitulate the location of the stations in Paris and environs after the establishment of the Tour Centrale station. Taking the lines in clockwise order they were: *Lille line: Montmartre* (on tower of the old church); *Ecouen; St. Martin-du-Tertre and Ercuis* (church tower); *Strasbourg line: St. Eustache* church tower, *Belleville and Gagny; Lyons/ Toulon line; St. Sulpice* church tower, *Villejuif* and *Athis-Mons; Bordeaux line: St. Sulpice* church tower and *Fontenay-aux-Roses; Brest line: Passy* and *Mont-Valérien.*

In the larger provincial cities and towns, stations are known to have existed as follows:

Calais (Citadel and Tour du Guet); Metz (Palais de Justice); Strasbourg (Cathedral); St. Malo (Cathedral tower); Brest (St. Louis church tower and telegraph director's house); Rennes (N. tower of Cathedral, telegraph director's house and St. Sauveur church tower); Nantes (N. tower of Cathedral and telegraph director's house); Dijon (Palais des États); Lyons (at St. Just, in the Rue Bourbon and at the old prefecture); Marseilles (at the Batterie de la Pinéole and at the Cimetière de l'Hôtel-Dieu); Toulon (Fort Lamalgue and the Tour de l'Horloge); Besançon (at 8 Place St. Jean and St. Pierre church); Orleans (Hôtel de Ville); Tours (Hôtel de Ville); Poitiers (Palais de Justice); Bordeaux (St. Michel tower); Narbonne (Archbishop's Palace); Montpellier (Observatory); Nîmes (Cité Foule); Perpignan (Citadel)

This list is incomplete as in some cities and towns there were more stations than those shown.

In April 1830 the Comte de Montureux proposed that the public should be allowed to use the telegraphs. An annual subscription would give the right to send messages, which would be separately charged for by syllables. Official despatches would still have priority. His useful suggestion was not adopted as the government wished the telegraph to remain exclusively political. In any case, the interruption to the service at night and in bad weather would have limited the value of the plan, and if the public had responded enthusiastically the system would soon have become impossibly congested.

Ferdinand Flocon became second joint administrator in 1834. For the *Calais-Boulogne* line he adopted a modification of the apparatus that had been proposed under the Empire by Durant, who became telegraphic inspector in Paris. The regulator was fixed and above it was placed a second, smaller regulator, also immobile.

During the 1830s there were several more efforts to perfect a night telegraph, as there had been when Napoleon meditated his descent on England. A trial at Montmartre in 1832 showed that a mixture of oxygen and hydrogen projected on to carbonate of lime gave a very good bright light, enhanced by painting the telegraph arms white. But the method was considered too risky and the supply of combustible to stations too difficult.

Trials were made between Paris and Tours using two lamps on the regulator and two on the ends of each arm — the invention of Dr Jules Guyot, who found that hydrogen and carbon gave a bright white light. The method was also tested satisfactorily between Montmartre and Écouen in 1839 but it was thought that the fixing and upkeep of the lamps by the operators would be difficult if not impossible in bad weather. Morice, a former director, was authorised to try another kind of lighting said to be cheaper than Guyot's but there was no lasting result of these ingenious attempts.

Throughout the history of the Chappe telegraph close attention was paid to the siting of the stations and in a number of places extra or alternative stations were erected to improve visibility. Perhaps the most interesting variations were those made in the 1820s and 1830s on the Paris — Lyons — Toulon line in Côte d'Or. Hence in 1820 a new station was built at *Bard* between *Pisy* and *Millery* and in 1822 the stations at *Bussy-la-Pesle* and *Fleury-sur-Ouche* were replaced by new stations at *La Chaleur, Remilly* and *Quemigny* on a new deviation.

Until 1824 there was no director between Paris and Lyons but in that year one was created at *Semur,* linked by courier with Dijon. In 1827 Magol, the director at Semur, suggested to the Prefect of Côte d'Or that the line should be deviated to pass through *Dijon,* an important administrative and military centre, as had been originally intended. Under Foy's administration the plan was adopted and on 4 May 1833 Magol's successor Kelsch was instructed to negotiate with the city council for the hire of rooms in the Tour Ducale for use as a telegraph station and accommodation for the director.

The *Journal de Carion* reported on 15 July 1835 that the Dijon station was operating. Some stations in the neighbourhood had to be abandoned and others opened on an amended route, and by 1837 the line passed entirely through Dijon.

In 1832 the Duchess of Berry, Louis XVIII's niece by marriage, tried unsuccessfully to raise Vendée against Louis-Philippe. While she was imprisoned in the citadel of Blaye, her warder General Bugeaud and Soult, then President of the Council and Minister of War, were in constant touch about her. As already stated, a branch telegraph line was even built from *Bordeaux* to *Blaye,* with seven stations; it functioned until 1845.

The Duchess of Berry was not the only exalted personage to trouble Louis-Philippe's government. In the evening of 31 October, 1836, the King and his cabinet were thrown into confusion by part of a message received in Paris by the Strasbourg telegraph. Dated the preceding day, it ran:

> This morning, about six o'clock, Louis Napoléon, son of the Duchess of Saint-Leu, who had in his confidence the colonel of artillery, Vaudrey, traversed the streets of Strasbourg with part of...

Fog had prevented the rest of the message from being sent and the King could only guess at the further exploits of the nephew of Napoleon who was destined to become Napoleon III.

In 1834 the Prussian Government sent Major O'Etzel (later General von Etzel) to study the French telegraph. He spent six weeks in France and submitted a report on it which is a mine of valuable information. He noted that the greatest distance between any two stations was 2½ leagues. In hilly districts they were often very close, as at Metz, where only 3,000 m. separated the main station in the town and the next station towards Paris. Between Bordeaux and Bayonne the stations were numerous because the vibration of the air in the Landes made transmission difficult over long distances.

At the divisional point of Metz, O'Etzel noted, each wing of the Palace of

Justice was surmounted by a telegraph, one the terminus of the line from Paris, the other the terminus of that from Strasbourg. The Director lived and worked in the central part of the horseshoe-shaped building, his office being placed to enable him to observe both telegraphs. Above the window of his dwelling stood a small apparatus repeating from or to the main apparatus. No news went direct to Paris from Strasbourg; it was decoded at Metz and transmitted thence afresh. The same procedure applied at the other divisional points.

From 10 to 15 stations, depending on the distance involved, were supervised by an *inspecteur*. The *inspecteurs* were chosen mainly from young men from the École Polytechnique and were paid fr. 2,000 – 3,000, plus travelling expenses. As well as looking after the technical equipment of the line, the upkeep of the apparatus and the structural condition of the buildings within his territory an *inspecteur* had to engage the station officials, whose appointments were confirmed by the *direction*, and discharge them, and make out the payrolls. The ablest *inspecteurs* were promoted to *directeur* or to headquarters.

The operators were locally recruited. There were two to each station. They relieved each other at midday, the duty being 24 hours on and 24 off. Each was paid 1fr. 25 a day. According to O'Etzel, only at one station in each district did the operators have to keep a journal, and received 75 c. extra for the duty.

As a rule French telegraph operators followed a second occupation, such as a handicraft, to bring in more money, a circumstance of which, as we shall see, Dumas makes use in *The Count of Monte Cristo*. While a married man was on duty, his wife would pursue this occupation. She was often familiar enough with the apparatus to be able to stand in for her husband if necessary. O'Etzel seems to have been struck by the fact that although the men were poorly paid the French telegraph was very efficient. He was led to point out differences in the French and German character. 'The Frenchman,' he wrote, 'is light-minded, has few requirements, turns with ease from one thing to another, and in the lower classes particularly is more active than the German. The latter is undeniably ponderous but thereby more solid.' As he observed, this difference in temperament meant that a German could never take on a hobby or secondary occupation at which he was likely to be disturbed. Conditions in the German telegraph service could therefore never be like those in France.

O'Etzel noted that on the Paris – Strasbourg and Paris – Lille lines, which he inspected, single signs (*signaux reglémentaires*) could be sent at an average speed of 12.18 leagues a minute in favourable weather. He recorded that such a signal had been sent over the new Prussian telegraph from Berlin to Koblenz and back, 145 leagues, in 28 min. that is, at 10½ leagues a min. The 'no news' sign had been sent there and back in only 15 min.

To demonstrate the speed of the French telegraph, observers had been stationed at the posts on the towers of Strasbourg Cathedral and Montmartre church. In good atmospheric conditions 30-40 signals were received in 15 min. or two a minute. In poorer light the frequency was 20-25 in the same period. In Paris, in dazzling sunshine, clear sky and light south-west wind, signals were received every one, two or more minutes from the Bayonne direction between 10 am and 1 pm. By the Prussian telegraph two signals had been passed per minute in clear air, but the average was 1 - 1½. French telegraphy was faster because only the divisional stations kept a record, whereas every station on the Prussian line did so.

The greater number of the signs in the Prussian system (8,910 against 2,304) meant that the same number of words could be made with fewer signs. O'Etzel noted the greater skill of the French *traducteur,* born of his longer experience, and he had learned that on some days up to six messages of 20-30 words had been transmitted. The Prussian telegraph, he reported, would become faster than the French once the operators had become more dextrous and the code book had been perfected.

The divisional system in France had the disadvantage that any time gained by the operators' skill was lost by the deciphering and reciphering at divisional points. On the other hand it was of value when bad weather interrupted the service. The whole line was then not necessarily put out of action. If bad weather or darkness caused a message for, say, Bayonne to be held up at Orleans divisional point, it could be sent on to Tours, the next divisional point, by messenger. Thence it would be telegraphed, if possible, to Bordeaux, the next point, or sent on again by messenger.

The Paris headquarters building housed the deciphering office (*Bureau de traduction*), the administrative offices and the living quarters of the administrator, translator and other officials. In the courtyard were a guardroom and stables for a cavalry detachment of some 20 horses. The cavalry were posted in front of the building and acted both as messengers for the rapid delivery of messages and as a guard.

The *administrateur-en-chef* received a salary of 15,000 fr. and enjoyed free quarters in the building. Under him were a first and second *administrateur adjoint* and the following services: *bureau de traduction* with *traducteur-en-chef, secrétaire du cabinet,* and clerks; *bureau du personnel; bureau de la comptabilité et des fonds; bureau de matériel.* The first office enciphered all messages leaving Paris and deciphered those coming in.

At each terminus of a line, as well as at the divisional points, was a *directeur,* with his *adjoint, secrétaire,* etc., who were concerned with the coding and transmission of messages. The *directeur* received between 5,000 and 6,000 fr. — a near contemporary French source says fr. 2,400-2,500 — and had free accommodation at the office.

Some additional administrative details from French sources may be given here to supplement O'Etzel's information.

In Louis-Philippe's reign, telegraph officials wore a silver-embroidered uniform similar to that of officials of the Highways and Bridges administration. Directors were allowed to have embroidery on collars, pockets and

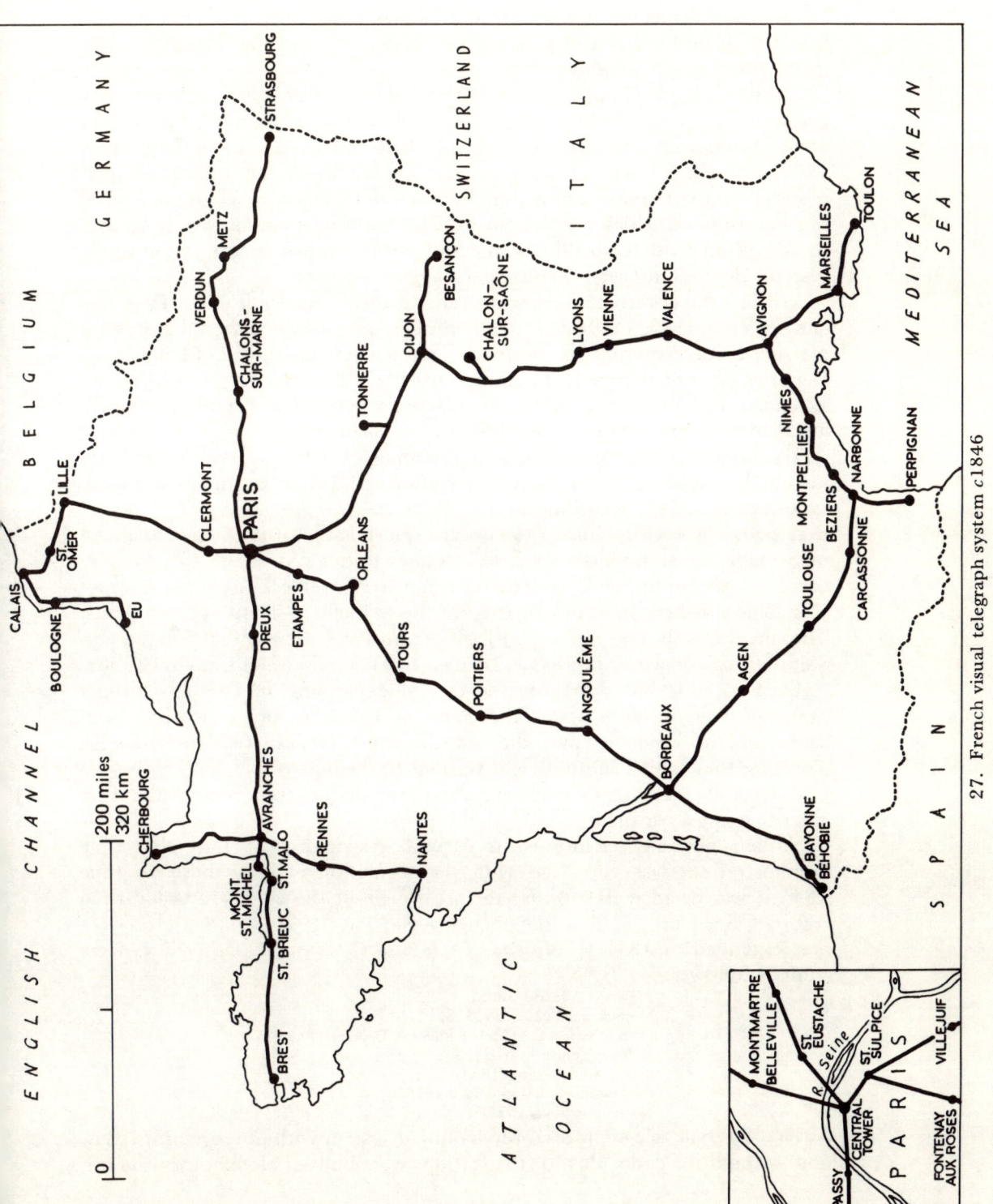

27. French visual telegraph system c1846

facings, and inspectors on collars and facings only. Operators wore a short jacket with uniform buttons showing a Gallic cock and the words *Administration des Lignes Télégraphiques.*

Although René Chappe had renounced direct association with the telegraph, he retained a keen interest. On 9 June 1839 he wrote to the Minister of the Interior to say that the telegraphs had cost fr.18 — 20,000,000 since 1792. He said he had succeeded in reducing still more the number of signs required and cut transmission time by one-third, if not by a half, and asked the government, which paid about fr. 1,000,000 to maintain the telegraph, to allow him to supervise the application of his improvements to the whole system. He does not seem to have persuaded it to do so.

In 1844 the government drew up a new project for a Paris — Le Havre line and for completing the Metz — Avranches peripheral connection, but in view of successful experiments with the electric telegraph the Chamber of Deputies did not debate it. As stated earlier, the Bayonne line was, however, extended to Béhobie in 1846, doubtless to form a link with the newly installed Mathé system in Spain, described in Chapter XIX.

The system in France was by then at its most widespread, with some 5000 km. of line and no fewer than 534 stations. It had even an international function. By 1844, according to *The Illustrated London News* of 6 July of that year, it was a vital link in the inward transmission of the Indian Mail. All mail made up at Bombay went by steamer to Suez, overland to Cairo, by canal to Alexandria and by steamer again to Marseilles. There, an abstract was made for the instant information of the French and British governments. The abstract was telegraphed to Paris. From Paris the abstract for London went by mail coach to Boulogne, steamer to Folkestone and train to London.

Even so, this was not quite the end of expansion. In 1850, two short branches were opened from the Paris — Lyons section, the first from *Massangis* to *Tonnerre* and the second from *Mary* to *Châlon-sur-Saône.* Together they added another eight stations to the network.

After 1847 despatches sent from Paris over all lines were preserved in files each covering a fortnight.

To aid military operations in Algeria, Alexandre, of the telegraphic administration, was entrusted with the formation of lines there. In June 1844 it was decided to organise them throughout the colony, a task carried out by César Lair. The first line opened was from *Algiers* to *Blidah* (1837). It was extended in 1847 to *Miliana, Tenes* and *Orléansville.* Other lines were built as follows:

        Algiers — Aumale (1848)
        Mostagenem — Médéa (1848)
        Oran — Mascara — Sidi-bel-Abbès — Tlemcen (1853)
        Dellis — Bougie — Sétif (1853)
        Sétif — Constantine (1853)
        Constantine — Batna — Biskra (1854)
        Bône — Guelma (1854)

The Algerian telegraph used the modified system with the regulator fixed, and a simplified code. Most of the stations were built as blockhouses and the clear atmosphere enabled some to be up to 12 km. apart. Mobile telegraphs carried on mules were also used in Algeria.

Foy constantly improved the system in France and extended the vocabulary. Flocon carried out successful experiments on the Narbonne — Perpignan section, a difficult one to work because of high winds. By modifying the mechanism he raised the speed of transmission above that of other sections.

Nevertheless, Foy was so impressed by the electric telegraph between Paddington and Slough on the Great Western Railway in England that a Cooke and Wheatstone apparatus was imported into France in 1842. After being modified by Louis Breguet it was installed on the Paris — Versailles Railway. In May 1845 the electric telegraph was opened alongside the Paris — Rouen Railway and in July 1846 authorised between Paris and Lille.

The optical system thereupon began slowly to disappear, although during the 1848 Revolution there was a brief revival under the administration of Flocon, who was devoted to it. Under the second administration of Foy, while the electric telegraph was being gradually but systematically substituted, messages might be sent by either means and the Foy-Breguet electrical apparatus was specially devised to reproduce the Chappe signals and so ease the transition. Only under Foy's successor Vougy, administrator-general from 1854, did Morse replace Foy-Breguet. Massie states that the last message from Bayonne to Paris by Chappe telegraph was relayed on 24 March 1853. Belloc records that the last Chappe station went out of service in 1856. Pamart says that the last general message sent by Chappe telegraph was the announcement of the capture of Sebastopol in September 1855. The equipment was sold or scrapped as the lines were abandoned.

The French sent out both visual and electrical telegraph apparatus for the Crimean war. An optical line of seven stations was installed along the Black Sea coast between Varna and Balchik, where troops embarked for the Crimea. It operated from 15 August to 15 November 1854.

The army telegraph service was subsequently divided. One section built an electric telegraph from Varna to Bucharest to link with the Austrian electric system. The other took the optical material to the Crimea where it was used to connect headquarters, strategic points, armies in the field, detached divisions and supply points.

The telegraph indicator arms were made of sheet steel instead of wood, for lightness. A post could be set up in less than 20 min. and transmission was rapid. For some time officials and operators camped in tents and not until November 1855 were the permanent stations given buildings. Only one operator was allocated per station and he had to keep watch 16 — 18 hours a day. More than 4,500 messages were sent.

During the Kabylie (Algeria) expedition of 1857 mobile visual telegraphs linked Algiers with the advancing French troops.

Chappe's invention had rendered immense service to the French state in peace and war and it is not too much to claim that it had played a notable part in conserving and consolidating national unity in troubled times.

As a young man, Victor Hugo had satirised the telegraph:

> Tandis qu'en mon grenier rongeant ma plume oisive,
> Je poursuis en pestant la rime fugitive,
> Ce maudit télégraphe enfin va-t-il cesser
> D'importuner mes jeux qu'il commence à lasser?
> La devant ma lucarne ! Il est bien ridicule
> Qu'on place un télégraphe devant ma cellule !
> Il s'élève, il s'abaisse ; et mon esprit distrait,
> Dans ces vains mouvements cherche quelque secret.
>
> Bon ! me dis-je à la France, il annonce peut-être
> Des Ministres du Roi qui serviront leur maître.

While visiting Mont St Michel in 1836, Hugo was almost struck by the movements of the telegraph and lost his hat! Perhaps the circumstance for ever warped his view of it, for his epitaph on it was also ironic:

> Tout se dit avec l'A.B.C.
> L'A.B.C. partout F.E.T.
> Longtemps par le sort K.O.T.
> Nous cesserons de V.G.T.
> Le télégraphe est A.G.T.
> De fureur il est R.I.C.
> Il ne peut supporter l'I.D.
> Que du monde il est F.A.C.
> Oui, malgré son R.E.B.T.
> Trop longtemps, il nous est R.S.T.
> Debout comme une D.I.T.
> Vieillard que le temps A.K.C.
> C'est une affaire d'S.I.D.
> Son F.I.J. est même O.T.
> De lui nous allons R.I.T.
> Car il est enfin D.C.D.

A more gracious adieu was sung by Gustave Nadaud in *Le Vieux Télégraphe*, of which the following are two stanzas:

> Que fais-tu, mon vieux télégraphe,
> Au sommet de ton vieux clocher,
> Sérieux comme un épitaphe,
> Immobile comme un rocher
> Hélas, comme d'autres, peut-être,
> Devenu sage après la mort
> Tu réfléchis, pour les connaître
> Aux nouveaux caprices du sort.
>
> Tu fus l'enigme de notre âge;
> Nous voulions, enfants curieux,
> Deviner ce muet langage,
> Qui semblait le parler des dieux,
> Lorsque tes bras cabalistiques
> Lançaient à l'horizon blafard
> Les mensonges diplomatiques
> Interrompus par le brouillard!

Perhaps the best-known literary reference is Dumas' description of the telegraph — 'the insect with the black claws and the terrible name' — the sight of which aroused such emotion in *The Count of Monte Cristo*. The Count, it will be recalled, visits the telegraph station on the ridge above the plain of Montlhéry, on the Bordeaux line, some 40 km. S. of Paris.

The tower, Dumas tells us, has three floors, the lowest containing gardening tools, the second being the operator's dwelling and the topmost the operating room. Monte Cristo works on the man's passion for gardening and induces him to accept a bribe of fr. 25,000 to ignore a message from the south and to send on instead to Paris false messages supplied by the count, who wishes to engineer a fall in Spanish bonds to ruin his enemy Danglars. (In truth, the operator would not have known the content of the message from the South and the Count, unless he had suborned divisional staff, would not have known the code).

The success of Chappe encouraged many would-be emulators. In 1797 Breguet and Béthencourt brought out a telegraph consisting of an arrow turning about a mast. The device was well publicised, though a criticism made was that the angles were difficult to distinguish, but money was too scarce for the government to subsidise the invention, particularly as by then the Chappe system was already established. However, in 1799 the Directory voted a sum to establish a *vigigraphe* between Paris and Le Havre. This device was the invention of a naval engineer named Moncabrier, a harbour official and a mathematician. A beam moved up and down one face, and a disc up and down the other face, of a kind of ladder, which had fixed beams at top and bottom. The *vigigraphe* had been devised as a ship-to-shore communication but Depillon's *sémaphore* was found to answer better. In the event, the Paris-Le Havre proposal was not carried out.

In 1820 Rear-Admiral de Saint-Haouen brought out a telegraph for day and night use. Three wickerwork globes, painted black, were hung from a yardarm 18 ft. long at the top of a 30 ft. mast. A fourth globe moved horizontally 2 ft. above the station house, to indicate thousands. Cords ran from the globes into the operating room. Globes 1, 2 and 3 represented units, tens and hundreds. As it was difficult to distinguish the positions of the globes, they were hoisted to take up different positions, without much improvement. Lanterns replaced the globes at night.

Louis XVIII is said to have watched a trial on Mont Valérien, from the Tuileries. Commissions of naval officers reported favourably and the Council of Ministers proposed the system in 1821 for use on the projected Paris-Bordeaux line. But a more extended test is believed to have failed.

Information is confused but it would seem that Saint-Haouen had persistence and influence for *Le Moniteur,* the official government organ, reports a trial that took place in October 1822 of a system of night telegraphy he had devised, presumably applied to the Chappe telegraph. The test was conducted on the Paris-Orleans section of the new line to Bordeaux.

At 21.13 hours Paris sent the preparatory signal to Orleans. At 21.20 a message announcing; 'His Royal Highness, Monsieur the King's brother, has arrived at Montmartre.' It was followed at 21.15 by 'The Prince asks what the weather is at Orleans this evening.' Orleans answered at 21.52 – 21.53; 'much wind.' Paris acknowledged this at 22.00, then at 22.07 sent; 'The Prince has withdrawn,' followed a minute after by 'The Prince has been very pleased with the service over the whole line.' He followed with the statement that he appreciated the 'impenetrable' character of the system and the utility of its application to the safety of commerce and navigation.

Abraham Chappe was stung to reply that the telegraph had been operating for 29 years and needed only a minute per word to send a message more than 100 leagues, twice or three times the distance between Paris and Orleans. 'When the government wished, the telegraph could be used with the same rapidity day or night,' presumably a reference to the fact that night transmission had proved feasible even if it had not been thought necessary to apply it. Indeed that year Chappe had been testing lanterns with reflectors — like those with which he had experimented at Boulogne in 1804 — at the Paris headquarters and on Montmartre church. The application was successful even though the arms had not been strengthened to take the extra weight, but no general application followed. In the event, nothing more was heard of Saint-Haouen's proposals.

Ennemond Gonon's telegraph comprised two masts, one 10.6 m. high, the other 8.5 m. high, each carrying two movable arrows worked by levers. The space between the masts was occupied by six movable valves or flaps, controlled by pedals. Gonon claimed his invention gave much more rapid communication than the Chappe system.

As well as by his tombstone, Claude Chappe was long commemorated in Paris by a handsome bronze statue erected in 1893 at the junction of the Boulevard Saint-Germain and the Rue du Bac. It showed Chappe standing in front of his apparatus, telescope in hand. The statue became a favourite rendezvous, especially as it was close to a Metro station, and Parisians would say: 'See you at the Chappe.' Unfortunately, this fine piece was removed to be melted down during the German occupation in 1942.

Some at least of the telegraph towers still stand, whole or in part. In the Paris region, for instance, the three-storey square tower at *Trou d'Enfer* (Bailly), north of Versailles and close to the Autoroute de l'Ouest, remains intact. The station at *Mont-Valérien,* on military territory nearer Paris, retains only its ground floor. Other stations known to have survived exist at *Mont-Dol* (Avranches — St. Malo section); *Bain de Bretagne* (Rennes-Nantes section); *Concoeur* (Dijon-Lyons section); *St. Georges* (Paris-Dijon section); and *Gradignan* (La Burthe) (Bordeaux-Bayonne section). The Gradignan station is a round tower about 8 m. in height and 3 m. in diameter; it became a water tower after the telegraph was given up.

Though many stations were demolished after the system was abandoned, others have fallen into ruin and still others have disappeared in more recent times, there are doubtless yet more intact or in reasonable condition and their discovery will engage the writer.

But pride of place must undoubtedly go to the reconstructed station at *Haut-Barr,* a high rocky spur some 6 km. from Saverne, Alsace. There, on 31 May 1968 Les Amis de l'Histoire des PTT d'Alsace inaugurated a most interesting and praiseworthy venture, the restoration to working order of the ruined round-tower type station. (The first Haut-Barr telegraph was built within the perimeter of the 12th-century castle. It was replaced within a few years by a new station about 200 m. distant. The course of the Sarrebourg-Strasbourg section proposed in 1794 was modified in 1798 and again in 1814 and 1824.) The society has created a faithful reproduction of the second station, complete with all its mechanical equipment. A former adjacent house for the operators has been rebuilt in the original style but on a slightly larger scale to serve as a museum of the Chappe telegraph.

The reconstruction is a fascinating addition to the tourist attractions of the region.

A postage stamp showing a Chappe apparatus was issued in 1972 — stamps depicting Claude Chappe had appeared in 1944 and 1949.

In France some of the Chappe sites have been used in connection with telecommunication (TV, telegraph and telephone) trials and one at least — *Mont Affrique* near Dijon — has been permanently adopted. The modern communication, be it noted, needs only eight instead of some 60 intermediate stations for relaying between Paris and Lyons.

CHAPTER XI

Italy and Switzerland

In Chapter X it has been shown how on Napoleon's orders the French telegraph was extended over the Mont Cenis into Piedmont, Lombardy and Venetia. Like the other lines extended beyond the bounds of France itself, the line to Milan and Venice ceased working in 1814. Dr. Augusta Lange of Turin has found a commemorative tablet in the courtyard of the town hall at Susa showing the names of two persons 'who from 1806 to 1814 have been at this telegraph post.'

In 1810 it had been decided to extend the Venice line still further, to cover the N. Adriatic coasts, in one direction N. to *Trieste* and in the other SE. to *Monte Santa Lucia* (Porto San Benedetto del Tronto) on the border of the Papal States. The main object was to protect the coast against the British navy. To this end telegraphs were also erected along part of the coast near *Genoa,* and also that near Naples, in the Kingdom of Naples ruled by Joachim Murat, Napoleon's brother-in-law. Records of the Direzione Belle Arti del Comune di Genova show that one of the Genoese stations stood on the hill of *San Benigno,* near La Lanterna, and communicated with *Savona* and *Sori.*

More detailed references have survived of the Adriatic line. In a little over 330 km. between Venice and Monte Santa Lucia there were 36 stations, some on sites which had been beacon posts since the first century AD. They included *Punta Ravenna, Rimini, Fano, Senigallia* and *Ancona* (Monte ai Cappucini).

In view of the predominantly defensive nature of the Adriatic and the other Italian coastal lines it may seem surprising that Chappe's essentially overland system was used instead of the French coast semaphore, arranged to form a continuous telegraph.

The masts of the Chappe stations along the Adriatic were 12 m. in height and 90 cm. wide. They carried the normal long indicator and two short arms. Each station had two operators — seamen were chosen where possible — whose duty period was dawn to dusk. As in France, the divisional system was used, but the divisional points averaged only five stations apart. The line was managed by a general inspector in Venice and divided into three sections each with an inspector and a sub-inspector: Venice — Trieste; Venice — Rimini; Rimini — Monte Santa Lucia.

It appears that the Adriatic system was given up even before the fall of the Empire in 1814. The end of the Napoleonic era and the departure of the British fleet laid open the Italian coasts to the attacks of Turkish and Barbary corsairs. In 1817 the harbourmaster at Ancona suggested to the Papal Secretary of State in Rome, that the Adriatic telegraphs should be reinstated to warn of approaching pirates. Although the idea was agreed in principle, the country, impoverished by war and blockade and weakened by famine, could not afford such a project.

The Naples telegraph seems to have continued in service after the end of French rule, possibly for many years. Ironically, its most dramatic use was against Murat.

The Naples State Archives contain a message dated 20 May 1815 from Adjutant-General Galleni at Monteleone to the Minister of War in Naples advising of the arrival of the 46 sail of the Anglo-Sicilian expeditionary force. On 28 September 1815 Murat and a small force landed at Pizzo, Calabria, in an ill-fated attempt to regain his kingdom. Jean-Paul Garnier in an article in issue No. 166 of *Historia* relates that the news was telegraphed to Ferdinand IV, who was attending the San Carlo Theatre in Naples. On 11 October Murat, his fate then sealed and a captive in the fortress of Pizzo, asked General Nunziate: 'If you were ordered by telegraphic message to put

28. Telegraph used by professors of Jesuit College of Schwyz, transmitting messages over Lake Lucerne from Brunnen to Seelisberg, 1847.

me before a military commission, would you do so?' The evening before, the general had told him that he had received instructions by telegraph: 'You will consign him to...'

In Switzerland the Chappe telegraph was, of course, well known and much discussed, but never tested or applied as such. Unstable weather conditions were thought to be too great a deterrent. Even the traditional fire signals on mountain peaks throughout the country were often handicapped by fog or mist, either in the valleys or in the mountains.

The State Archives of Schwyz contain a pencil drawing showing 'how the professors of the Jesuit College acquaint their pupils and the people of Schwyz with the new invention.' The drawing was probably made just before the religious war of 1847. The reference is to an imitation of the Chappe system by Johann Nepomuk Schleuniger, professor at Lucerne in 1846-47. It was used during the 1847 war under the name of 'Jesuit Telegraph' but with indifferent results.

CHAPTER XII

Netherlands

We have seen that in 1809 the Paris — Brussels telegraph was extended via Antwerp to Sas and Flushing. In the same year a line was opened from *Amsterdam* to *Ossendrecht* near Antwerp, the headquarters of King Louis Bonaparte's Marshal Dumonceau. It ran from the Westerkerk in Amsterdam via *Aalsmeer, Langeraar, Waddinxveen, Nieuwekerk, Rotterdam, Klaaswaal, Willemstad, Standdaarbuiten, Steenbergen* and *Bergen-op-Zoom*. A station at *Woubrugge* seems to have been substituted for that at Langeraar soon after the line came into use, and an additional station was built at *Ierseke* on South Beveland, on a branch from Bergen-op-Zoom.

A second line followed, from Amsterdam to the palace of *Het Loo*, with intermediate stations at *Muiden, Naarden, Eemnes-Buiten, Nijkerk, Putten, Garderen, Elspeet* and *Niersen*.

The apparatus consisted of a wooden cross hanging from a yardarm and furnished with five discs. The system lasted only until Louis Bonaparte's kingdom was absorbed into the French Empire on 9 July 1810. In that year there was a proposal for a similar telegraph to link The Hague and Brussels through Rotterdam, Willemstad, Steenbergen, Bergen-op-Zoom, Antwerp and Malines.

Chappe Line

In October 1810 Abraham Chappe went to Amsterdam to supervise the extension of the Paris-Lille-Antwerp line to Amsterdam, with stations between Antwerp and Amsterdam at: *'sGravenwezel, Hoogstraaten* (Meerle), *Bavel, Dongen* (near Breda), *Sprang, Heusden, Veen, Fort Loevestein* (on the Waal), *Leerdam, Hagestein, Houten, Utrecht* (St. Jacob's tower), *Westbroek, Vreeland* and *Ouderkerk*. The Amsterdam station was at the Weesperpoort.

One French source lists Chappe stations in Holland at Kudelstaart, Vianen, Heukelom, Bommel, Crevecoeur, Dieren, and Gelder. This selection presents a mystery as, although the first five are on a reasonably direct line SW. and S. of Amsterdam, there is a Dieren well to the NE. of Arnhem — such a line, with sharp changes of direction, would seem to serve no obvious purpose, except perhaps an ephemeral military need. In any case, the Amsterdam — Crevecoeur section roughly parallels the Amsterdam — Fort

Loevestein section of the Amsterdam — Antwerp line.

This venture was also short-lived. Some of its stations, certainly that at Amsterdam, were destroyed as symbols of French rule after French troops were recalled on 15 November 1813 — even the coast signal stations are said to have suffered likewise at the hands of francophobes, though the reference may apply only to the easternmost extension of the French coast semaphore at Flushing and not to the yard-and-ball stations of Dutch origin thence to Den Helder. In any case the Lille — Amsterdam line, like the other Chappe lines beyond the borders of France itself, did not outlast the Peace of Paris.

In May 1815, in the newly-established Kingdom of the Netherlands, a telegraph was ordered to be installed between The Hague and Tournai, via Brussels, because of the threat from Napoleon. In view of the massing of the Allied armies the order was rescinded.

Lipkens's Telegraph

The next and last telegraph in the Netherlands operated in the 1830s. In August 1831, because of the uprisings in the southern provinces which led to Belgian independence, William I ordered A. Lipkens, Chief Engineer to the Ministry of Home Affairs, to establish a line within 14 days between the Residency at *The Hague* and the army in North Brabant, where the Prince of Orange had his headquarters. Resting only in his coach and sometimes rousing his workmen in the middle of the night in his haste to finish the job in time, Lipkens succeeded in completing the section between The Hague and *Breda* in only 11 days.

By Royal Decree of 16 December 1831, First-Lieutenant G. Jooss was appointed telegraphic director. The line was extended to *'sHertogenbosch* and placed under the control of the Topographical Bureau of the Ministry of War.

The stations between The Hague and 'sHertogenbosch were: *Delft, Overschie, Rotterdam, West Barendrecht, Puttershoek, Willemsdorp, Terheijden, Breda, Dongen, Loon-op-Zand* and *Helvoirt*.

The Lipkens telegraph, which Jooss declared to be 'certainly the simplest which has come into operation up to the present day', consisted of six discs of wickerwork, 110-120 cm. in diameter. They were attached to spars and could be moved to present either their face or edge. Of the 63 signs possible, 10 were reserved for numbers, 26 for the letters of the alphabet, and the remaining 27 for special messages. The signbook was divided into groups of 999 numbers, signified by the letters of the alphabet, except H, which showed that the preceding sign betokened a number.

Each station had a *seinmeister,* paid three florins a day, and his assistant, who received one florin. They were expected to keep watch by telescope every five minutes at least from 15 min. before sunrise to 15 min. after sunset. The preparatory signal was first given by the station at The Hague, to which it was repeated back, and this was followed by the time signal.

The favourable results of this telegraph in military service induced the

Director-General of Marine to propose its use in Zeeland, where communications were often held up in winter. In the summer of 1832, Jooss was instructed to select sites for a line from Breda to *Antwerp* and to *Flushing*, forking at Bath. He established them at *Zevenbergen, Fijnaart, Steenbergen, Bergen-op-Zoom, Bath, Lillo, Antwerp* (Citadel), *Waarde, Kruiningen, Heinkenszand* and *Arnemuiden*. There were also two stations at *Koudekerke* and *Domburg* in Walcheren, communicating with Flushing. The line came into use on 1 January 1833 between Lillo and Flushing. Antwerp was never served as the citadel there was ceded by the Dutch in 1832.

Although Jooss experienced difficulty in finding buildings high enough on which to erect the telegraph, the apparatus was so simple that the drawback was not so great as had been feared. There was little delay in transmission, as the following table shows:

The Hague to:	seconds	no. intermediate stations
Breda	30	7
Bergen-op-Zoom	38	11
'sHertogenbosch	50	11

Remarkably, the telegraph was only out of service because of bad weather for 20 days a year on average. No ships could approach Walcheren or the Wester Schelde without the news being known at the Residency within a few minutes. Jooss maintained that the telegraph was quicker than either the French or the Prussian which was then just coming into use.

After the Netherlands had adopted the 24 articles of the London Conference of 1839, the army reverted to a peacetime footing and the fortresses were demilitarised. The telegraph, being solely for national defence, became redundant and was given up by Royal Decree on 24 September 1839.

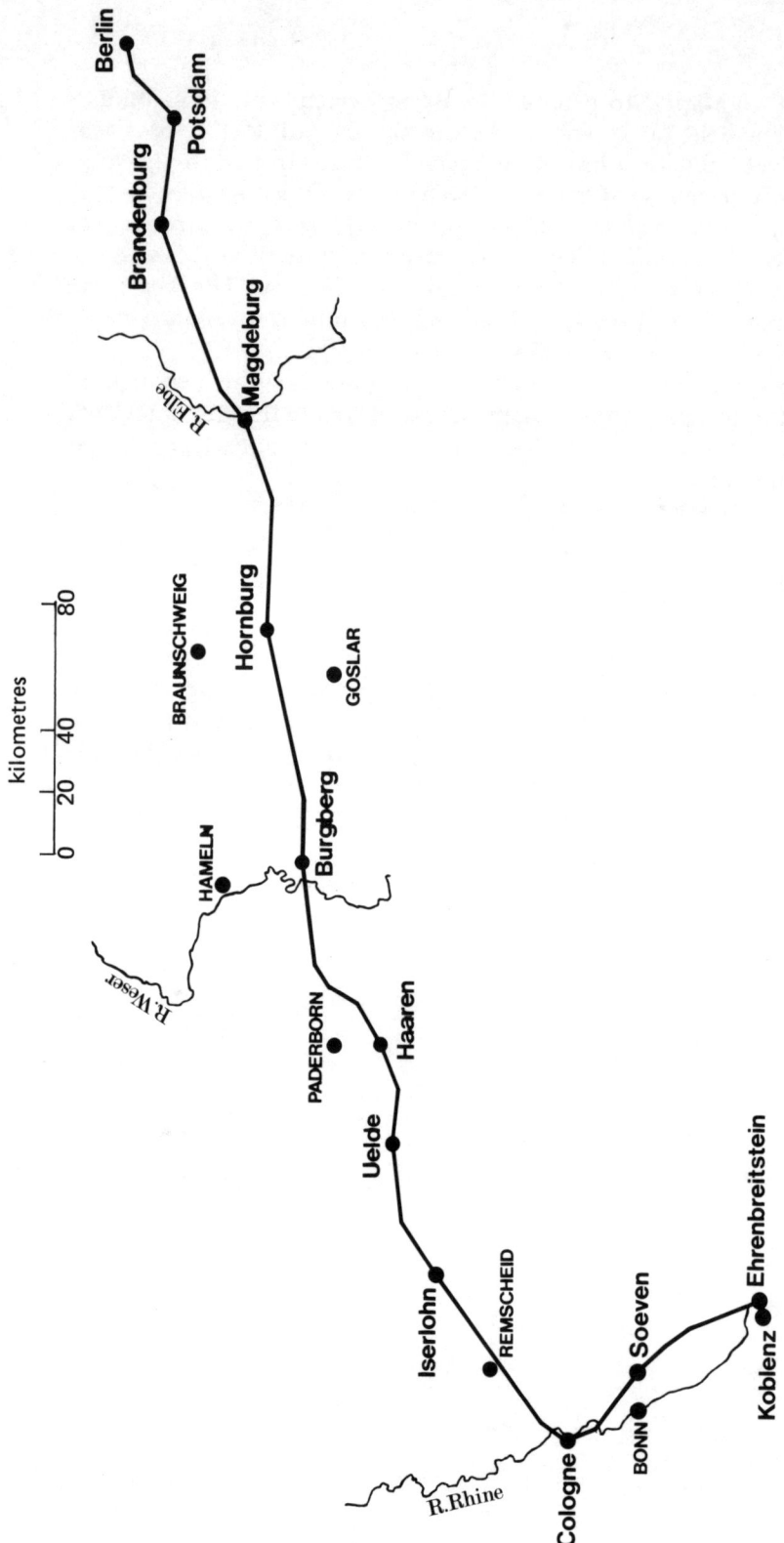

29. Berlin – Koblenz telegraph route.

CHAPTER XIII

Germany

In the German states of a century and three-quarters ago there was no lack of interest in improving communications but political disunity made all experiments local and transient.

On 22 November 1794, Professor Böckmann of Karlsruhe telegraphed a distance of 1½ leagues, sending rhymed good wishes to the Margrave Karl Friedrich of Baden on his birthday. Two lines ran: 'O Prince, see here what Germany saw not yet, How today the telegraph sends thee good wishes.'

No record has apparently survived of Böckmann's apparatus, though we know that he tried to modify Chappe's telegraph, of which details were brought from Paris to Frankfurt-am-Main by a former member of the Bordeaux *Parlement*. It has been noted that two working models of the French telegraph were made at Frankfurt and sent to the Duke of York in England. The fact may account for suggestions that there was a telegraph in Frankfurt in 1798.

Meanwhile, on 30 October 1794, two-and-a-half months after Chappe's telegraph had proved itself by announcing the fall of Le Quesnoy to the Convention in Paris, Senator Günther of Hamburg proposed to the Hamburgische Gesellschaft zur Beförderung der Künste und nützlichen Gewerbe that a telegraph should be established between Hamburg and Cuxhaven for rapidly transmitting shipping and commercial news. The Referent, Domherr Dr. Meyer, warmly supported this sensible plan but nothing was done. In December 1796 Meyer told the Gesellschaft of a visit he had made that year to study the French telegraph. The Gesellschaft, which lacked the funds to finance such a scheme, contented itself with bringing it to the notice of commercial interests in Hamburg.

Schmidt's Telegraph

Although the Napoleonic Wars and their aftermath retarded technological progress in Germany, the idea of a Hamburg-Cuxhaven telegraph was not forgotten, but it remained a paper scheme for 40 years. In 1836 Johan Ludwig Schmidt of Altona floated a company to carry out the long-deferred project. Little time was lost and the line was ready in February 1837. The first terminus was in *Altona*, which communicated with *Blankenese*. The line crossed the Elbe and had stations at *Hohenwedel vor Stade*, near Hechthausen (Telegraphenberg), *Fahlenberg* (now Deutsche Olymp), *Otterndorf* (church tower) and *Cuxhaven,* where the terminus was a turret erected on the warehouse in the Old Harbour. In 1838 the line was extended from Altona to the *Baumhaus* in Hamburg. An additional station was later erected on the Holstein bank of the Elbe.

The accompanying illustrations indicate the type of apparatus. Normally, the arms, 1 m. in length, were mounted on separate masts, but at Hamburg, and possibly also elsewhere, they were apparently formed into a cross, as shown.

In 1839 Schmidt proposed to the Senate of Bremen a plan for a similar telegraph between Bremen and Bremerhaven. The Bremen authorities were enthusiastic but the governments of Hanover and Oldenburg, through parts of whose territories the line would have to pass, for long withheld consent. Eventually the line was completed and opened throughout at the beginning of 1847.

From a station on the Rabeschen Haus, by the Domhof in *Bremen,* the line ran by way of *Oslebshausen, Vegesack, Vorbruch, Elsfleth, Brake* and *Dedesdorf* to *Bremerhaven.* It crossed the river Weser twice. A station was also erected at *Solthoeren bei Wremen* for the benefit of shipping in the outer Weser and was linked with Bremerhaven by a branch passing through *Debstedt.* A link line was also built from Bremerhaven to the Hamburg-Cuxhaven line at Hechthausen, via *Elmlohe, Bederkesa* and *Lamstedt,* making through communication between Bremen and Hamburg possible. Communication was established between Hamburg and Brake on 13 July 1846 to enable Schmidt to send birthday greetings to the Grand Duke of Oldenburg. Schmidt seems to have entertained the ambitious idea of extending his telegraph from Bremen to the Ems and possibly into Friesland but no line was built.

It was unfortunate for Schmidt that the electric telegraph began working between Bremen and Bremerhaven at the same time as his line. No doubt the bureaucratic delay was responsible for this duplication, but some accounts say that the object was to afford a comparison between the two modes.

An article in a Hamburg journal on 15 June 1848 contained a protest by local bodies against the passage of electric wires across their land, on the score of danger to persons and damage to property. The protests were backed up, if not instigated, by Schmidt, who alleged that a telegraph wire stretched in the air attracted to it all the electricity in a storm, with resulting ill effects on the weather. The wires, he said, were too weak to carry lightning to earth and those who tried to apply electricity to telegraphy had no knowledge of physics!

At the end of 1848 the electric telegraph came into use in the lower Elbe. The competition at both places soon put Schmidt out of business and he lost all his means. Only a few of his company's shares had ever been taken up, as the merchants in both Hamburg and Bremen had proved lukewarm. The Hamburg-Cuxhaven line closed down in 1850.

Plate 66: Telegraph at Naples during Murat's rule.

Plate 67: Stone in the yard of the Tower Hall at Susa, west of Turin, with the name of the two operators of the Susa telegraph (Paris-Milan-Venice line).

Plate 68: Detail from hand-made map of Amsterdam, 1812, showing Chappe-type telegraph on the Weesperpoort.

Plate 69: Netherlands: Lipkens-type telegraph on the Residency at the Hague, 1831.

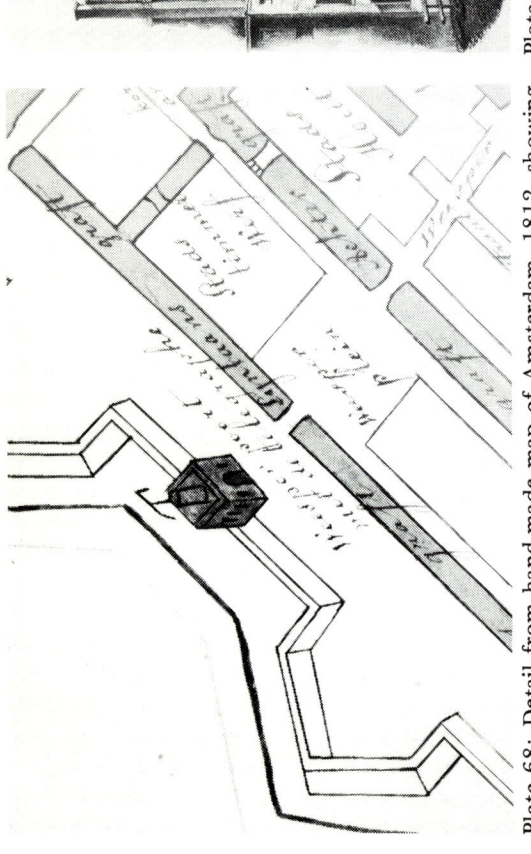

Plate 70: Germany: Johannis-kirche, Magdeburg, showing telegraph on roof.

Plate 71: Telegraph on south pavilion of the castle, Koblenz; terminus of line from Berlin.

Plate 72: Telegraph inspectors outside Dahlein church telegraph, near Berlin.

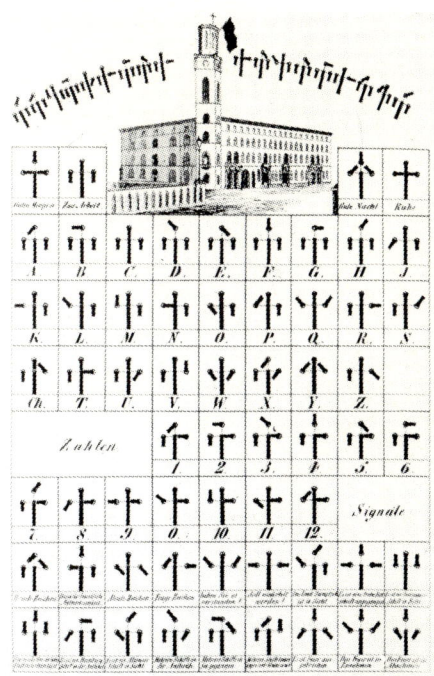

Plate 73: Hamburg telegraph and code.

Plate 74: Bremenhaven telegraph.

Plate 75: Cologne-Flittard reconstructed telegraph station.

Plate 76: Telegraph station at Old Harbour, Cuxhaven.

Plate 77: Denmark: telegraph on Kronborg Castle.

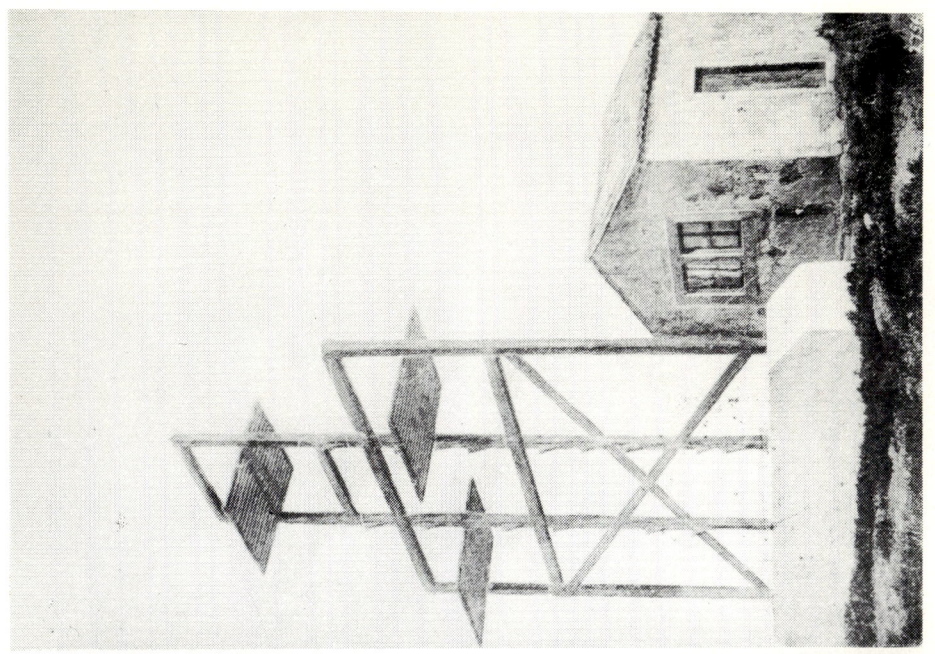

Plate 79: Portugal: shutter telegraph in use between Cabo Carvoeiro and Berlengas, near Peniche, as late as 1889.

Plate 78: Representation of telegraph with Great Belt in background.

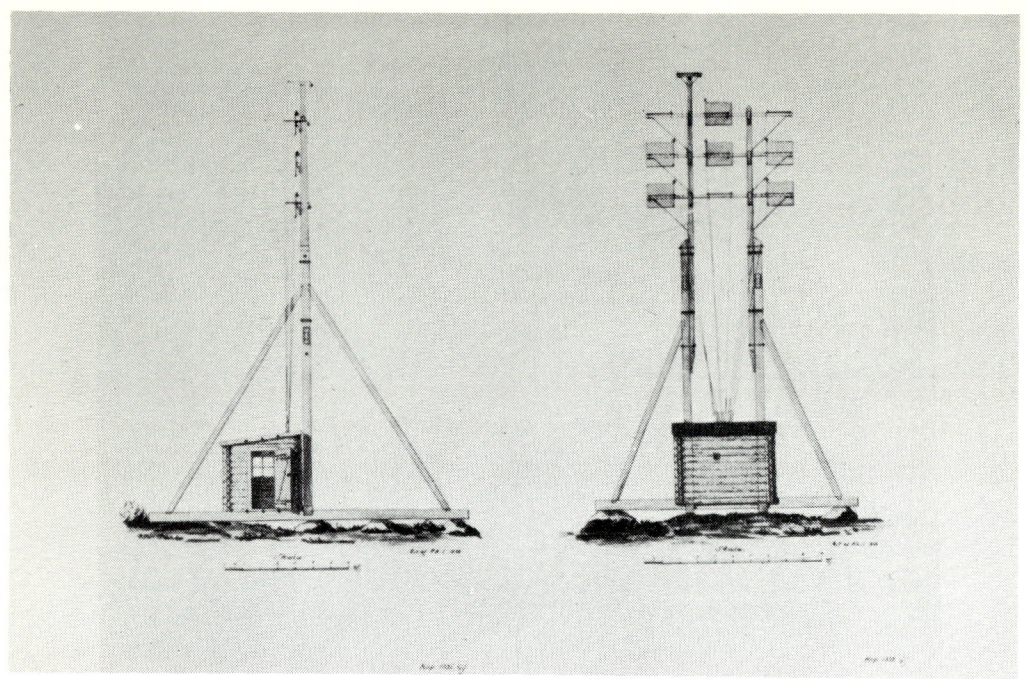

Plate 80: Sweden: dimensional drawing of west coast telegraph station, 1838.

Plate 81: Sweden: Mosebacke telegraph.

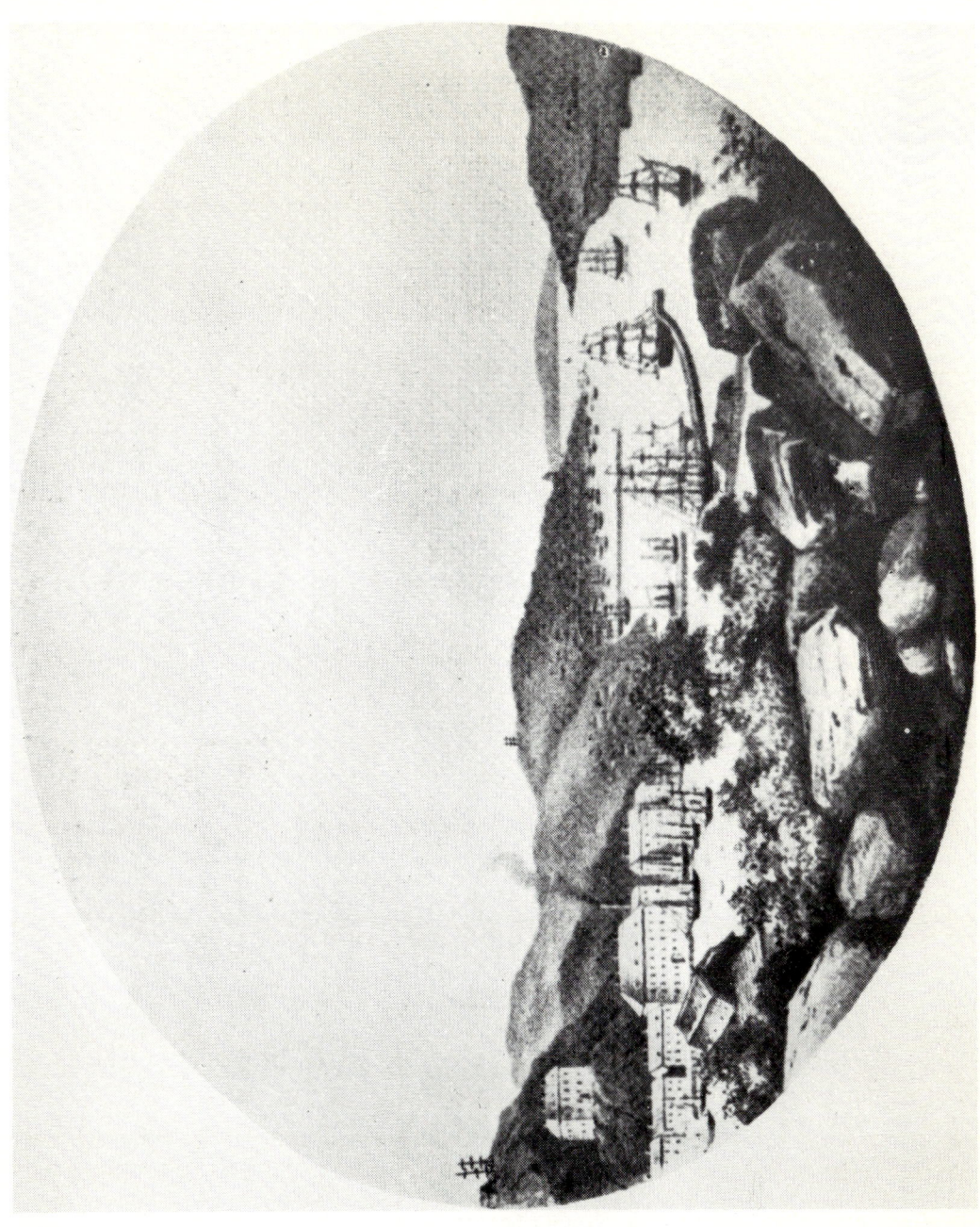

Plate 82: Sweden: view showing (left) Osterhällen and Stigbergsåsen telegraphs in 1855.

Plate 83: Russia: (left) telegraph on the Winter Palace, St. Petersburg (now Leningrad); (right) sketch of Russian telegraph tower showing lanterns fixed to arms for night operation.

Plate 84: Russian telegraph at Otchakov, near the mouth of the Dnieper, during the Crimean War.

GERMANY 161

Pistor's Telegraph

In the meantime the Prussian Government had established a far longer and more important system, for military purposes, between Berlin and Koblenz, seat of the military governor of the Rhine Province. A Royal decree of 27 June 1832 sanctioned the appointment of a commission, chaired by General-Leutnant von Krauseneck, to discuss the question of telegraphy. Another decree, of 21 July 1832, authorised construction.

A noted mechanic and optician named Carl Pistor, who was a *Geheime Postrat* (Postal Privy Counsellor) devised the system used, which he modelled on Watson's first telegraph and brought out in 1830. Von Krauseneck supervised construction and Major O'Etzel, whom we have already met in the chapter on the French telegraph, superintended the operation. Freund, founder of one of the first machine works in Berlin, supplied and erected all the equipment to Pistor's specification. Frauenhöfer of Munich and English opticians provided the telescopes.

One of the most ambitious telegraphs to be built in continental Europe outside France, the line was about 550 km. in length and included 61 stations originally.

The officials, who wore military uniform and were selected from men just entering military service, were part of a telegraph section under the control of the Army General Staff. Two men were allotted to each station.

The stations were, on average, 1½-2 German miles (10-14 km.) apart and erected on hills or, in four cases, on conspicuous buildings. Those on hills were substantially built of stone and comprised a square tower and adjoining living quarters. Some towers were apparently of four, others of two storeys. If the station was remote from a town or village, the accommodation was ample enough for two families. In addition to obvious items like telescopes, the operating room equipment included the homely feature of a Black Forest clock, and a spittoon.

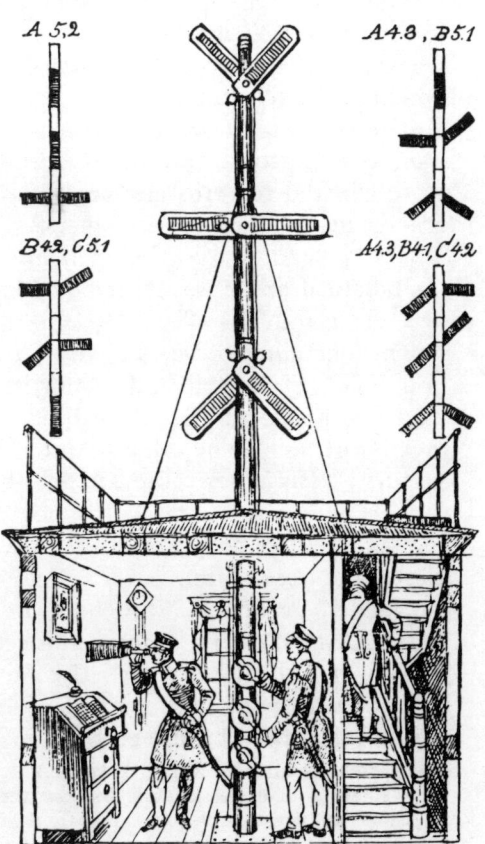

Station der preußischen optischen Telegraphen-linie zwischen Berlin und Magdeburg (um 1835)

30. Section of operating room of Berlin-Koblenz line station.

The wooden tapered mast, of pine or spruce, rose six or seven m. above the tower roof, which was railed round. It carried three pairs of wooden arms, each 1.56 m. in length and 0.47 m. wide and painted black. They incorporated sheet iron slats.

The arms were worked by cords or chains passing over pulleys and down the mast to a wheel and crank in the operating room. The lowest pair of arms was the 'A' pair, representing hundreds, the middle pair 'B', for tens, and the topmost pair 'C', for units.

By placing the arms at 45°, 90° or 135° to the mast, 256 angle positions could be given. The code made possible 4,096 abbreviations and combinations. Some accounts say that a message took at most only 15 min. to travel the length of the line, though this probably applied to the simplest communication only. A more normal transit time seems to have been an hour for a message of about 30 signs.

The instruction manuals were particularly detailed. Each station kept a journal in which the signs received were recorded, along with details of the weather and light, and statements of which operator was at the machine and which at the telescope.

The vocabulary was divided into classes, as for names of places and rivers, names of persons, and about 1,000 words in alphabetical order. Numbers were allotted to verbs and certain syllables, for the months, weeks and days, and even for punctuation marks! For instance, the signs to be made for *Abend* (evening) were 'A' — 4 and 1, 'B' — 4 and 1 and 'C' — 1. Next in alphabetical order was *Abscheid* (departure), signified by 'A' — 4 and 1, 'B' — 4 and 1 and 'C' — 2.

The decoding officials at the terminal stations were expected to add the courtesies of speech and honorifics that the telegraphed message would exclude. For example, 'Der König befiehlt' (The King commands) would be given out as 'Seine Majestät der König haben allergnädigst zu befehlen geruhtet' (His Majesty the King has been graciously pleased to command).

The telegraph operated for not more than six hours a day. An engineer named Treutler proposed to use lanterns for night signalling but officialdom fiercely opposed him.

Opened as far as Magdeburg in 1832, the line was completed in the autumn of 1833 and entered full service early in 1834. The *Berlin* terminus was on the platform of the Alte Sternwarte (Astronomical Observatory) on the northern wing of the Academy (later State Library), between the Unter den Linden and Dorotheen Strasse. Telegraphy was yet another use for the Academy, a hive of activity which the author Gutzkow pictured as:

> A Pantheon of all the Arts and Sciences. The Muses all round and Mars in the centre. The home of savants and artists, with a race track for the cavalry horses, an exhibition of works of art, meetings of the Leibnitz Society, the Royal Prussian coach-house and headquarters of the Uhlans.

Voltaire had proposed for it the witty inscription, *Mulis et Musis*.

Gutzkow amusingly wrote of the Berlin telegraph that it continually threw its arms together over its head as if it were surprised at something. Berliners soon christened the telegraphs *Hampelmänner* (jumping jacks).

GERMANY 163

The first station out was on the church in the village of *Dahlem,* now a suburb of Berlin. It communicated on the W. with a station at *Schäferberg,* a site now marked by a modern telecommunication tower. Thence the line went by way of *Potsdam, Brandenburg* and *Ziegelsdorf* to *Magdeburg,* where a station was erected on the Johanneskirche. From Magdeburg it continued along high ground forming the N. foothills of the Harz and then the N. part of the central highlands of Germany, passing a little to the N. of Goslar,

31. Telegraph on Garrison (now St. Pantaleon) Church, Cologne.

Burgberg near Holzminden, then S. of Paderborn, Soest and Iserlohn to *Cologne*.

At *Cologne*, where the station was on the Garrison (now St. Pantaleon) church, the line changed direction sharply to follow a course east of the Rhine, passing near *Siegburg*, to *Ehrenbreitstein*, where it crossed the river to the terminus on the castle, now destroyed, at *Koblenz*.

The section between Magdeburg and Holzminden crossed parts of the Duchy of Brunswick and the Kingdom of Hanover. In the summer of 1842 an additional station was erected at *Berg Heber,* on Brunswick territory, to

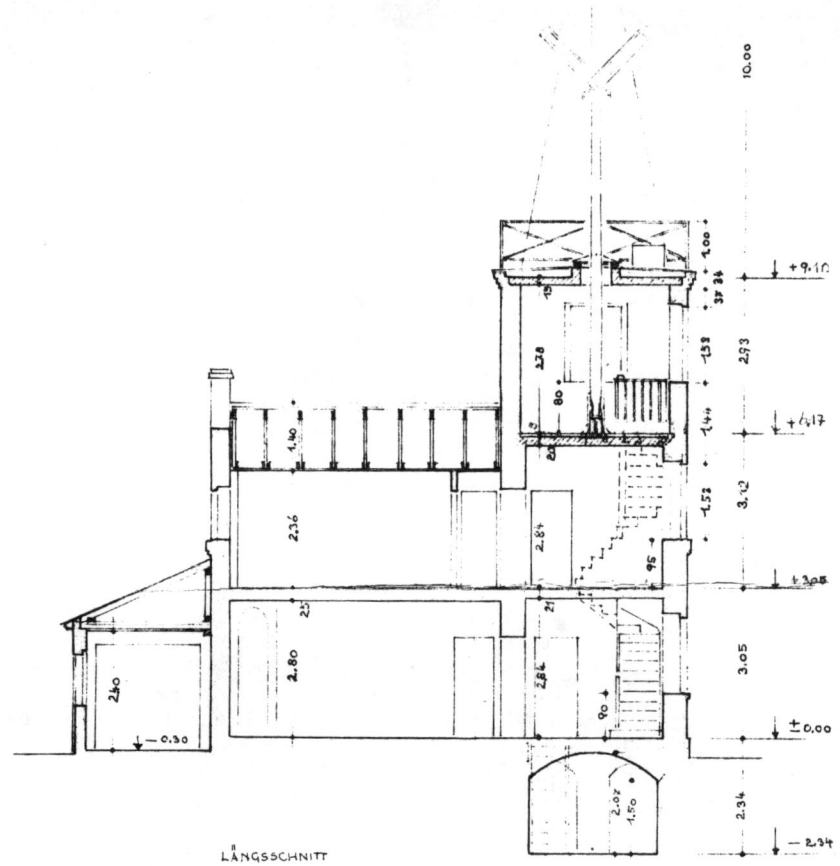

32. Reconstructed telegraph station at Flittard (Cologne).

ensure that communication was not affected by fog, in view of the impending visit of the King of Prussia to the Rhineland. This station brought the total to 62, of which five were in Brunswick and two in Hanover.

Fortunately, some, at least, of the stations in the German Federal Republic have survived, some intact, others in part. At the station at *Köln-Flittard* — now No. 152 Egonstrasse — the Cologne city conservator has restored the house and tower and mechanism. The tower of station No 28, at *Burgberg* near Holzminden, is classified as a technical memorial by the Conservator of Lower Saxony. Herr Horst Drogge of Göttingen erected a new mast and arms on it in 1968 for a film on the history of the German post office. At *Soeven*, south of Cologne, both the house and tower have survived and are occupied.

In 1849 *Frondenberg* telegraph near Iserlohn incurred the displeasure of rioters, who regarded it as an informer. According to Uhlmann-Bixterheide's *Chronika van Iserlaun*, a party of men armed with muskets climbed the hill and threatened the operators. The senior operator managed to prevent the mast from being cut down but the arms were broken off and cast into a ditch at the foot of the hill.

Berlin businessmen vainly tried to be allowed to use the line for commercial purposes. It remained predominantly military, though not, apparently, wholly so. To break up the length of the line and increase its capacity an intermediate divisional point was established at Cologne in 1836. This not only simplified and speeded transmission but also enabled news from Prussian embassies in western Europe to be sent by telegraph between Cologne and Berlin.

After Werner von Siemens had successfully proved his needle telegraph, the first electric telegraph in Prussia went into service between Berlin and Cologne in May 1849, displacing that section of the optical telegraph. The Cologne — Koblenz section lasted until 20 October 1852, when it also succumbed to the electric system.

CHAPTER XIV

Sweden

The long history of optical telegraphy in Sweden has been thoroughly chronicled in N.J.A. Risberg, *Den Optiska Telegraphens Historia i Sverige, 1794-1881,* (1938). This book is the third part of a monumental work on Swedish telegraphs.

Risberg's work goes into the smallest details, even including such items as an inventory of the tools allowed to a station, the uniform, pay and pensions of the telegraphists, and a schedule of punishments for misdemeanours. Comprehensive lists of authorities and a full bibliography, including the works of ancient authors, are included. The present author acknowledges his indebtedness to this meticulous history for much of the information in this chapter.

The visual telegraph in Sweden certainly merits consideration on account of its scope and long service, but it had a rather aimless existence, fulfilling no special function and not connecting essential cities or other centres of communication as, for example, in France and England. It is true that at one period of its existence there was an ambitious plan to link Stockholm across country with Gothenburg, but it never materialised, partly because of the advent of the electric telegraph. Unlike the Chappe telegraph, which at one time stretched over much of Napoleonic Europe, the Swedish telegraph had no political object. Unlike the English organisations it provided no special service for the navy, then a negligible force in Sweden.

The system was confined to three separate regions: in and around Stockholm; round Gothenburg; and round Karlskrona, chief naval port at that time. (Under an abortive scheme of 1811, a line would have extended from Karlskrona right round the S. coast, along the whole E. shore of the Sound and to the N. again on the W. side as far as Kullen. This would have embraced about 30 main towns and had many intermediate stations, one of them at Hälsingborg where a telegraph had been erected in 1801 but not used. One obvious advantage would have been a link with the Danish system, with perhaps further international possibilities).

The Swedish telegraph performed a useful service in keeping up communication in some parts of the conglomeration of islands and islets that fringe the coast. Particularly was this so with the lines that connected places around the Stockholm archipelago and up and down the coast, and with islands which might otherwise have been cut off.

At its zenith, as well as linking *Stockholm* with *Gävle* in the N. and with *Landsort* in the S., the Stockholm system sent out branches to extremely isolated places. An important special branch crossed the channel from *Grisslehamn* to a rocky spot called *Signilsskär* and thence to *Ekerö* in the Åland Islands. It was useful until the Russian invasion in 1808, when those islands and the mainland of Finland were occupied.

News of the wonderful success of Chappe's telegraph had reached Sweden by the autumn of 1794. Sweden claims a slight lead over England in following France's lead. In England the telegraphs were to operate almost solely in the service of the Admiralty and the shipping interests, and there were to be at least three important practitioners, Murray, Popham and Watson, each with his version of the art. In Sweden the aims, eventually at any rate, were to be more social and commercial and only one name is significant, Edelcrantz.

Edelcrantz's Telegraph

In England the shutter gave way to semaphores of various designs. In Sweden Edelcrantz began with a rather crude semaphore, and almost at once switched to the shutter system which survived until 1881 without much alteration. Its longevity may be accounted for by the cost of installing a network of under-water cables to link the innumerable islands.

Abraham Niclas Clewberg-Edelcrantz was born on 29 July 1754 at Åbo (now Turku) in Finland, then part of Sweden, where his father was a university professor. Educated at home, he later had a varied literary career, culminating in his being created a member of the Swedish Academy and, in 1800, a Knight of the Order of the North Star. By that time he had dropped the Clewberg.

In 1796 he published a treatise on telegraphy entitled *Afhandling om Telegrapher och forsök til en ny inrättning däraf* which gained him renown. On the title page he describes himself as the King's 'hand-secretary' and archivist. Risberg tells us that Edelcrantz travelled abroad from the end of 1801 to May 1804. He certainly gained international prestige. Chappe l'Aîné wrote appreciatively of him, incidentally spelling his name throughout as 'Endelerantz'. The English Society of Arts elected him a member and referred to his telegraph model in its *Transactions*, in which he is called both 'Chevalier' and 'Sir'.

Edelcrantz's first machine consisted of a tall post carrying cross-pieces rather like the indicator of the Chappe 'T' telegraph, save that there were two, one above the other, each pivoted at the centre. Each arm had only four settings: vertical, horizontal and diagonal in each direction. For instance, vertical indicated numeral 1, horizontal 3 and the diagonals 2 and 4 respectively. Hence both arms horizontal signalled No. 33. At a distance the contrivance looked rather like a ship's mast with yards, 'running free' or braced round. As there were only 16 signals possible, the telegraph was restricted.

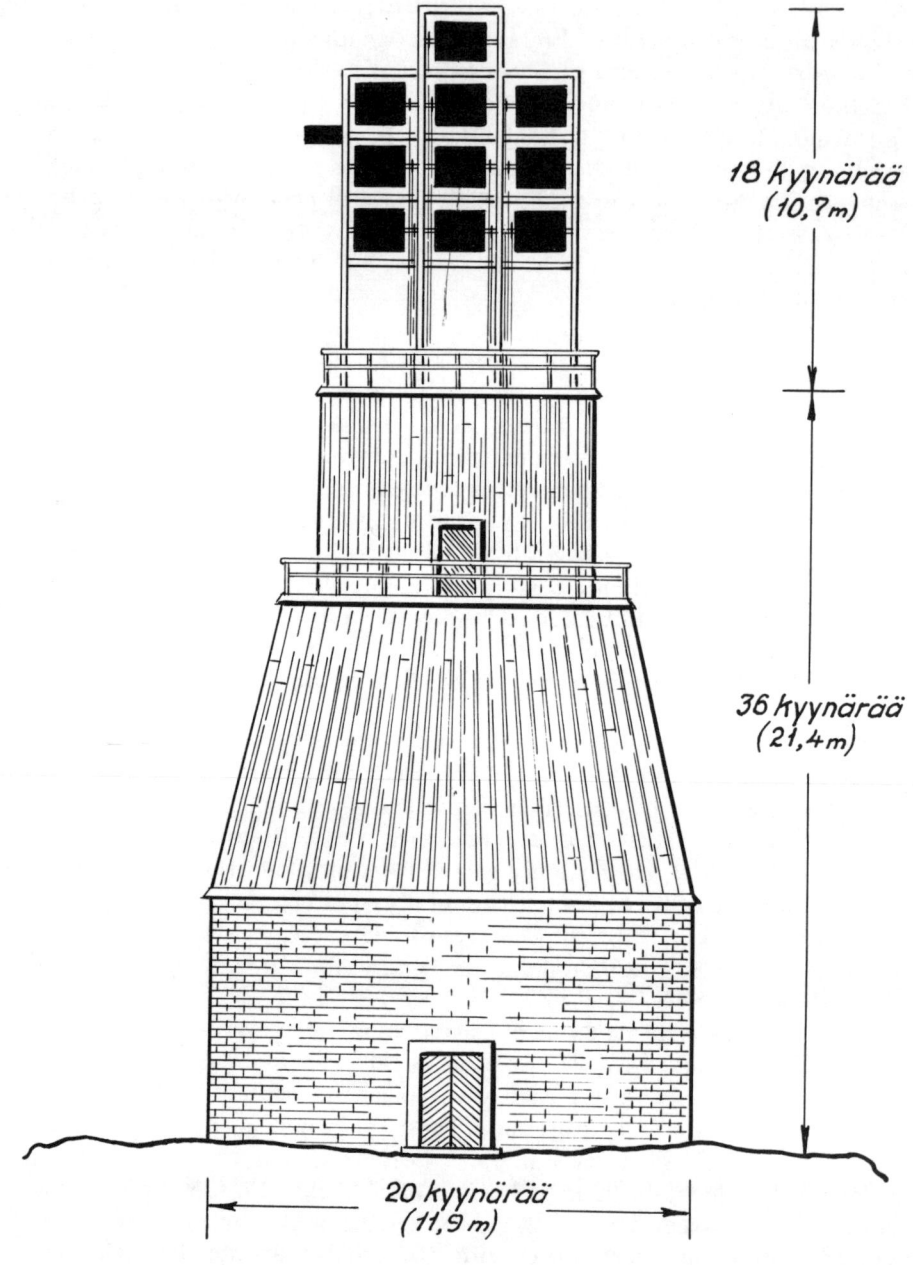

33. Diagrammatic sketch of Signilsskär station, Sweden.

Edelcrantz first tried out his invention at the end of October 1794. He built three stations, the first on the roof of the castle in *Stockholm,* the second on the Trane mountain towards *Stora Ässingen* and the third on the Götiska tower at *Drottningholm.* To commemorate King Gustav IV's birthday on 1 November a rather doggerel poem in praise of him was telegraphed from Stockholm to Drottningholm.

Almost at once Edelcrantz, dissatisfied with the limitations of his first machine, was at work on a more elaborate and comprehensive successor. Much experimentation took place. In July 1795 a station was set up on the tower of St. Catherines Church, Stockholm. This station remained the Stockholm headquarters until a separate telegraph building was erected in 1808 in Tullports Street. Although the churchwardens and parish council protested, the St. Catherine's station was used for much trial communication with the royal castle and fortresses in the neighbourhood of Stockholm. Edelcrantz refers in his book to the trials at Drottningholm in August and October 1795, at which the King, the Regent and the whole royal household, as well as several foreigners, including Englishmen, were present.

The new telegraph was a 'trellis' carrying 10 flaps, arranged three by three, the tenth being a kind of top-board. The flaps, light in weight, differed from the English shutters in that they did not open and close in a frame but were carried on spindles and simply turned to show their face or edge. Of the nine main flaps the topmost had the value 1, the middle one 2 and the lowest 4. The topmost and middle together indicated 3. When open they were invisible at the distant station. Nine flaps could make 512 signals. The addition of the tenth raised the number to 1,024.

Telegraph stations built before 1808 had no operating room. A report of 15 February 1808 stated that at *Grisslehamn, Signilsskär* and *Ekerö* the operators could not attend all day long because the machines were exposed to the weather. The King was therefore petitioned that three under-officers or other able persons should be ordered to keep watch with the telegraphers to maintain constant communication across the Ålands Sea. Later that year orders were given to build a 'despatch room' or 'observatorium' at each station. This room was built with one wall against the mast. A stove was not allowed because in cold weather it might dim or mist the glass of the telescopes and hinder communication. As a concession the windows of the observatory were made very small indeed. With their heavy timber struts holding up the framework and the primitive huts for the operators the Swedish shutter stations were not unlike those of England, though the accommodation seems to have been even more cramped.

During the war of 1808-9 against Russia the telegraphs around Stockholm played an outstanding part in coast defence, after which they fell into decay. In 1813-14 the campaign against Norway focused attention on the W. stations. They in turn became disused thereafter, save for the three linking *Gothenburg, Karlsten Castle* and the naval station at *Nya Varvet.*

Because of Sweden's impoverished state during the years of peace that followed, the telegraphs were abandoned. They were not reorganised until 1830, when relations with the Russian-Prussian-Austrian coalition began to be strained. The system then gained a new military defence status. Edelcrantz had died in 1821 and the telegraph administration had been placed under the Engineer Corps the next year. In 1834 it was transferred to the Topographical Corps.

At its zenith, between 1794 and 1810, the E. region had more stations than at any other time. There were 10 local stations in the neighbourhood of *Stockholm,* 30 more to the N. as far as *Gävle,* including the branch to the *Åland Islands* and that to *Söderarms Beacon.* Two more lines extended E. to remote *Korsö* and S. to *Landsort,* adding another 15 stations. Other important points served were *Dalarö,* a seaside resort some 30 km. SE. of Stockholm, and *Vaxholm* 19 km. NE. By 1836 the numbers had been reduced to three Stockholm local stations, nine to the N. as far as *Arholma Beacon* and 11 in the S. and E. group as far as *Landsort* and *Sandhamn.*

At *Gothenburg* there were two local stations and five more in the neighbourhood. The *Karlskrona* group comprised four stations.

In 1839, when there was a threat of war, lines were proposed from *Stockholm* to *Karlskrona, Gothenburg, Malmö* and many other towns. The line to *Karlskrona* via *Söderköping* was to follow the E. coast through *Västervik* and *Kalmar. A* line linking Stockholm with *Gothenburg,* via *Öreborg* and either *Askersund* and *Karlsborg,* or *Mariestad,* thence via *Lidköping* and *Vänersborg,* would have needed about 50 new stations.

The Gothenburg scheme commended itself for political and commercial reasons. Steamers were beginning to ply between Gothenburg and Hull in 1840. Taking advantage of the telegraph 'which on the English side they had already learnt how to establish,' a message could have been sent from Stockholm to London in 2½-3 days. The line to *Malmö* could have run either from Karlskrona via Kristianstad, needing about 20 stations, or round the other way via Gothenburg and along the Halland coast through Hälsingborg and Landskrona.

Carl Fredrik Akrell began his career as a copperplate engraver but subsequently became a lieutenant-general and Chief of Telegraph Works. As general-adjutant he was appointed Chief of the Topographical Corps on 8 March 1834. He considered the Swedish telegraph satisfactory but wished to test it against another. He chose Pasley's Universal Telegraph 'which in England had gained widespread application.' This observation was hardly correct. By 1834 Pasley's telegraph had been adopted by the British Admiralty only for harbour work and only later in the century did it begin to enjoy a deservedly wide vogue. It is surprising that Akrell did not try a well-established type, such as Chappe's, Popham's or Watson's.

The trials took place during August 1834 between Karlskrona and Drottningskär. Akrell submitted his findings to the King in October 1835. He reported that the arms of Pasley's semaphore were 'not more than two ells long' and needed a man's strength to operate. The machine, he said, averaged about six signs a minute. (Incidentally, the speed of Pasley semaphoring in the Royal Navy reached ultimately 20 signs a minute!).

In like conditions, observed Akrell, Edelcrantz's flaps could be managed by a 'half-grown boy' and the speed was 16 signs a minute. He remarked that even in a moderate wind the arms of the semaphore tended to 'bind' against the mast and became almost immovable. This was an admitted defect of semaphores not shared by shutter machines, in which the boards were balanced.

Akrell added that because of the country in which the Swedish lines had to operate, the station distances were much greater than in England and France, and the boards were sometimes necessarily larger and heavier, which amounted to an additional complaint against Pasley's apparatus. Finally he pointed out that the Edelcrantz shutters could signal no fewer than 1,024 combinations whereas Pasley's machine could achieve only 28 signs. The comparison exemplified the conflict between those who liked a telegraph that used a cumbersome vocabulary (of some 40,000 words or phrases) and those who wanted one based merely on the letters of the alphabet and spelling. The Swedes preferred a vocabulary.

Akrell omitted to mention an advantage of machines like Popham's and Pasley's, that they could be trained round if necessary. Such a provision would have been useful in Sweden where there were many places at which a branch required two shutter machines at right angles to each other. Notable were the two machines at *Mosebacke,* which in certain directions obstructed each other, and those at *Dalarö* and at *Grisslehamn,* where the forward transmission to the *Åland Islands* was at right angles to the main line.

In most countries government visual telegraphs were never allowed to accept private or commercial messages, but in Sweden it was decided quite early that an income could be derived by allowing the system to be used for such traffic at a fixed tariff. The decision was sensible as the lines were not busy and could be usefully employed, especially at remote places such as Landsort where many intervening islands would have to be crossed or a boat journey needed to get a message through. The service began in May 1834 and Sweden claims to have possessed the first telegraph in the world to be managed by a State for private and commercial ends.

Risberg records the periods and hours of telegraph working, which varied much over the years. In 1836 the lines were open for seven months of the year and for three hours in both the forenoon and afternoon. In 1858 all stations were open the whole year round. In summer the hours were 8 to 11 am and 4 to 7 pm, and in winter from 10 am to 2 pm. On Sundays the hours were 8 to 12 in summer and 10 to 12 in winter. In 1867 the yearly record showed the following numbers of hours when conditions ruled out telegraphy: *Dalarö-Sandhamn,* 252 hours; *Dalarö-Landsort,* 360 hours; *Nya Varvet-Marstrand,* 404 hours.

By that time the electric telegraph was making headway. The *Stockholm-Dalarö* line was ready in 1859. The station commander in Stockholm was discharged as being redundant and a commander was appointed instead at Dalarö to supervise the remaining lines thence to Landsort and Sandhamn. The visual stations in the Stockholm region were given up between 1858 and 1876. Those in the Gothenburg and Karlskrona areas lasted a little longer but all went by 1881.

CHAPTER XV

Denmark

A bone turner named Dicksen made the first experiments in telegraphy in Denmark. He carried them out on 9 December 1794 between the Round Tower and the Vesteport in Copenhagen, using an apparatus patterned on Chappe's. Nothing came of them but in 1796 Lorens Henrich Fisker, a naval captain, submitted to the Danish Admiralty a plan for a coast telegraph of his own design. In 1799 he drew up a new scheme and in 1800 set up a telegraph at Quintus Fortress, Copenhagen, for training officers and men. Fisker, son of a vice-admiral, became a commander in 1804 and a rear-admiral in 1809.

Fisker's Telegraph

Fisker's apparatus resembled that of Edelcrantz. It comprised a mast with five cross-trees, to which were fixed 18 movable flaps of sailcloth. When horizontal the flaps were at rest. When vertical they represented figures, the combinations permitting 1 to 42,221 to be signalled. The apparatus was improved and simplified over the years, and by about 1850 a framework with only five flaps seems to have been used.

A line was laid out in 1800 between *Maglebylille*, just S. of Copenhagen, and *Nakkehoved* on the N. tip of Sealand, with intermediate stations at *Quintus, Fortunen, Sandbjerg, Islandshöi, Hestensbakke, Kronborg Castle* and *Bodebakken*. Apparently, by the time of the Battle of Copenhagen, 2 April 1801, only Quintus and Kronborg stations were ready for use.

Extensions followed rapidly and lines were set up in Sealand between *Copenhagen* and *Korsør*, and from *Nyborg* across Funen into *Slesvig* (later German Schleswig) and to *Fredericia*, Jutland. For some reason the stretch across the Great Belt between Korsør and Nyborg — with a station midway on the island of Sprogø — was at first equipped on the Edelcrantz system. Fisker's system replaced it there in 1808 after the Sprogø station had been destroyed by a British warship.

By Royal Decree of 23 October 1801 the telegraphs passed to the control of the Danish Post Office. During the war years 1807-14 the system is said to have been extended by the following new routes: *Kalundborg – Samsø; Helgenaas – Aarhus; Aarhus – The Skaw.*

Lists show that, in addition to places mentioned, there were stations at one time or another at the following: (*Copenhagen*) *Valby, Rødovre, Høje Taastrup, Egegaarden, Ledreborg, Stiftrup, Outrup, Ebberup, Slots Bjergby, Baegdrup, Høyerup, Galgebakken, Trebjerg, Bøjden, Sønderborg, Sønder Højerup, Kieler, Warteberg, Hüttner, Slesvig, Tolker, Stierberger, Büttler* and *Hüggeberger*. There were also stations on the island of *Langeland* and at *Nakskov*. In June 1849 additional stations were built in Funen, at *Østrupgaard, Ravnebjerg* and *Baaring*.

At some periods the telegraphs were not used to their full extent, although across the Great Belt and the Little Belt (between Funen and Jutland) they were constantly operated. The Great Belt telegraph was used to transmit news during the Napoleonic period when the British fleet hindered the mail service between Sealand and Funen, and this service was kept up subsequently.

The Fisker telegraph was seldom used after the electric telegraph was inaugurated between Helsingør (Elsinore) and Hamburg in 1854. It came into use again for a short time after that when the cable under the Great Belt broke, and finally ceased working in 1862.

CHAPTER XVI

Norway

A Danish Royal Commission was set up in 1802 to consider establishing telegraphs along the south coast of Norway. Lines were surveyed from Kristiania (now Oslo) south-eastwards to Fredrikshald (now Halden) near the Swedish frontier, and SW. to the cape of Lindesnes, at almost the southernmost point of Norway, west of Kristiansand. The project was not carried out for economic reasons and flag signalling had to suffice for the next few years.

In autumn 1807, Prince Christian August, Commander-in-chief in Norway, ordered Captain Ole Ohlsen and First-Lieutenant G. Hagerup to travel along the coast from the Swedish border to Stavanger and arrange for coast defences, linked by flag signal. The first signal line was ready in a few months. Messages took 75 min. to pass between *Fredriksvern* (now Stavern) on the Skagerak and *Kristiania,* with 2 min. delay at each station.

Ohlsen, who directed the system, was not satisfied with it. At the end of 1808 he proposed to introduce a visual telegraph of his own pattern, a simplification of those of Edelcrantz and Fisker. The General Command accepted his plan and his telegraph was soon brought into use in three command districts.

Ohlsen's Telegraph

Ohlsen's apparatus comprised a mast with two cross-trees, each carrying three movable flaps. Of the 12 resulting positions, the flaps on the upper arm represented the figures 0 to 5 inclusive and those on the lower arm, 10, 20, 30, 40, 50 and 60.

The telegraphs were first installed on the E. side of Kristianiafjord from *Kristiania* (Akershus) via *Jelφy* to near *Fredrikshald* and from Jelφy along the entire S. coast through *Larvik, Langesund, Arendal, Kristiansand,* and *Mandal* to *Hitterφy* (Flekkefjord). The W. section continued the communication through *Stavanger* and *Bergen* to the island of *Fedje.* The N. section ran from *Kristiansund* via *Skibnes* and *Munken* to *Trondheim* (Kristiansten Fortress). One source extends it in one direction to *Stadlandet* and in the other to *Folda.* The Bergen station was at the fortress of Bergenhus.

In all this remarkable coast telegraph was more than 1300 km. in length. It had at least 175 stations, most of them on hills near the coast or on islands, and at an average of 6-8 km. apart. The system rendered far better service than the flags and Beutlich's *Norges Sjøvaebning 1750-1809* acknowledges its importance to the country's coast defence. The Arendal flotilla was summoned by telegraph to the action at Lyngør, when the Danish frigate *Najaden* was destroyed by a British man-of-war.

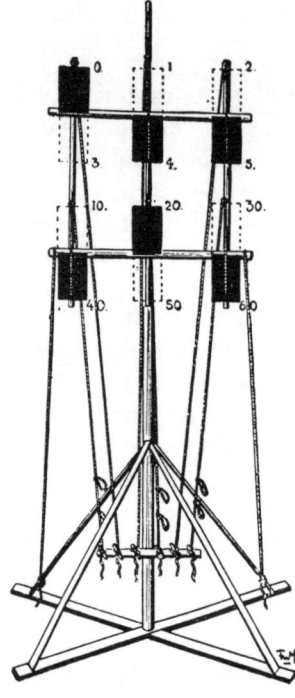

34. Captain Ohlsen's Norwegian telegraph.

But the telegraph proved a costly installation, often at the mercy of wind and weather, particularly in the rigorous Norwegian winter. When war and the fear of war alike were over in 1814 its value dropped sharply. On 18 November of that year, only days after the centuries-old union of Norway and Denmark had given place to a Swedish-Norwegian dual monarchy, the telegraph was ordered to be given up and the staff discharged.

In 1848, the year of great political turmoil in Europe, Norwegian military circles discussed the need for a telegraph between Kristiania and the coast. In a resolution of 24 December 1848 King Oscar I of Sweden and Norway urged the Norwegian Parliament to examine a possible link, passing through Horten on the western side of Kristianiafjord. Second Lieutenant of Marine Carsten Tank Nielsen looked at possible sites for stations between Nevlunghavn, south-west of Larvik, and Kristiania, and between Horten and the Swedish frontier. He proposed 18 stations on the first line, compared with the 32 on the same section of Ohlsen's S. line. The Naval Department reported its findings on 9 August 1849. It is believed that the Edelcrantz system was intended.

The plan was not pursued, partly because of cost and partly because Nielsen declined a commission to study the Swedish telegraph on the grounds that 'the day of the optical telegraph would soon be over.' Nielsen became a great enthusiast for electrical telegraphy and on 1 January 1856, a year after the Kristiania-Drammen electric telegraph was opened, became the first Director of State Telegraphs in Norway.

CHAPTER XVII

Finland

When the Crimean War broke out in 1853 Tsar Nicholas I ordered a visual telegraph to be built from the Russian naval base of *Kronstadt* along the N. shore of the Gulf of Finland to *Hangö* (Hanko), passing through *Inonmemi, Pitkapaasi, Helsingfors* (Helsinki) and *Porkkala*, with a branch to *Viborg* (Viipuri). The main object of this line, the first in the Grand Duchy of Finland, was to relay intelligence of the movements of the Anglo-French fleet which had been sent to the Baltic to bottle up Russian shipping and neutralise troops who otherwise would have gone to the Crimea.

The line was ready at the beginning of April 1854. The stations were sited on hills near the sea and were 5-11 km. apart. The system was a variant of the Edelcrantz devised by a Captain Ramstedt. The apparatus could display 54 signs, or 93 if a ball was added. The ball was hoisted as a preparatory signal, after which the first number group was shown. A civil telegraphic organisation was set up under military supervision and employed 482 persons.

The line was given up after the Anglo-French fleet had left the Baltic for the winter. But when the fleet returned at the beginning of 1855 a new telegraph line was laid out from *Hangö* W. via *Åbo* (Turku) to *Nystad* (Uusikaupunki). Branches ran to *Hiitinen* and from Åbo to *Korppoo*.

Experience in 1854 had shown that stations on cliff tops were too vulnerable and those on the new line were built farther inland. A British roving expedition was ordered to attack and destroy the telegraph station at *Jarsö*, in the parish of Kyrkslatt, about 45 km. W. of Helsingfors. The station was sited so high above a narrow creek that the guns of the warships could not be trained on it until sailors had landed and gathered wood to raise them to the necessary height.

The only important action of the second Allied fleet — commanded by Rear Admiral Dundas and Rear Admiral Penaud — was the destruction of the fortress of Sveaborg, which protected Helsingfors. HMS *Cossack*, when off Hangö Head, sent a cutter with a flag of truce to land three prisoners who had been captured on merchant vessels. The officer-in-command, the ship's doctor, three stewards, one displaying the flag conspicuously, and the prisoners went towards the telegraph station to communicate with the

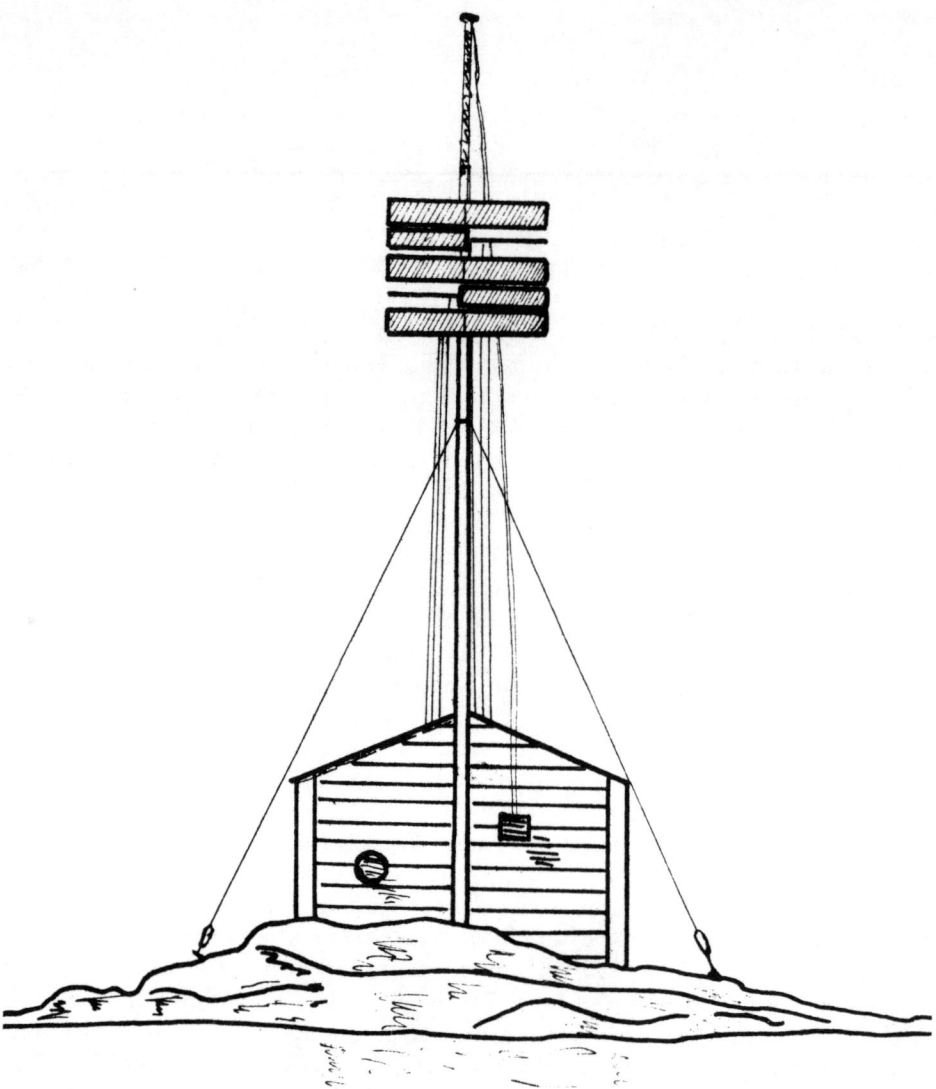

35. Ramstedt's telegraph used along the Gulf of Finland during the Crimean War.

operators. Russian troops suddenly rose from an ambuscade and fired on them, also attacking the boat's crew. Dundas vainly sought redress for the incident.

A watch was kept at the stations throughout the 24 hours, with two men on duty. The stations were manned whatever the weather. The Russian historian Borodkin wrote that the duties of the operators were neither easy nor pleasant, with the enemy squadrons delighting to single out the 'poor little huts' for attention. The stations were barely furnished. A small table, on which lay the entry book, a rough bed and a teapot, 'always with a broken spout', made up the inventory. For all that it seems that morale was good.

When war ended in the spring of 1855 the second line was likewise given up. Both lines cost 147,500 silver roubles, which the Russian Imperial Government charged to the finances of the Grand Duchy. In the same year, on 1 June, an electric telegraph was inaugurated between St. Petersburg, Viborg and Helsingfors, with a branch to Åbo.

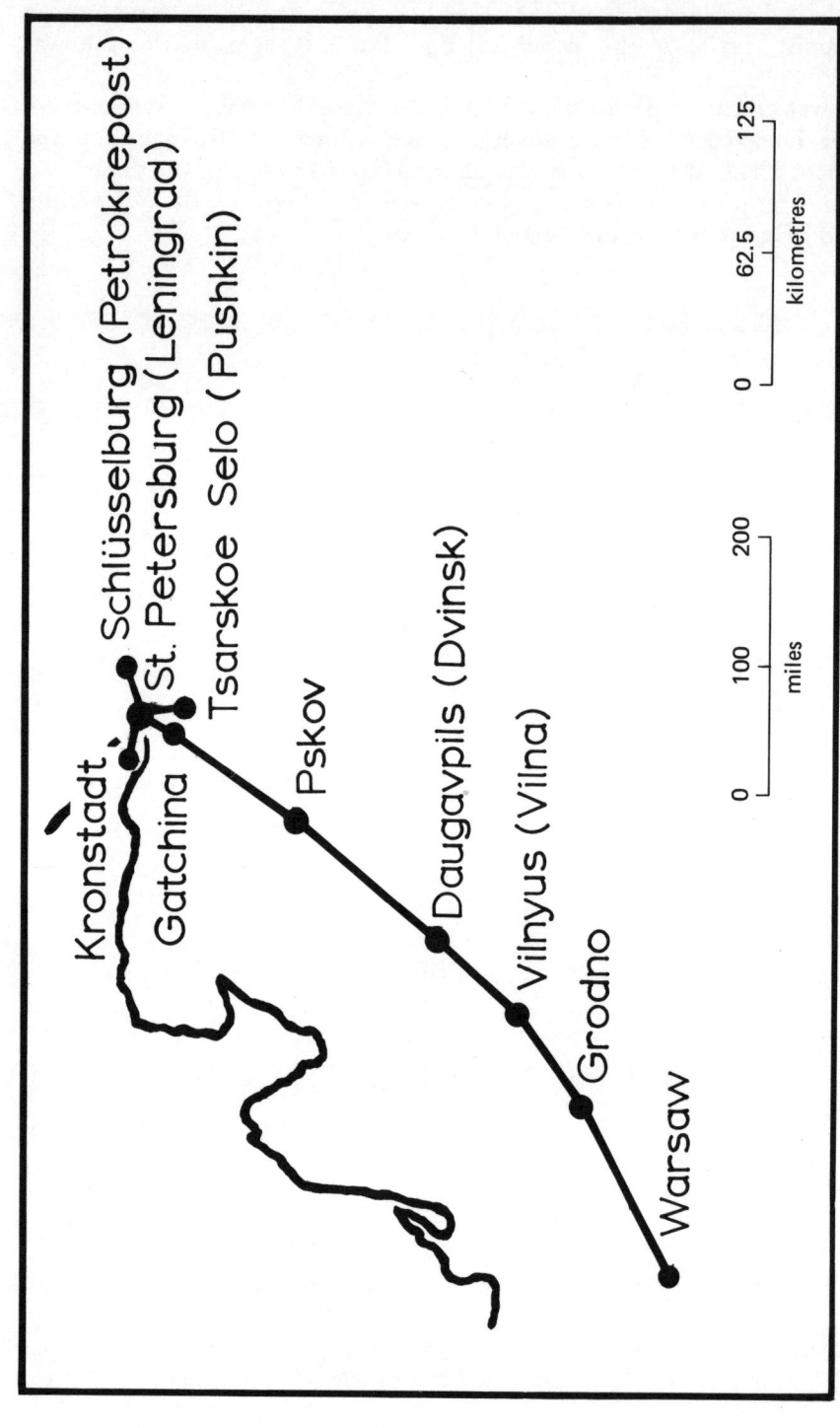

36. Russian telegraph routes.

CHAPTER XVIII

Russia

After abortive attempts, telegraphy was established in Russia under Tsar Nicholas I. The first line was opened in 1824 between *St. Petersburg* and *Schlüsselburg* (now Petrokrepost) on Lake Ladoga. It was followed in February 1834 by a line from the city to the naval base of *Kronstadt*, on the Gulf of Finland, with eight stations. In 1835 two other lines were opened, one to the imperial palace of *Tsarskoe Selo* (now Pushkin) the other to *Gatchina*. Finally, in March 1839, a much more ambitious project was completed, nothing less than a trunk line between *St. Petersburg* and *Warsaw*, about 830 km. in length and comprising 148 stations. It operated until 1854.

Tsar Nicholas is said to have sent the first message of 30 words after first practising with a model and when the acknowledgment came correctly from Warsaw, to have granted Chatau, the inventor, the Cross of St. Vladimir and a pension of 10,000 roubles. The cynical may suspect that the message was sent only a short distance along the line, to ensure success, as the reply came in only 10 min. On 23 April 1839 *The Times* reported that on the 12th the first official despatch sent over the Warsaw line had been received in Berlin and related to the health of the Tsaritsa.

Chatau's Telegraph

The system was a modification of the Chappe system by Pierre-Jacques Chatau, who had worked with the Chappes before emigrating to Russia. He wrote *Le Télégraphe de Jour et de Nuit* (1842). His apparatus comprised two long arms set side by side and centrally pivoted. Each arm bore a wide square at its outer end. A code, on the Chappe model, provided for sentences, words, parts of words and letters.

Chatau seems to have developed a night telegraph for use on the projected Warsaw line in 1835. Plans, bearing an official annotation 'Most Highly Approved', show a tapering tower, about 17 m. in height, with a lantern room at the top. A light source could be shown or obscured by means of a sliding panel. There is no record that the system was adopted.

182   THE OLD TELEGRAPHS

37. Russian telegraph tower of the type used on the St. Petersburg-Warsaw line.

Taliaferro P. Schaffner says in *The Telegraph Manual* (1859) that the towers and houses of the Russian telegraph

> were constructed for permanency and beauty. They were neatly painted, and the grounds were beautifully ornamented with trees and flowers. I have seen these stations, situated on eminences along the routes mentioned, every five or six miles, and the towers were in height according to the face of the country, and sufficiently high to overlook the tall pine so common in Russia.

The towers were up to 20m. in height and the apparatus added another 3m. Each station had a bedroom, kitchen, two storerooms, cellar, yard, garden and well. Of the telegraph establishment of about 2,000 men at least four men were posted to each station — some stations had six men.

The terminal station in *St. Petersburg* was a specially erected tower, still preserved, on the roof of the Winter Palace. That in *Warsaw* was a tower on the Royal Palace in Castle Square. The French divisional system of operating was used, the Warsaw line being split into sections, with intermediate divisional points at *Pskov, Daugavpils* (Dvinsk), *Vilnyus* (Vilna) and *Grodno*.

It is likely that the Chatau telegraph was also used elsewhere in Russia, as during the Crimean War *The Illustrated London News* published a view of *Otschakov,* at the mouth of the Dnieper, showing a Chatau machine on a tall wooden tower.

As stated in Chapter XVII, the Russians operated a coastwise telegraph along the Gulf of Finland during the Crimean War, when the French Army used a simplified form of Chappe telegraph in the Crimea.

CHAPTER XIX

Spain and Portugal

Spain

The first optical telegraph in Spain of which a record can be traced was devised in 1805 by Don Francisco Hurtado, a lieutenant-colonel of engineers. It was applied for military purposes in the province of Cadiz, where it is said to have given excellent service. No details have apparently survived. Hurtado also invented a field telegraph in 1829.

The constantly disturbed state of the country hampered the general application of telegraphs but in 1831 Don Juan José Larena, a naval Lieutenant, inaugurated a telegraph linking the Royal palaces in Madrid, Aranjuez and San Ildefonso. He is believed to have adopted a system not unlike Murray's but with much smaller shutters and capable of both day and night operation. It was suppressed by a Royal Order of 14 May 1837, which decreed the establishment of a national system, at the same time acknowledging that Larena's telegraph had given valuable service and been useful as an experiment.

Another Royal Order, dated 1 June 1837, entrusted the Ministry of Highways, Waterways and Bridges with the work. Nothing more was done until 1844, when an Order of 1 March announced details of the financing of the project and the siting of stations. On 4 March 1844 the Director General of Public Works issued orders as to the stations.

Four possible systems, three by Spanish inventors, were examined. The choice fell on that of Don José María Mathé, a Colonel on the General Staff.

A Royal Order of 16 June 1845 approved plans for a line linking *Madrid* with *Irún,* close to the French frontier, and ordered construction to begin. Work went ahead rapidly. The line passed its first message from Irún to Madrid on 2 October 1846 by announcing the entry into Spain of the Duc de Montpensier and the Duc d'Aumale, fifth and fourth sons of Louis-Philippe, preparatory to the marriage of Montpensier to Doña María-Luisa, sister of Queen Isabel II.

When the line was in regular operation despatches from Paris could be received in Madrid in six hours. They were conveyed by hand over the short distance between the French telegraph terminus at Béhobie and Irún.

Mathé seems to have infused great energy into the construction of the network and the efficient working of the system. The Irún line was followed by lines from *Madrid* to *Zaragoza* and *Pamplona,* and to *Valencia,* with a branch from *Tarancon* to *Cuenca,* opened in 1849. The Valencia line was extended almost at once to *Barcelona* and thence to *La Junquera,* near the French border with Catalonia. The network was completed with a line from *Madrid* to *Cadiz,* with a branch to *Badajoz,* in service in 1850. There is said also to have been a link with Portugal, possibly by an extension W. from Badajoz.

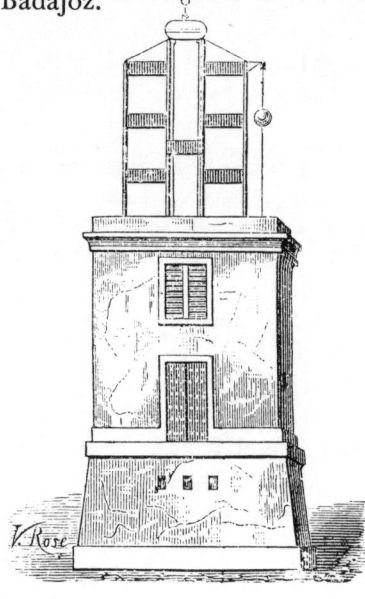

Figura 6.
38. Mathé's telegraph as applied in Spain from 1846.

Mathé's telegraph had a frame with six fixed shutters, arranged three by three, and another fixed shutter placed centrally between the two sets. An oval shutter could move up and down the centre of the frame, taking up different positions in relation to the fixed shutters. To augment the power of the apparatus, a ball could be moved up or down on one side of the frame. The moving parts were actuated from inside the station through gearing and a counterpoise. The towers were squat, square buildings of stone.

Although it seems to have functioned efficiently, the system was introduced too late to enjoy more than a brief life, as it gave way to the electric telegraph as early as 1855.

It has not been possible to trace the precise routes of the lines or the sites of the stations. It seems likely that a number of the substantial towers remain. At *Lérida* for example, there are, or were, the remains of three towers, two on the Barcelona side, at *Bell-Lloch* and *Tossal de Mollerusa,* and one on the Madrid side, at *Mariola.*

Other telegraphs known to have operated in Spain include a line for naval use opened in 1848 between *Bilbao* and *Portugalete,* with an intermediate station, and a military telegraph introduced in Catalonia (and the Philippines) the same year. In addition the Spanish Army made much use of optical telegraphs, including heliographs, throughout the 19th and well into the present century.

Portugal

The first telegraph in Portugal was installed in 1809, the year in which Wellington for the second time freed the country from French occupation. It seems unlikely that it was constructed under French auspices for Soult and his forces then held only the north of the country, but the system was apparently the Chappe or a modification. Telegraphic tables of the period show that numbers were used for the letters of the alphabet and for proper names.

On 5 March 1810 control was transferred from the navy to the army engineer corps, with an establishment of a director-general, nine other officers, 17 corporals, 28 lance-corporals and 64 privates.

A regulation of 8 December 1828 relates to a 340 km. line between *Lisbon* (Penha de França) and *Oporto*, with intermediate stations at *Apelacão, Monte Cerres, Monte Gordo, Bõa Vista, Santarem, Alviela, Golega, Atalaia, Tomar, Ceiras, Alvaiazera, Monte de Vez, Volta do Monte, Coimbra, Agrela, Buçaco, Boialvo, Montedo, Vila Nova dos Fusos, A Branca, Santa Luzia, Souto Redondo, Murado, Canelas* and *Alto de Bandeira*.

The installation was later reorganised and a similar system introduced in Madeira and the Azores. A link is said to have been made with the Spanish national telegraph of 1846-55. If so, it may have been formed by an E. branch from Santarem connecting with an extension from Badajoz and would have permitted intelligence to be transmitted between Paris and Lisbon, by a route of well over 1,600 km., albeit over three separete systems. Such a through connection, if it existed, would have been by far the longest to have been established by a combination of visual telegraphs. The rapid introduction of electrical telegraphy in the 1850s would doubtless soon have effaced the memory of so notable a link.

The main Portuguese system lasted until 1855 but as late as 1889 a report refers to two visual mechanical telegraphs still operating, one a semaphore across the mouth of the Tagus, between *São Julião da Barra* and *Bugio*, the other, a three-shutter apparatus, near Peniche, on the coast north of Lisbon, linking *Cape Carvoeiro* with the isle of *Berlenga*. In addition, the Portuguese army, like the Spanish, employed optical telegraphy until recent times, including a complete heliograph network which covered the whole country.

CHAPTER XX

The British Army in the Peninsular War

The British Army occupied the famous Lines of Torres Vedras, commanding the routes to Lisbon, between October 1810 and March 1811. As early as 15 June 1810, Wellington wrote to the British naval officer in command at Lisbon (Admiral the Hon. Sir George Berkeley, with his flag in *Barfleur*) for copies of Popham's *Telegraphic Signals or Marine Vocabulary* and the services of officers of Marines or steady midshipmen, or even Marines NCOs or sober signalmen, to work telegraphs for the army. Berkeley told his flag lieutenant, John Keith, to superintend the erection and manning of posts within the Lines.

The system seems to have been by flags of Popham's Code, using Popham's signal book. Letters went from Berkeley to the Admiralty asking for stocks of both to be replenished as the army had taken the spares. On the other hand, Oman's *History of the Peninsular War* refers to masts and yards, with balls suspended from the yard. The only signals used were the arbitrary words and sentences of the vocabulary and there was no method of spelling except as laid down in the code. This was not wholly satisfactory. In her *Wellington—the Years of the Sword* Elizabeth Longford quotes from Wellington's MSS. and Ward's *Wellington's Headquarters, 1809-14,* the phrase from the code book 'I have sprung my bowsprit but cannot fix it at sea' as being hardly suited to military needs! Nevertheless, as Ward points out, the copies of the vocabulary that have survived show that such local place names as *Sobral, Alenquer, Alhandra and Santarem* had been added.

There was a hitch in supplying the code books. Elizabeth Longford mentions that General Sontag of the King's German Legion reported that the midshipman who put up the telegraph in the town of Torres Vedras had received no book and so could not explain the operation, and that his colleague at Sobral had the same difficulty.

Oman lists stations at *Alhandra, Monte Agraça, Nossa Senhora de Socorra, Torres Vedras* and *Redoubt 30,* but Marshall-Cornwall in *Marshal Massena* says there were nine.

Ward mentions that at this time Wellington was in telegraphic communication with the Spanish garrison in besieged Badajoz, 130 miles away, but does not say what system was used. *Through to 1970,* published by the Royal Signals to mark the golden jubilee of the Royal Corps of Signals, says that Gamble's telegraph was used in the Peninsula 'in more forward areas' but gives no more details. Presumably Gamble's portable telegraph is meant.

On 4 April 1811 Wellington asked that the naval officers lent to help with the Torres Vedras telegraph should return to their normal duties.

After Wellington had carried the war into France in 1813 and was investing Bayonne, a telegraph was set up on a divisional basis to link army headquarters at *St. Jean-de-Luz* with *Vieux Mongerre* via *Guéthary, Arcangues, Château d'Urdains* and *Château de Larralde.* Posts were from three to five miles apart. Communication began on 18 December 1813. Contemporary notebooks in the War Office Library show that signals were made at a single mast by a flag and balls hoisted at the mainmast or yard. They were referable to a numbered dictionary, limited to about 600 words and phrases. It was not intended that mast and yard should be carried on the march. Instead, tent poles were recommended — 1½ poles for a mast and another pole for a yard.

Ward says that the short period that the St. Jean-de-Luz telegraph operated is of particular interest in that the divisional system then established was the ancestor of the modern divisional signals. A divisional telegraph officer was appointed. An NCO and three privates, who received extra pay, were entrusted with the poles, flags, balls and dictionary, and had to carry them wherever the division moved. The dictionary was designed for the use of the 'unlearned' (Wellington's expression).

It is evident, declares Ward, that the telegraph sections of the divisions that were instituted on 12 January 1814 were intended to become an integral part of the divisional organisation as much as Artillery, Engineers, Staff Corps, Medical Department, Commissariat and Staff. Although circumstances did not call them into operation after most of the army left the Bayonne district in February 1814, their short life deserved recognition as yet another example of the means for independence with which Wellington equipped his divisions.

CHAPTER XXI

Egypt

It is said that the fame of the French telegraph induced the Turkish ambassador in Paris to ask for a model of it to be sent to Constantinople, where nobody could make it work. It was a different matter in Cairo, where the ruthless Albanian-born Viceroy, Mohammed Ali, bent on at least outwardly westernising Egypt, engaged an Armenian named Abro to construct a telegraph on the French plan, and with French material between Cairo and Alexandria. Abro and Coste, the Viceroy's engineer, chose the sites.

The line was opened about 1823. It had 19 stations in the 100 miles, as follows: *Cairo Citadel; Boulak Fort; Abu-el-Gheyt; Ziffet-Chalahan; Faraounyeh; Serou; Menouf; Nadir; Bechtamy; Zaiyat; Byban; Gizayr-Yssa; Telbany; Damenhur; Karaouy; Burket-Gheytas; Leryoun; Beydah;* and *Alexandria.* Murray's *Handbook to Egypt,* 1847 edition, says that the monotony of the banks of the Mahmoodeeh Canal [another of the Viceroy's improvements] 'is not relieved by the telegraphs rising at intervals above the dreary plain'. One would have imagined the reverse.

In 1835, Lieutenant Waghorn, the East India Company's agent in Egypt, initiated the Overland Route whereby travellers and mails between the United Kingdom and India were conveyed from the Mediterranean to the Red Sea by way of the Nile to Cairo and then the desert to Suez. Messrs. Raven and Hill, the owners of a Cairo hotel, soon began to compete with Waghorn. A telegraph opened between Cairo and Suez in 1839 benefited Raven and Hill's clients, for the shipping intelligence it transmitted enabled them to stay until the last moment in Cairo and so avoid spending more than a night in squalid Suez.

The telegraph towers on the Suez line were close enough to the caravan route to serve as shelters for travellers, and at one or two of them at least drinking water was kept in porous bottles. It is not certain whether the line was used for any but shipping business, but it seems unlikely as Suez then had no other importance. Probably the line was owned by the Egyptian Government and leased to the shipping interests.

The Chevalier de Negrelli summed up the telegraphic state of Egypt in the 6 September 1856 issue of the *Railway Times,* as follows:

The chief lines are from Alexandria to Cairo, and from thence to Suez, and along the course of the Nile to Upper Egypt. On the line from Alexandria to Cairo the aerial telegraph has now been supplanted by the electric telegraph, and this line is now being extended to Suez ... If ever there was a country where telegraphic communication was most advantageous, that country is Egypt. The usual communciations are frequently either broken or suspended by the desert, or by the periodical inundations of the Nile. Formerly, the Bedouin tribes of the desert took advantage of these inundations to rise against the Government and to pillage the country; for they knew that they could retire before the news of the insurrection could reach Cairo. The days of brigandage are past. The camp of Saidieh, near Cairo, is in connection with the lines of telegraph ... At the first news of disturbances, the Government is enabled to send its flotilla to the south ...

The electric telegraph to Suez doubtless followed the course of the Cairo-Suez railway then under construction and itself following the road. *The Scotsman* of 26 September 1855, reporting that three Egyptians had been undergoing training with the Electric Telegraph Company in Edinburgh, stated that a line of electric telegraphs across the Isthmus of Suez was projected and would be of great advantage in telegraphing the arrival of the Overland Mail on either side of the isthmus.

It seems possible that some of the former telegraph towers remain. The well-informed French journal *La Vie du Rail* referred in 1955 to the Cairo-Suez railway as running only a short distance from the desert road, 'toujours jalonnée des hautes tours du télégraphe Chappe qui servaient d'abris au temps où des caravanes acheminaient la malle des Indes par l'Overland Route.'

CHAPTER XXII

South Africa

A semaphore telegraph was laid out in Cape Province in the 1840s. It was installed by one Henry Hall, who described it in an article in the November 1859 issue of *The Cape Monthly Magazine*. Details of the installation were given by the late Professor Percival R. Kirby in the December 1960 issue of *Africana Notes and News*. The following account is taken from Professor Kirby's article, by kind permission of the editor.

Hall, born in Dublin in 1815, was apprenticed to a firm of builders and government contractors. In 1839 he won first prize in a competitive examination held by the Royal Engineer Department, in which he became Foreman of Works. He served in Ireland with the Department until 1842, when he was ordered to the Cape. During the six years before Hall's arrival in South Africa, the Royal Engineer Department there had been commanded by Lieutenant-Colonel Griffith George Lewis. He planned and built fortifications found to be necessary after the Sixth Kaffir War (1834-5). Hall carried on some schemes which Lewis could not complete. They included the superintending of the building of many of the most important military posts, towers and signal stations on the line of the Great Fish and Kat Rivers. He and Lewis planned a system of signal towers on the frontier of Cape Colony.

Before the installation was begun, there had been a signal station on Woest Hill, about a mile S. of Grahamstown, a high point from which the horizon is clearly visible in most directions. It had been used during the war of 1834-35 but had later been abandoned. It would probably have made an admirable headquarters for the new system planned but unfortunately Lewis selected instead Fort Selwyn, on a hill only 300 ft. high and just above where the Drostdy Barracks was situated. The view of the E. was extensive, and the hill chosen for the most important link in the chain of stations was clearly visible from it during fine weather. But in the early morning Grahamstown is frequently covered by mist, which often effectually obscures the hill on which Fort Selwyn was built.

Lewis proposed to establish two telegraph lines connecting *Grahamstown* with *Fort Beaufort* (and, later, with Tarka Post, although this was never carried out) and *Fort Peddie* respectively, with an auxiliary line to *Bathurst*. The 'key' position for both lines was to be at *Governor's Kop*, 2700 ft.

above sea level, and nine miles E. of Grahamstown, from which magnificent views of the surrounding country are gained in all directions.

Towers on the N. line were erected at *Governor's Kop, Gras Kop* (a hill near Fort Broen), *Botha's Post Tower* (between the Kat and Koonap Rivers), *Dans Hoogte* (above seven miles S. of Fort Beaufort) and finally at *Fort Beaufort*, a little to the NW. of the present town of that name. Those on the E. line were placed at *Governor's Kop, Fraser's Camp, Piet Appel's Tower* (near Trompetter's Drift on the Great Fish River) and finally, *Fort Peddie Tower*. The auxiliary line, for which towers were never built, was to have received signals from Fraser's Camp at a tower to be erected at Mount Donkin, at Cawood's Post and thence to Bathurst.

The towers were about 30 ft. high, constructed so as to be readily defended, with quarters for telegraphists and others. Each cost about £500. The towers of the N. line, from Grahamstown to Fort Beaufort via Governor's Kop, received their semaphore masts. Unfortunately the nature of the terrain had made the selection of sites most difficult and it was found that after the towers had been built only in a few instances were the arms silhouetted against the sky as they should have been. In addition, the telescopes supplied from London were not sufficiently powerful. Worse, the phenomenon of refraction caused the signals on hot days to assume fantastic shapes and made them thus indecipherable.

Hall, who personally tested the telegraph between Dans Hoogte and Fort Beaufort, only seven or eight miles, has recorded that he found great difficulty in reading from Fort Beaufort the signals made at Dans Hoogte, even with the aid of a telescope, whereas from Dans Hoogte he could decipher those made at Fort Beaufort with the naked eye.

The towers on the N. line had received their apparatus before the outbreak of the Seventh Kaffir War of 1846 and were to be put in operation then. Those on the E. line were completed but were not fitted with apparatus. Hall tells us:

> in one month after the outbreak of war all these towers were in ruins, abandoned by us or burnt by the enemy; and their blackened walls are to this day [1859] seen crowning the crest of every commanding eminence along the old colonial boundary of the Fish River.

The annual cost of maintaining staff at each station was at least £2,000 and the original outlay on towers and their equipment was about £5,000. The towers, each manned by only a few men, became almost completely isolated during enemy invasions of the Colony and their occupants were soon short of food and water. Advertisements in the *Graham's Town Journal* in 1844 reveal that every gallon of water consumed by the men detailed to live in the towers had to be supplied by contractors, who undertook to refill the tanks at stated intervals, never less than every second day. The contractors could not, of course, fulfill their obligations during a campaign.

The system selected was Pasley's, or a close imitation, as it was a Royal Engineers' commitment. In the illustration reproduced as Plate 85 — from a painting in 1850 by Thomas Baines which hangs in Grahamstown City Hall and entitled *Graham's Town from Selwin's Battery* — the semaphore mast is shown mounted on the Fort, with the regulator vertical and retracted. The lower indicator is half hidden by the picture frame. In this position no message could be sent but the presence of a member of the signal corps, who is looking through a telescope, suggests that he is deciphering a communication from Governor's Kop which a second soldier, with his back to the beholder, is possibly transcribing.

The author says that the fact that the semaphore is shown in the painting seems to indicate that it had not been destroyed or removed when the system collapsed at the beginning of the 'War of the Axe', that the apparatus at Governor's Kop had also been spared and that both were still in use in 1850.

When Baines entered Grahamstown for the first time in March 1848 he noted in his journal that 'the semaphore of Selwin's Battery appeared upon the heights on our right hand.'

Professor Kirby says that the 'signal towers remain as a monument to what was an honest if misguided attempt to apply a modern invention in an area that was neither suited to nor prepared for it.'

Plate 85: South Africa: Grahamstown, Cape Province, about 1850, showing Fort Selwyn telegraph at extreme right, from a painting by Thomas Baines.

Plate 86: India: The Alambagh, near Lucknow, in 1857, showing telegraph on roof.

Plate 87: Tasmania: telegraph at Mount Nelson.

Plate 88: Tasmania: preserved telegraph station in Prince's Park, Hobart.

Plate 89: United States of America: New York Harbour, telegraph at Sandy Hook in 1856.

Plate 90: Model in New York city museum showing the telegraph on Staten Island and barrel signals being hoisted.

CHAPTER XXIII

St. Helena

In 1802 Colonel Robert Patton, who had been the East India Company's Military Secretary in Bengal, was installed as governor of the company's island of St. Helena in the South Atlantic.

The St. Helena Consultations of 30 May 1803 contain a report from Patton that his new 'telegraph' system

> is found completely to answer the intended purpose by conveying information in a simple, easy, clear and distinct manner without a possibility of error that may not be immediately corrected.
>
> I was led to prefer the sort of Telegraph I have adopted composed of a frame of wood and balls, because our stations are generally upon heights having the sky behind which renders such objects distinctly visible...With only four balls, one hundred different signals can be distinctly made, in the numerary manner, in that the system adopted is a species of communication in cypher – a Ball hoisted above the frame indicates an hundred, the progress in this manner being unlimited.

On 25 July 1803 Patton produced a book of instructions. He proposed to supply 'good glasses' to each station and set out a system of 'approbation money' to each NCO in charge of a station. 'Celerity and accuracy' were his watchwords. All signals were to be answered and repeated. Each NCO was to have an assistant from the guard.

The instructions refer rather confusingly to both telegraph posts and outlook stations.

> When signals are made from the Outlook Stations or any of the Telegraph Posts to Headquarters, a red broad pendant must be hoisted and kept flying while the signals are making and when they are to be made; and then struck. When back signals are made from one post to another, or from Headquarters to any Telegraph Station, a white broad pendant must be hoisted.

Every station had to keep a log of all signals made. An outlook station operator had to write down the words he intended to convey, with the numbers or signals by which he expressed them, but at posts transmitting or passing signals only the numbers answered and repeated had to be entered.

A Sergeant Weager, later Sergeant-Major of Artificers, who had made a model apparatus and improved the working, was named as Telegraphic Superintendent.

For night operation the wood and balls were replaced by a system of candles in lanterns. When a vessel or vessels were sighted, a gun was to be fired. Other stations would acknowledge also by gunfire, whereupon the first station would indicate the number of ships by suspending lights vertically or obliquely. Three oblique lights, for example, expressed six ships and a perpendicular light added signified seven ships. Lights in the form of a triangle or square indicated the approach of boats.

It is not clear how many telegraph posts were erected, nor whether the sites were changed from time to time. A map by Colonel James Cocks, dated 30 November 1804 shows nine stations, mostly disposed near the coast but with one near the centre of the island on a site identical with that of Halley's observatory of 1677, recently excavated.

Sir Home Popham visited the island in 1806. According to an island historian, G. C. Kitching, he 'contributed to the St. Helena telegraphs by introducing numerary signals' but this statement seems to conflict with Patton's own words. It is more likely that he merely made a few improvements.

HMS *Northumberland* carrying Napoleon Bonaparte and his suite anchored off St. Helena on 15 October 1815. Gilbert Martineau says in his *Napoleon's St. Helena*:

> The rusty pulleys of the aerial telegraph installed by a previous governor were cleaned; and an office was set up at Longwood [the house allocated as the residence of Napoleon] making it possible to communicate immediately to the authorities the information that:

> All is well with respect to General Bonaparte and family
> General Bonaparte is unwell
> General Bonaparte is out, properly attended, beyond the cordon of sentries
> General Bonaparte is out but within the cordon of sentries
> General Bonaparte has been out longer than usual, and is supposed to have passed the sentries not properly attended
> General Bonaparte is missing

The last signal, a blue flag, was never used!

There is some doubt whether a post was set up at Longwood. Kitching argues that Sir Hudson Lowe, the governor, never referred to it, nor was it mentioned in reports. But, as W. G. Tatham, Hon. Archivist of St. Helena, observes, it is likely that there was a station though independent of the earlier posts and used simply to report on Napoleon's activities and health to Lowe at Plantation House and to Jamestown, the capital. It would, of course, have been redundant after Napoleon died in 1821.

St. Helena 1502-1938, by Philip Gosse includes the following extract from 'Sally Phil's Poem', a doggerel published in the *St. Helena Herald* in 1854:

> At Longwood a Captain was put to look out,
> To see what Napoleon was alway about;
> He was ordered to see, Bonny every day twice,
> A duty it was, I should say far from nice:
> At Longwood the signals were plac'd on a pole,
> And passed to the Governor, through the post at High Knoll,
> The purport of which, was the Governor to tell,
> That Buonaparte was, both safe and quite well.

CHAPTER XXIV

India

As early as 1788 Sir Archibald Campbell, Governor of Madras, had studied a plan to establish a signal system to give warning of an invasion of India but no action seems to have been taken.

By 1815 the value of a telegraph system to establish more rapid communication within the sub-continent had become clearer. The Bengal and Bombay Presidencies were in communication over a plan proposed to the Government of Bombay by William Boyce, proprietor of the Bombay Tavern. Boyce had been trained as a land surveyor under Richard Lovell Edgeworth in Ireland and assisted Edgeworth in his telegraphic work there. He planned to use an improved version of Edgeworth's telegraph (*See* Chapter VIII).

Boyce wrote:

> The machinery is so simple as to be perfectly manageable by a boy of ten years old... I have effected improvements which render it infinitely superior to any telegraph whatever. When an overland despatch arrived or any news of importance by sea, it could be communicated to the Supreme Government in half an hour, and an answer received back in the next half-hour.

One line, Boyce suggested, would run from *Bombay* to *Calcutta* via *Poona, Hyderabad, Ellore* and *Cuttack* and have 75 stations. The other, with 133 stations, would also link Bombay with Calcutta, running along the coast to *Mangalore,* thence via *Seringapatam* and *Bangalore* to *Madras* and along the coast to Calcutta.

In recommending the plan to the Supreme Government in Calcutta, the Bombay Government observed: 'a man better qualified than Mr. Boyce to superintend the establishment will hardly be found.'

Asked to comment, Colonel C. Crawford, Surveyor-General of Bengal, wrote:

> Mr. Boyce... has... overlooked some obstacles that, had he travelled much in the peninsula, he would have found very difficult to overcome... First... the Pindaris, or any other lawless tribes; for, although the towers may be musquet proof, and sufficiently strong to prevent the tower being carried by force, yet how are they to procure water or provisions? If the tower must be protected, then only one man can go on the errand, whose fate could easily be guessed... In many places... the distance from any supplies and, what would be still worse, not being able to procure water, or of its proving bad, would prove... very difficult to overcome... In most of these jungles it would often happen... that... the whole complement of men would be... all down with fevers at the same time.

Nevertheless, in August 1816 Boyce was asked to superintend an experimental line between *Fort William* (Calcutta) and *Barrackpore.* It is not certain whether this line was ever built. In August 1817 the Military

Department appointed a committee to establish an experimental line between *Fort William* and *Nagpur*, with the probable extension of the system in mind. The members of the Committee included Lt. Colonel C. Mackenzie, Madras Engineers and first Surveyor-General of India, as president, and six other officers, including the Adjutant-General and the Quartermaster-General. Their report of 20 September 1817 recommended a line from *Calcutta* to *Chunar,* following the line of the new military road through *Hazaribagh* via *Ranganathpur* and *Sherghati*. Chunar, on the S. bank of the Ganges about 16 miles S./SW. of Benares and some 400 miles from Calcutta, was the European settlement for the invalid or veteran battalion of the East India Company army and for long the N. frontier post of the company's jurisdiction.

The Bengal Military Despatches for 1 September 1819 advise of the appointment of a committee to superintend the establishment of the *Calcutta-Chunar* telegraph. Major H. Faithful was named as Secretary, at a salary of 600 rupees a month, and Captain-Lieutenant George Everest, Bengal Artillery, and Lieutenant R. B. Ferguson, Ramghur Battalion, as Surveyors at 500 rupees a month each. Everest, who was knighted in 1861, had been appointed Chief Assistant on the Great Trigonometrical Survey of India in 1817. He was later, as Surveyor-General of India, to give his name to the highest peak in the world.

There is record of an allowance, equal to the salary of a Barrack Master, granted in 1825 to Major W. D. Playfair, Superintendent of Military Roads, while engaged on building the telegraph towers, in which Ferguson had assisted him.

The Bombay Despatches of the East India Company contain a resolution dated 1 April 1818 to suspend further consideration of Boyce's project for a Bombay-Calcutta link in view of likely physical obstacles. What the obstacles were is suggested in an inter-departmental letter of Bengal dated 20 December 1820. The writer doubts the practicability and advantage of Boyce's plan, by this time said to be for telegraphs throughout India. Because of haze and frequent fogs, particularly in Lower Bengal, telegraphic communication would require many stations, which would have to be built high enough to clear forest and jungle. If left unguarded, the stations would be liable to be destroyed by marauders.

The same writer regrets that without consultation with his department a line had been authorised between Chunar and Calcutta

> with the avowed intention of extending it through Bundlecund and Saugor to Nagpore [Nagpur]. And, whatever may be the result of that experiment, which you will of course report to us, with a statement of the whole expense attaching to it, we positively prohibit any further progress in the plan without our previous sanction.

There was no objection to a donation of £1,000 to Boyce for his labour and ingenuity.

Meanwhile the survey had been going ahead. Boyce by now seems to have rejected his planned improvement on Edgeworth's system. The Bengal committee asked Lieutenant-Colonel W. Lambton, Superintendent of the

Trigonometrical Survey, for his opinion of the method Boyce now proposed and on routes to form an all-India network. In Boyce's method, he stated, there were

> four balls which, being numbered from 1 to 4, are singly or by combination sufficient to express the nine units. For applying this system to written language it was only necessary to form a vocabulary of all words and forms of words that can be useful, together with names of persons and places, and then to number them in such a manner as to exclude o or zero ... The apparatus for communicating them is composed of a mast and yard, to the latter of which are affixed four pullies serving to raise and lower the balls, which are from four to six feet in diameter, according to the distance at which they are to be seen, and made of the lightest bamboo work covered with black canvas.

The committee stressed the key position of Bombay and the importance of safe and rapid communication with W. India.

> On the expediency of establishing one general line of communication between the three Presidencies ... the different lines would naturally meet both to the east and west of Berar.

Lambton replied stating the advantages of a route by Nagpur, in view of the large military stations in the W. provinces and in the Deccan.

The survey party headed by Everest reached Sherghati in mid February 1818. At the end of that month the committee asked Lambton for his opinion on the power of telescopes 'which you would deem sufficient to ensure a view of telegraphic spheres of 6 feet in diameter throughout the year at the distances of 7, 9, 12 and 14 miles.' Lambton answered that if the balls could be seen sufficiently distinctly at a distance of one mile without the aid of a telescope, a magnifying power of 7 would be proper for viewing at a distance of 7 miles.

The survey reached Chunar in May 1818. Everest and Ferguson stayed there during the hottest weather and rainy season and studied their maps and reports. In June Everest reported to Lambton. Throughout the plains, he wrote,

> there is a peculiar vapour in the hot weather which affects the atmosphere at a less height than 100 feet above the surface of the earth, and causes so great a divergency in the rays of light that telescopes of large magnifying powers are of little use, and in such situations the telegraphic distances have seldom been greater than 7½ miles, whilst in the hilly tracts 18 miles has not been too great.

The survey closed its work in October 1818, after experimental signals had been exchanged between some temporary stations. The line was probably completed about 1821. Captain H. E. Gilbert-Cooper, Bengal Infantry, was Telegraph Superintendent from 1822 to 1824, being succeeded by Captain C. T. G. Weston, Bengal Infantry. Both were styled Director of Telegraphic Communications.

The line cost Rs. 54,593 up to 1 November 1818, of which Rs. 31,250 could be considered as an extra charge to Government. The annual cost of maintenance was put at Rs. 27,216, plus about Rs. 9,000 as the salary of the Committee secretary.

Costs seem to have caused concern quite early. A report of 30 April 1828 records that the Government made no decision on measures proposed by the Military Board 'relative to the repair and the charge of the bungalows and telegraphic towers on the military road from Calcutta to Benares.' Between

1818 and 1828 total expenditure was five lakhs of rupees. The idea of extension to Nagpur, let alone elsewhere, seems to have been abandoned quite soon. In April 1828 Sir Charles Metcalfe recommended to the Governor-General the discharge of the telegraph establishment,

> being for my own part strongly impressed with the belief that the expense of the telegraph is not repaid by its services, and also, that, if the institution be maintained, the expense will progressively and largely increase.

When Watson died on 27 May 1828, the Governor-General

> though inclined to avail of this opportunity to abolish the establishment, yet he deferred the matter till the receipt of the final report on the expenses of the telegraph.

A Lieutenant Beatson, Adjutant of European Invalids at Chunar, who had been appointed Acting Superintendent of Telegraphs, reported that total costs need not exceed Bengal Rs. 2000 a month.

A letter of the Public Department of 18 February 1829 reads:

> Sir Charles Metcalfe has likewise observed that the whole expense incurred on account of the Telegraph is to be regretted, its services not being likely to counterbalance the outlay ... as yet, for about ten years, it seems to be nothing better than an expensive plaything. We were of this opinion from the beginning ... and the sooner the Establishment is abolished, the better.

With its great length and 45 stations the Chunar line was the most ambitious telegraph completed up to that period outside France. Almost all the stations seem to have been specially built towers of brick. Constructional costs must have been heavy, even in view of the low rates payable to native labour. Some towers were 100 ft. high and of necessarily massive construction. After the line was given up, some towers found a new use for a time in the service of the Survey of India during the great trigonometrical survey carried out about 1830.

One tower still standing is at *Nibria,* near Mahiari, five miles W. of Howrah. It is a round, tapering tower 88 ft. high and with five storeys. There is a window at each landing of the staircase, which has collapsed.

Each station was manned by five to seven Indians. The total paybill for the line was Rs. 992 a month.

The service seems to have worked efficiently when conditions favoured it. For example, a message was sent from Chunar about noon, arrived at Calcutta before 1 pm and the reply reached Chunar before 5 pm. About this time it was stated that it was very rare for a message sent all the way to arrive unintelligible. A drawback about which Watson complained in May 1828 was the lack of a second telescope at each station. The one telescope was liable to damage through being constantly turned round in a crowded room. In the same letter Watson also said that unusually dry weather was raising clouds of dust along the Benares road and so impeding telegraphy.

Although regular transmission ceased on 1 September 1828, the line was kept on a caretaker basis until 1830. One man remained in each tower 'as a nucleus for the education of others qualified for prompt employment in case of some emergency should it become indispensable to re-establish the line to Chunar.' Captain J. N. Jackson, Bengal Infantry, was appointed Superintendent for the caretaker period. The telescopes and other equipment were sent to Fort William arsenal, Calcutta.

A telegraph to link *Calcutta* with *Saugor Island* was proposed at least as early as 1823. On 21 March of that year the Bengal Public Department concluded that from the erection of two navigation beacons on the River Hooghly, apparently on the E. bank, the 'projected telegraph between Calcutta and Saugor, by the mercantile body of the community, has been abandoned.' It was agreed that the government could not spend upwards of Rs. 24,000 to make the beacons subservient to the plan of the telegraph.

Nevertheless the line was certainly operating by 1833, when Charles L. Smartt published a signal book entitled *The Hoogly River Code adapted for the general use of the semaphore stations on or near its banks, in aid of Conolly's Vocabulary and Marryatt's* [sic] *Code, and intended to facilitate communication between Saugor* [Sagar] *Island and Calcutta*. Another signal book, published in 1845 and for use by the Bengal Pilot Service, gives the stations as under:

East Bank
1. Calcutta Exchange (E)
2. Fort William
3. Akra
4. Budge Budge (Budj Budj)
5. Moyapoor (Achipur Semaphore)
6. Roypoor (Royapur)
7. Fulta
8. Hoogly Point (E)
9. Diamond Harbour (E)

West Bank
10. Jiggercolly (Jigerkolli)
11. ?
12. Bellary (Balari Semaphore)
13. Sundea (Sandia Island)
14. Gungra (Gangra)
15. Kedgeree (Khijiri Semaphore) (opposite Saugor (Sagar) Island) (E)

Stations marked (E) had European superintendents, the remainder Indian. Modern spellings appear in parentheses. Conolly's system, as described on page 117, was used.

Semaphore telegraphy rendered notable if limited service during the Siege of Lucknow in 1857. The Residency, in which Brigadier Inglis and his force were besieged after 1 July 1857, stood on a site a little higher than the city. The arrival of Sir Henry Havelock and Sir James Outram's relief troops on 25 September brought forces sufficient to prolong the defence but not to raise the siege. In fact, the relief force itself was besieged by the rebels and its communications were cut with the Alambagh, a large house with a walled enclosure south of the city. The force had left its baggage train there, guarded by 300 men.

Those in the sorely-battered Residency decided to try communicating with the Alambagh, though the distance was about three-and-a half miles and there was often a haze over the intervening city. The means taken are well recounted in Michael Joyce, *Ordeal at Lucknow*, from which some of the following detail is taken.

Martin Gubbins, Financial Commissioner in the Residency, turned up particulars of the construction of Popham's semaphore in the *Penny Cyclopaedia*. A machine of Popham type was made and erected under the supervision of Lieutenant Henry Moorson, 52nd Light Infantry, then DAQMG, and half-caste lads from the Martinière College were deputed to work it.

On 31 October Outram concluded a despatch to McIntyre at the Alambagh as follows:

I enclose a plan for telegraphic communication, your share of which Sibley [McIntyre's second-in-command] will, I hope, be able to construct, as I know he is a great mechanic. Ours will, I hope, be ready in a couple of days, and you will be able to make it out from the top of your house. A second set of apparatus should be got ready to send with the relieving column, for the purpose of being placed on the top of the Martinière.

The evening before the day on which we purpose telegraphing to you, a bonfire will be lit on the highest point of our position [the Residency roof] to enable you to know exactly our whereabouts. A similar illumination on the top of the Alambagh will be proof to us that our signal has succeeded.

We shall signal at twelve, noon, of each day, the time best suited; for the enemy annoy us least at this hour, and our signallers consequently will incur less danger. Even should our signals fail from your being too far from us, still do not delay in having two sets of telegraphic apparatus prepared; for so soon as we establish one set of apparatus at the Martinière, and yours also is ready, the signals will be carried on without difficulty...

In his reminiscences General Francis Maude recounted amusingly the first trials and errors of the Residency semaphore:

One afternoon, when sufficient time had elapsed for the despatches to have arrived at the Headquarters Camp at Alambagh, Moorson and I ascended the Residency Tower and began to work the semaphore, assisted by the Collegians. We noticed, with some amusement, that the enemy, thinking they had the key in their hands, maintained complete neutrality and did not fire a shot at us, although, in spite of the sandbags, our party presented a fair enough mark for musketry. The exact distance between the two semaphores was estimated by us to be exactly three miles and we had a very powerful telescope, every movement was clearly visible. We laid our code on the coping of the parapet and began to send a message to the Alambagh. After some delay, we were delighted to see the arms of their semaphore in motion and we, each of us in turn, noted the message which was sent. But, to our intense disappointment and confusion we could make nothing intelligible of it. The first four letters were a complete puzzle. 'M Y Y R' I think it ran; and the following signals were equally mysterious. So again we had recourse to our own semaphore and the Martinière lads vied with one another in their alacrity in working the arms. A pause, and the same cryptogram was revealed by the telescope. By the time we had repeated the dumb show once more it was too dark to distinguish their movements and we went, with rueful countenances, to report our ill success to the Brigade Mess.

Conning the signals again, a lucid idea came into my head and I asked Moorson how it would be if the workers of their semaphore were standing on a different side of the instrument to the one on which we stood? Or, in other words, supposing they were reversing the movements of the arms. The meaning of this will be best understood if a person will move his right arm up and down after the manner of a semaphore and while doing so turn to the right about; or, if the spectator will walk round to the other side of him. The position will of course then appear as though he was working with his left arm. 'A happy thought,' exclaimed Moorson, and we forthwith proceeded to develop the new Key. The first four letters did not appear to throw much clearer light upon the puzzle. We read them as 'G-o-o-n'. Our discomfited faces made the members of the Mess party laugh heartily at our expense. But the mirth was on our side a few seconds after when another flash of intelligence enabled us to spell out: 'Go-on-we-are-ready.'

Then warm and sincere congratulations were showered upon us and at day-dawn the next morning we were both again perched up and reversing our own code and reading theirs reversed, without the slightest hitch. Soon every detail as to the strength and intended movements of Campbell's [Sir Colin Campbell, C.-in-C. of the British Army in India] force was made known to us and ours to them.

It is unfortunate that the Residency party had no details of Pasley's telegraph. Its indicator would have avoided the confusion of left hand and right hand.

After preliminary announcements on 12 and 14 November of the intended movement from the Alambagh, the signal 'Advance to-morrow,' was received at the Residency just before dark on the 15th, a little after a semaphore had been erected on the Martinière College.

The Residency was successfully evacuated and its occupants transferred to the Alambagh. Michael Joyce, reporting the events of 25-26 November, states

> Mr Gubbins had a last look at the old position from the top of the Alambagh house. Through the glasses the enemy could be clearly seen crowding the roof of the Residency building, where they had already smashed the semaphore.

This was not quite the end of semaphoring, or attempts at it, during the siege. On the evening of 8 February 1858, the Commander-in-Chief wished to communicate between the camp at Bibiapore, near the Dilkusha Palace, and that at Chinhut, on the Faizabad Road, which were separated by the Gumti River. An officer and three men of the Royal Engineers were deputed to construct the apparatus. Two complete double-arm semaphores were completed on the 8th, one erected on the roof of the Dilkusha and the other sent to Outram's camp, to which the officer (Lt. P. Stewart) went next morning to superintend operations if required. But the service was not used as the semaphore could not be erected on the left bank of the river, where the only prominent buildings were still held by the mutineers. As the Royal Engineers were in charge, it seems likely that Pasley's rather than Popham's system was chosen.

39. Surviving tower near Calcutta.

CHAPTER XXV

Australia

New South Wales

Flag signalling was used early in the life of the colony of New South Wales. From 1790 news of the arrival of vessels was signalled by flags from a post at South Head to Port Phillip, about five miles away. In the late 1790s the communication was extended inland to Parramatta, the country residence of the Governor. Warned in this way of the advent of a ship, the Governor could reach Sydney before the mail had been brought ashore.

The *New South Wales Post Office Directory* (1832) lists intermediate flagstaffs at Bedlam and One Tree Hill between Port Phillip and Parramatta. In that year tenders were called for the erection of a 'small stone house' to accommodate the signallers at South Head. At Port Phillip the unfinished fort was used as the station.

Some time in the 1830s the flag signalling between South Head and Port Phillip gave place to a telegraph, probably of Popham or Pasley type, and the Parramatta section may have followed suit. Certainly as from 1 June 1833 four 'telegraph masters' are listed, one each at *South Head, Port Phillip, Bedlam* and *Parramatta*. They lived on the premises. The arrangement continued until the Parramatta line was given up on 1 November 1842.

Flag signal stations were also in use at Moreton Bay (Brisbane), Newcastle, Port Macquarie and Melbourne by 1840 and remained in use until at least 1853.

A pictorial atlas of Australasia published in the 1880s refers to a plan to link Sydney and Melbourne by telegraph as early as 1845, but no official record of such a project is known. Such a line would certainly have been a semaphore system as the electric telegraph was not introduced into New South Wales until 1854.

By far and away the most important visual telegraph in Australia was developed not on the mainland but in Tasmania (Van Diemen's Land until 1853).

Tasmania

In 1811 Lachlan Macquarie, Governor-in-Chief of New South Wales and Van Diemen's Land, instituted a ball-and-flag system to announce the arrival of ships to Hobart (then Hobart Town). He himself chose the site for the outer signal station on a hill about four miles S. of Hobart and named it Mount

Nelson, after the ship which had conveyed him from Sydney.

The system was displaced about 1828 by a visual telegraph between *Hobart* and *Mount Royal,* some 30 miles S. and commanding D'Entrecasteaux Channel. The Hobart terminal was at *Mulgrave Battery*. Then came *Mount Nelson,* followed by *Mount Louis,* above Pierson's Point, at the entrance to the Derwent River.

As soon as a ship appeared off the coast, Mount Louis reported to Mount Nelson whether it was from the SW. or SE. and what type it was. When the pilot at Pierson's Point went aboard he signalled Mount Louis the name of the port from which the ship had sailed. The intelligence was then telegraphed to Mulgrave Battery, which hoisted a flag to tell the people of Hobart that a vessel was entering the Derwent and followed it with a ball to denote the type of ship. Later, one or more of 30 flags was raised to indicate the ship's port of origin.

The apparatus consisted of two arms, one above the other on a mast — possibly a variant of Pasley's system. Each arm revolved and used three positions on each side of the mast. Certain positions of either or both arms denoted certain numbers, but there was no orderly rotation of the arms, and the code was limited and hard to memorise. The mechanism was cumbersome and easily damaged.

The apparatus was improved about 1830, when a third arm was added to simplify the code. The top arm then represented units, the middle one tens, and the lowest hundreds. Each arm could take up a position indicating 1, 2 and 3 on one side of the mast, and 4, 5 and 6 on the other side. There was no provision for making 7, 8 and 9.

Lieutenant-Governor Colonel George Arthur, Governor of Van Diemen's Land between 1824 and 1836, set up the notorious penal colony of Port Arthur, at the southernmost tip of Tasman's Peninsula. Captain Charles O'Hara Booth, 21st Fusiliers, was its fifth commandant, being appointed in 1833. Under his enthusiastic direction a most efficient telegraph network was established to make sure that convicts did not escape from the peninsula and to convey messages between *Port Arthur* and *Hobart* — it also took over the Mount Royal line. In addition it relayed to the out-stations in the peninsula subsidiary to Port Arthur and used as probation stations.

M.C.I. Levy suggests in his book *Governor George Arthur* that Arthur may have been advised by Peter Mulgrave, who, as we have seen, had been concerned with the telegraph system of the Channel Islands before going out to Van Diemen's Land. Mulgrave rose to be Superintendent of Police in the colony and died at Launceston in 1847.

Booth's Telegraph

Booth accompanied his men to seek suitable sites on the hills and cliff tops of the peninsula. Smoke from bush fires in summer and fogs in winter made it necessary at times to change the position of stations. When the chosen site was a wooded hill, it was cleared of all but one tree, whose top was then lopped. A large tenon was cut and a substantial frame to carry arms and gear was bolted to it. The whole was supported by substantial rigging. Cross-trees or a yard were added for displaying the three flags used in addition to the arms.

Booth's semaphore, as finally evolved about 1837, had three pairs of single arms. They did not revolve but each had three movements on its side of the mast and each pair had three movements of its two arms in conjunction. The top pair represented units, the middle pair tens and the lowest pair hundreds. It was thus possible to count up to nine. The number 10 was denoted by making '1' on the tens (middle) pair. A 'millesimal' or 'thousands' pennant, chequered blue and white, was used to signify thousands. If this pennant was run up to half yard, the number made by the semaphore arms was read as thousands. If it was run to yard and another number made on the semaphore arms, the latter number was added to the first number shown. Thus the operator could display numbers up to 999,999.

The other flags used were the Blue Peter as preparatory and the tricolour to signify that the number shown by the arms was a number only and had no reference to the code. In the code book each letter of the alphabet had a number. Grouped under each letter were words beginning with that letter, each having its own number, and under headings for quick reference were also numerous orders, questions, answers and information of all kinds. As every bay, hill, cliff, river, road and township in the neighbourhood had its number, it was seldom necessary to spell out a signal.

On a clear day a message could be sent from Port Arthur to Hobart and a reply received within 15 min. passing through five stations each way. A sharp lookout was kept at all stations from dawn to dusk. No satisfactory system of night signalling was devised until the Morse lamp code was introduced. As the hempen ropes used to work the arms were found to perish or vary in length in different temperatures, chains of iron rods, about 2 ft. in length, with eyes forged at the ends and coupled by S-hooks, were substituted.

Marcus Clarke gives a graphic description of this system in his melodramatic novel *For the Term of His Natural Life,* written in 1874, soon after the penal colony had been given up. He likens Tasman's Peninsula in form to an ear-ring with a double drop, with Eaglehawk Neck — a sandy strip 450 yd. across, guarded by soldiers and chained dogs — the link between the drops.

Clarke writes:
At the Coal Mines [west of the Neck] was the northernmost of that ingenious series of semaphores which rendered escape almost impossible. The wild and mountainous character of the peninsula offered peculiar advantages to the signalmen. On the summit of the hill which overlooked the guard tower of the settlement was a gigantic gum-tree stump, upon the top of which was placed a semaphore. This semaphore communicated with the two wings of the prison — Eaglehawk Neck and the Coal Mines — by sending a line of signals right across the peninsula. Thus, the settlement communicated with Mount Arthur, Mount Arthur with One-tree Hill, One-tree Hill with Mount Communication, and Mount Communication with the Coal Mines. On the other side, the signals would run thus — the settlement to Signal Hill, Signal Hill to Woody Island, Woody Island to

Eaglehawk. Did a prisoner escape from the Coal Mines, the guard at Eaglehawk Neck could be aroused, and the whole island informed of the 'bolt' in less than twenty minutes. With these advantages of nature and art, the prison was held to be the most secure in the world.

The semaphore plays an exciting part in the story when absconding convicts rush the Signal Hill station and force the operator to send misleading messages to the settlement — a pardonable 'crib' from Dumas perhaps!

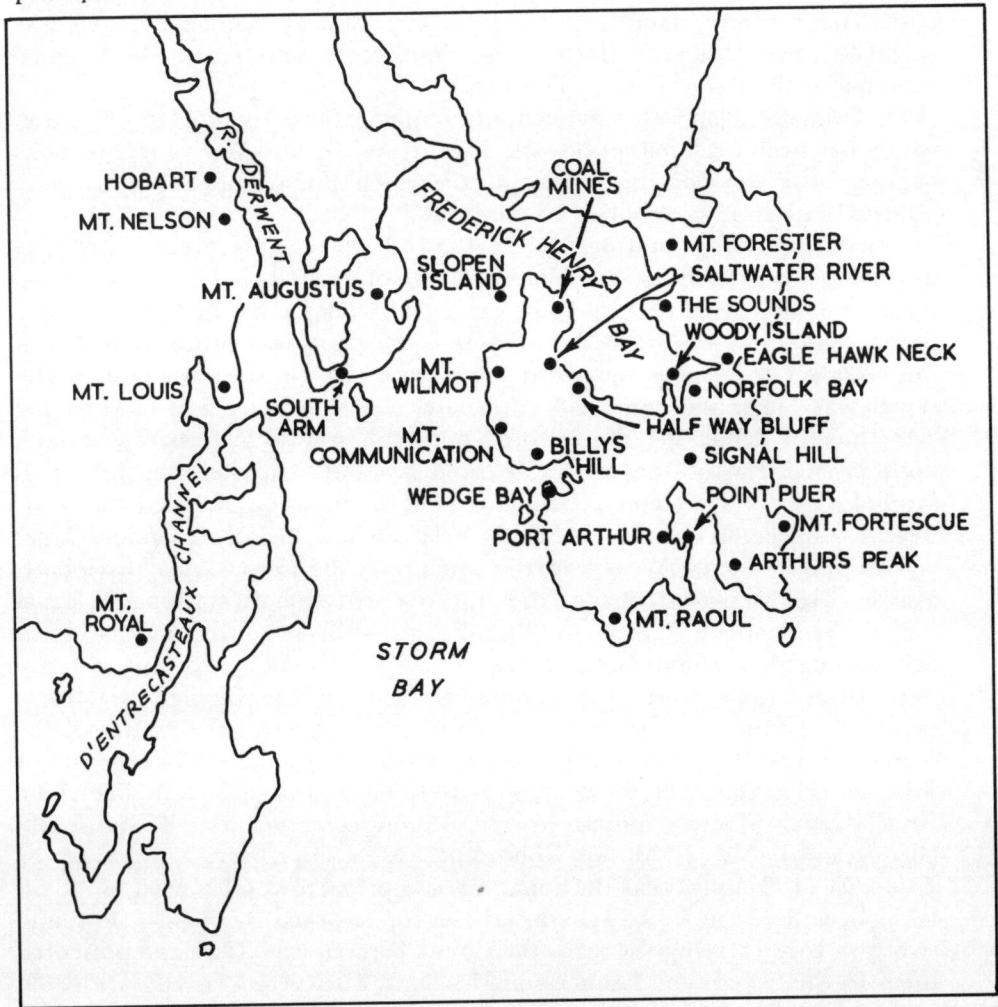

40. Map of telegraph stations in Tasmania.

The accompanying map shows the intercommunicating system in the peninsula and the connection with Hobart. At one time there were at least 21 stations on Tasman's and Forestier's Peninsulas. Their call signals included: 2–Eaglehawk Neck; 556–Mines; 3000–Mount Nelson; 9000–Impression Bay.

The system was long lived. Statistics for 2 May 1873 show that operators were still employed at *Mount Arthur, Mount Fortescue, Wedge Bay, Mount Communication, Slopen Island, Salt Water River, Signal Hill, Norfolk Bay, Woody Island, Mount Augustus* and *East Bay Neck.*

Largely on the initiative of Mr. W. E. Masters, the Hobart City Council has preserved the *Hobart* terminal station with its apparatus restored and in workable condition. The building is one of the historical showpieces of the city. The stations stands in Battery Park, Castray Esplanade, and was originally the Mulgrave Battery guardhouse. It superseded the original terminus at the top of Battery Point Hill.

In Tasman's Peninsula, some stones remain of the *Mount Arthur* station which has been declared an historic site. At *Salt Water River* a tree remains bearing near the top iron fittings, to which the semaphore arms were attached, driven into or bolted to the trunk.

The *Mount Nelson* station, like that in Hobart, remains intact. It is still in use, being controlled by the Marine Board of Hobart, but for many years communication with Battery Point has been by telephone. Just before it was dismantled about 1890 the Mount Nelson telegraph was photographed with the arms set to display the word 'forgotten' (343 in the code book). The signal was 'answered' some 50 years later, on 9 January 1940, when the ceremony of dedicating the restored telegraph station in Hobart was held, both in memory of the former telegraphers and as a cherished historical symbol of the State. Veterans who attended the function included Mr. W. A. Begent, a telegraph operator at Mount Nelson when the last semaphore signal was displayed there. As part of the ceremony, the Lord Mayor of Hobart, assisted by Mr. Begent, made from the restored Hobart station the signal '2095' ('remembered'), after the Blue Peter had been hoisted and had been acknowledged by Mount Nelson.

In 1825 the northern seat of administration in Van Diemen's Land was moved to Launceston from George Town, its outport at the mouth of the Tamar estuary. The need then arose for a telegraph to advise Launceston of ships entering the river. Four stations were erected on a 30-mile line: *Low Head* (George Town), *Mount George, Mount Direction* and *Windmill Hill* (Launceston). Mr. J. Moore Robinson, FRGS, in *Historical Brevities of Tasmania* (1937) says that the apparatus comprised a wooden mast, up to 90 ft. high, with a pair of 16 ft. arms at the top, worked by chains. He found two iron tops carrying the arms on Mount Direction in 1926 and presented them to the Hobart and Launceston Museums. The system lasted until 1869, when a cable was laid from Victoria under the Bass Strait to Low Head.

There is a suggestion that in the 1830s semaphores were erected on hills along the main road to carry messages between Hobart and Launceston, and that the passage of stage coaches was signalled in this way. In 1836, James Ross, a Tasmanian editor, wrote that there were by then nine telegraph stations in Tasman's Peninsula, with two more needed to complete the communication thence to Hobart. He visualised extensions to Launceston and even Sydney!

CHAPTER XXVI

Canada

Prince Edward, fourth son of George III and future father of Victoria, was appointed Commander-in-Chief, Nova Scotia, in 1794. He began to make Halifax the strongest fortress outside Europe and soon developed a telegraph system, the first in North America, if one excepts the smoke signals of the Indians and the baskets, barrels and flags used during the American War of Independence.

In *Halifax: Warden of the North* Thomas H. Raddall gives an entertaining account of the Prince's telegraph. Unless otherwise stated, the following references and quotations are from that work.

> In the meantime [writes Raddall], indeed while the Admiralty was still playing with the shutters, Edward's engineers had devised a system of visual telegraphy, using a code of flags and large black wickerwork balls and drums for signalling by day, and combinations of lanterns by night. There were signal posts of this sort near Chebucto Head, at York Redoubt, on Citadel Hill, and at the Dockyard. Haligonians became accustomed to the sight of baskets and bunting running up and down the flagstaff on the hill, a constant flutter from morn to night, for Edward was pleased with his new toy and kept it busy. The outposts reported everything that happened, even the petty misdemeanours of the soldiery, and the commander-in-chief bombarded them with questions, orders, regulations, and punishments.
>
> ...Edward soon decided to remove his charmer [his mistress, Madame St. Laurent] from the jostle of the town to a more secluded love-nest on the wooded shore of Bedford Basin...Finally, so that Edward could remain in constant touch with the citadel and outposts while spending pleasant hours in the company of his *chère amie*, a signal station was installed on the hill behind the lodge — a unique collaboration of Venus, Mars and Mercury.
>
> ...Daily, hourly, Prince Edward awaited the French invasion...[he] drew across the Atlantic a stream of troops, cannon, munitions, and gold for the fulfilment of his dream. Amongst other things he had determined to extend his telegraph along the Great West Road to Annapolis, the old capital of the province, where he maintained an alert garrison in Fort Anne. Engineers surveyed the route, and companies of sweating redcoats began to hack long avenues through the forest to provide a view from station to station along the way.

After leave in England, Edward, by then Duke of Kent, returned to Halifax in the summer of 1799. Raddall tells us that in the Duke's absence

> the telegraph system had been extended to the Annapolis Valley as he had directed, and when he arrived off Chebucto Head in the frigate *Arethusa* in September the news was signalled fifty miles to Windsor in something like twenty minutes...One of the Duke's new ideas was the enlargement of his telegraph system beyond all previous imagination. In view of his new responsibilities as Commander-in-Chief of North America he had determined to carry it around the Bay of Fundy to St. John, thence up the St. John Valley to Fredericton, and eventually to Quebec. The engineers and the axe-wielding redcoats fell to work at once. Edward's enthusiasm had taken no account of the Bay of Fundy fogs, which prevented visual signalling for weeks on end, and although the chain

was completed as far as Fredericton it does not seem to have been operated beyond Nova Scotia. There, however, he enjoyed it thoroughly. To test it on one occasion he made a mid-winter journey to Annapolis and stayed some time at Fort Anne, receiving a stream of reports from Halifax and sending a stream of orders back in return.

A visitor to a Halifax barrack in the snowy January of 1800 was surprised to see preparations for a flogging — the troops drawn up in the barrack yard, the naked culprits fastened to the triangles and shivering in the bitter air, the tallest drummers in place with their long-tailed cats, the others waiting to drown the cries with the roll of their drums, and the officers and the surgeon standing by. He was surprised because he knew that the Duke, who insisted on confirming all courts-martial and allotting the punishments, was at the moment in Fort Anne one hundred and thirty wintry miles away. But the explanation was simple. The Commander-in-Chief was about to flog his men 'by telegraph'!

The Duke, whose expenditure in Nova Scotia had appalled both the King and the Government, was recalled in August 1800. After the Peace of Amiens, the fleet was withdrawn from Halifax, the garrison reduced, and Edward's telegraph abandoned save for the stations along Halifax Harbour. By the time Sir John Coape Sherbrooke became Governor in 1811 the military decline of Halifax was complete.

> ...the Duke of Kent's great telegraph chain, abandoned outside Halifax in 1802, had fallen to pieces, and the long avenue cut through the forest towards Quebec was choked with another growth of trees and bush. The corps of expert signallers had vanished.

The desirability of sending rapid intelligence of ships in the St. Lawrence River to Quebec City brought the telegraph to Quebec early in the 19th century, only a few years indeed after the Duke's abortive attempt to link Halifax and Quebec by basket telegraph. A line was established in 1808 between *Isle Vert* (L'Ile Verte), below Rivière du Loup, and the capital, 120 miles, along the S. bank of the river. The system used is not known. A report dated 4 June 1828, from the Deputy Quarter Master General's office to the Military Secretary in Quebec, in reply to a *Report upon the Telegraphic Establishment of the River St. Lawrence* called for by the Treasury in London, shows stations as follows: *Brandy Pots;* (indecipherable); *Kamouraska; St. Denis; Rivière-Ouelle; St. Roch; L'Islette; Cap St. Ignace; St. Thomas; St. Vallier; St. Michel; Ile d'Orléans; Cape Diamond* (Quebec City).

In 1817, the stations between Brandy Pots and St. Roch inclusive, and in 1822 those between L'Islette and St. Michel, were given up, leaving only Ile d'Orléans and Quebec. Although the duties of Watt, the Telegraph Superintendent, had decreased in consequence, his long and faithful service, first for 16 years as Sergeant in the 49th Regiment and latterly in the telegraph service, were commended. The D.Q.M.G. further reported:

> With regard to the necessity of keeping up the Telegraph Establishment I beg to report that it is my opinion that so far from doing away with it it ought to be increased and kept up at least as far as the station below the Traverse over the River St. Lawrence, the difficult navigation of which river and the frequent accidents to Transports and Shipping generally and the impossibility of procuring assistance from the shore obliges the Master of the vessel to send to Quebec for it, the most expeditious means of communicating such intelligence being by Telegraph, as the Ship's boats are incapable of pulling against the Tide which runs at the rate of from 4½ to 5 knots an hour. Had the Telegraph Stations been further extended last year the greatest advantage would have been derived by the Shipping, several vessels having been lost by the impossibility of making known their distress in time to obtain relief.

> In addition to the above I beg to state that the Service would derive great benefit by the Telegraph Establishment as the only means of stopping the desertion which frequently occurs by soldiers getting on board vessels leaving the Port, and Quebec being a Garrison and fortified place under a Telegraph communication to a certain extent indispensably necessary.

Lord Dalhousie, the Governor, added a note to the report that he considered the telegraph indispensable and that it should not be reduced but one more added to its present establishment. A statement with the report shows the establishment of the Quebec City station as Superintendent (Watt), Second Class Superintendent and four 'labourers'. The Ile d'Orléans station has a second class superintendent and two 'labourers'. All but Watt are shown as being employed only between 25 April and 24 November, when the river was open to ships.

Watt's zeal was again praised in a letter from the D.Q.M.G. to the Military Secretary on 17 January 1832. He is called a 'scientific man, who adds to his other duties that of regulating the time of the Garrison, with the assistance of an excellent chronometer partly constructed by himself.'

The St. Lawrence telegraph seems to have lasted until the end of the 1830s, and some closed stations must have reopened. On 12 June 1844, the D.Q.M.G., Montreal, wrote to the Military Secretary asking for disposal instructions for stores left at St. Michel, St. Vallier and one other station (indecipherable) in the autumn of 1839 for the next season's service but never used because the whole line had been discontinued.

CHAPTER XXVII

United States of America

The first optical telegraph in the United States began working in 1801. It was patented by Jonathan Grout, Jr., a lawyer of Belcherstown, Mass., and intended to carry intelligence to Boston and Salem merchants of the arrival of vessels at Martha's Vineyard, an island off the coast of Massachusetts, west of Nantucket Island. The information about it which follows is based largely on a paper entitled *Early Visual Telegraphs in Massachusetts* given before the Bostonian Society by William Upham Swan on 16 April 1929.

Swan believed that Grout might have been inspired by the articles on both the Chappe and Murray telegraphs in the *Gentleman's Magazine* and preferred the Chappe system, or a version of it, because the description was clearer and it seemed simpler, though the description of Grout's apparatus has been lost.

The first message passed on 21 October 1801, between *Martha's Vineyard* and *Cohasset* near Boston. The vessel communicated was appropriately named *Mercury*. The line was completed through to *Boston*, 72 miles, in November, Grout announcing the fact in the *New England Palladium* for 17 November. The Boston terminus and office of the venture were first at 112, Orange Street and after December 1802 at 75 State Street. The route was as follows: *Martha's Vineyard* (West Chop), *Wood's Hole, West Falmouth, Pocasset, Bournedale, Cedarville, South Plymouth, Chiltonville, Duxbury, Millbrook, Marshfield, Scituate, Cohasset, Weymouth, Dorchester* and *Boston*. The name Telegraph Hill still describes seven of the sites.

The *Salem Gazette* of 14 September 1802, contained the following advertisement. (Incidentally, the French form *telegraphe* indicates the Chappe inspiration of Grout's project).

TELEGRAPHE

Merchants and others concerned in Navigation are respectfully informed, That the subscriber will recommence the operation of his Telegraphe by the first of October next. All persons who may wish to obtain by the Telegraphe, or by the Telegraphe and by the Mail, first intelligence of arrivals at the Vineyard or of arrivals at foreign ports — or who may wish to pass orders directing a vessel at the Vineyard to sail from thence to any particular port — or to wait there for further orders — or who may wish to learn the contents of a cargo — or whether a friend is on board of a particular vessel here, &c &c. may be accommodated.

The terms are lodged (for the convenience of all concerned) at the Post Offices in Boston, Salem, Newburyport, Portsmouth and Portland, and will be lodged at the Post Office of any other port, if desired. Agents are appointed in all the ports named above, to accommodate such as may wish for immediate intelligence from the Vineyard.

The terms contain different rates of fees. — If a man applies for first intelligence of an arrival at the Vineyard, by the 20th September inst. or three weeks before the day of such arrival, he shall have such intelligence at the lowest rate. An applicant has nothing to pay until the Proprietor or his Agents shall announce first intelligence of an arrival or other first intelligence desired.

Masters of Vessels will enter their arrivals at the Vineyard free of any expense, that the Proprietor (and owners, if *they* please) may have immediate knowledge of such arrivals.

JONATHAN GROUT, Patentee

Boston, September 14, 1802.

Grout's father, a wealthy landowner, may have helped to finance the enterprise. Business does not seem to have been brisk, for we find Grout raising his rates, originally $2 — $100, in 1804 and again in 1805. As the line operated only from October to May, it speaks much for Grout's pertinacity that he was able to induce two men to stand watch for even a few hours a day in a New England winter for the small remuneration which it was likely he could offer them.

Grout seems to have given an efficient enough service but he was poorly supported and his line lasted only until 24 April 1807.

In 1814 Samuel Topliff took over the commercial newsroom in Boston called the Exchange Coffee House, which his employer Samuel Gilbert had established in 1810. He soon made it one of the most famous in the world. It became the recognised meeting place of Boston merchants, who so staunchly backed Topliff, in contrast to their neglect of Grout, that in 1820 he was able to set up a harbour signalling system between Fort Independence and Long Island Head, using a mast with yard and three black balls.

Parker's Telegraph

John Rowe Parker, a Bostonian with musical interests, gave up his music shop in 1822 to devote himself to telegraphy. He substituted a semaphore for Topliff's signal post at *Long Island Head*, and later extended the line to *Boston Light*. He made his Boston headquarters in a building on *Central Wharf*, not far from Merchants Hall, to which Topliff's news room had moved.

Parker seems to have had much more success than Grout. The gross receipts from his subscribers for the first four years were $6000, or $2000 more than outgoings, not counting the 'arduous personal services of the conductor.' The conductor was presumably Joseph Pope, of Hull, Mass., who was appointed to take charge of the 'masts and semaphore' in 1825, when the outer station was moved to *Point Allerton* (Nantasket).

The *Long Island Head* station was operated by the light keeper, who stated in the July 1826 *Columbian Centinel:*

> there will not be any accommodations for Frolickers as the keeper does not wish to make a tipling-shop of any house belonging to the United States government. And that no liquors can be obtained, as his whole attention will be devoted to his duty to the government and to the Marine Telegraph.

The Long Island Head station was later moved to *Rainsford Island,* the port quarantine establishment, and another semaphore was erected on *George's Island.* According to Swan, the *Point Allerton* station was moved to *Hull Hill* as early as 1827, though the author of a lecture on the telegraph delivered before the Boston Marine Society on 5 February 1833 did not record the change. Pickering said the line was 12 miles in length, a signal passed from end to end in 2 min. and the operators were on duty from sunrise to sunset.

Parker must have kept in close touch with development in the Old World, for his apparatus was of a sophisticated type, with affinities to Pasley's final design. A single, lofty, tapering mast carried three arms, one above the other. That at the top of the mast was shorter than the other two. It was merely an indicator to denote which of three codes was being used. The arms could take one of six different settings. Several thousand combinations of four numbers each were possible.

By that time all American ships had numbers which were signified by flags when nearing a station that used a code devised by James Maud Elford, an English-born teacher of navigation in Charleston.

Parker, a strong advocate of coding rather than spelling, divided his 'numerical dictionary' into three parts, the first listing the numbers in the Elford code, the second an appendix of his own which included the numerals, phrases and so on of the Liverpool-Holyhead line, and the third particular to 1,200 vessels belonging to Boston and other American ports. The indicator showed which code was to be used. If it displayed '6' followed by '4', the Elford book was intended.

As from the first the Elford numbers were known at the Holyhead signal station, a close connection was established between Parker and his telegraph and the Liverpool shipping interests. It is pretty certain that the Liverpool Dock Company was well aware of the utility of Parker's system when in 1827 it commissioned Barnard L. Watson to survey a Liverpool-Holyhead semaphore line. (See Chapter V). Watson may therefore have been shrewd enough to adapt Parker's system for his own purposes, seeing that it was already well known to masters of American and British vessels on the Liverpool-Boston run.

Parker's telegraph was kept busy enough. Swan says that between October 1824 and October 1825 its first year of service, it relayed news of the arrival of 799 vessels. By 1833 the total had increased to 2104 in the year.

By 1841 Parker had a second town terminus, at the new *Exchange,* in addition to that at the Central Wharf. He wrote on telegraphic subjects, his work including *A Treatise upon Telegraphs Embracing Observations upon the Semaphoric System, with a Few Remarks upon the Utility of the Marine Telegraph Flags.* The first edition came out in 1838. Parker refers to the existence of harbour semaphores at *Baltimore, Charleston* and *Portland* (Maine) – the Charleston telegraph was the work of Elford.

It is possible that by then he had rationalised his service for he refers only to one outer station, at *Nantasket*. He describes the apparatus then as comprising a 60 ft. mast with three movable arms of board or copper, 6 — 10 ft. in length and 1 ft. wide, and worked by chains and wheels.

Parker claimed that no telegraph was simpler or faster than his. He advocated a nation-wide system which 'would be a sublime attempt at an approximation of time and space, and would be actually worthy of the high and enterprising character of our nation.'

Advancing years forced him to sell the undertaking as from 1 July 1843. It was bought by Joseph Pope and F. A. S. Brown, the operator on George's Island. They in turn sold it to John T. Smith and Robert E. Hudson. Smith became manager. Pope at least stayed on for when the magnetic telegraph was extended from Boston to Hull in 1853 he became a telegraphist there. The visual system seems to have been given up in or about 1856 when wire began to be strung up along Cape Cod to Highland Light.

Colles's Telegraph

New York City lagged behind Boston in applying visual telegraphy but the war of 1812 at length provided the spur.

A harbour telegraph was authorised on 14 March 1812. A mast and yard were erected in *Brooklyn Navy Yard* and another on Signal Hill, *Staten Island*. By 30 June preparations were reported to be 'making with all practicable expedition,' by Captain Chauncey and the officers of Government. A report of 14 May 1814, quoted by I. N. Phelps Stokes in his *The Iconography of Manhattan Island*, said that at noon Staten Island showed four black and two white balls, indicating the approach of four ships of the line and two frigates. (On 20 March 1813 New Yorkers had been alarmed by a signal from Staten Island that a British squadron was approaching, only to find that the news was false.)

Christopher Colles, a septuagenarian Irishman — described in the 1786 directory of New York as a fig-blue manufacturer — had by this time perfected his new 'Numerical Telegraph'. It is not certain whether he had already been associated with the official yard-and-ball communication.

In a pamphlet introducing his numerical telegraph Colles quaintly says:

.... eighty-four letters can be exhibited by the machine in five minutes, to the distance of one telegraphic station averaged at ten miles, and, by the same proportion a distance of twenty-six hundred miles in fifteen minutes, twenty-eight seconds.

On 3 July 1813 Colles announced that he had completed 'two of these important instruments', one at the top of the *Custom House*, the other on *Governor's Island*. He proposed on 9 July to 'exhibit a number of accurate and conclusive experiments'. He stressed that the machine was cheap,

perfectly accurate and capable of conveying any unexpected intelligence which can be written, with a celerity exceeding common belief, and also that it can convey registered pre-concerted sentences, orders or other intelligence with equal celerity, and at the same time, with perfect accuracy and privacy.

On 21 July he experimented with 164 letters from Custom House to Governor's Island, sending eight per minute. The results prompted him to petition the Common Council of the city for funds. In August he was granted a subsidy of $100. But support was discontinued on 13 December, probably because the Embargo Acts of the war had stopped shipping and so ended the need for a telegraph.

The Minutes of the Common Council record that the 'signal poles' were re-established on 20 March 1815, but the reference is probably to the yard-and-ball system which must have operated concurrently with Colles's venture.

On 17 January 1816 merchants and others met at the Tontine Coffee House, where a previously appointed committee reported Colles's telegraph to be superior to all others examined, including the English – presumably the shutter – and the French. This committee or another recommended it to the Common Council again, and merchants and others were urged to subscribe. The committee reported on 12 August 1816 that the Common Council had taken no definite action towards paying Colles $150 for his services.

This evidence suggests that Colles's venture never really got going or at least functioned only intermittently, and so far there is no indication how long either it or the yard-and-ball system lasted.

Thompson's Telegraph

We are on firmer ground by 1821, when on 11 June Jeremiah Thompson, representing an association of merchants and shipowners, was authorised to set up a telegraph of his own at the *Battery*. The system was first tried on 23 June. In his book *The Rise of New York Port* Robert Greenhalgh, who lists the Black Ball as one of the shipping firms, quotes the series of inaugural messages. First came a signal that no vessels were sighted, then 'light winds from the eastward' and 'foggy at sea.' By then a schooner came into view and the information 'Lady Tomkins is here' was relayed, followed by 'We have done.'

It is said that there was 'scarce a single misunderstanding' of the messages by the boy on duty at the Battery. The apparatus was described as having 'an upright and a centre, which a boy of 12 can manage.' It seems to have been a species of semaphore, though the description is confusing, with its reference to an alphabet in four parts with a distinct representation for each division, so that only six motions were needed to show the 24 characters used.

Directing the telegraph was Samuel C. Reid, who had been a noted privateer in the war of 1812 and who became harbour master and warden of the port in 1843. In Goodrich's guide *A Picture of New York* (1828) the telegraph is listed as one of the attractions of the Castle Gardens at the Battery.

Other American Systems

Like that of Colles, the Thompson telegraph was short lived. On 12 November 1827 the Superintendent of the Lighthouse Service authorised the President of the Merchants' Exchange Company of New York to erect a two-arm telegraph at *Fort Wadsworth* on Staten Island and on the *Sandy Hook Lighthouse* grounds and on the *Navesink Highlands* on the New Jersey coast. They were to communicate with a terminal station on the cupola of the company's building opened on 1 May 1827 in *Wall Street*. An extension from Sandy Hook to the *Navesink Highlands* on the New Jersey coast was sanctioned in 1829.

The system required a five-part dictionary. The first part covered all questions and answers relating to vessels, stations and compass points, and occupied the numbers 1 — 1000. Part 2 contained nautical phrases, Part 3 all essential words and Part 4 names of vessels. Part 5 contained the names of countries, ports, cities and harbours. Each arm could take up six positions.

After the building in Wall Street was burnt down on 16-17 December 1835 the company made *Holt's Hotel* the terminal and tried without success to transfer the enterprise to the government. In spite of such checks and the fact that by the mid 1840s the electric telegraph had reached Coney Island, whence quicker and more complete information could be relayed, the line seems to have continued for many years. An illustration in *Leslie's Weekly* for 20 September 1856 shows it as still in service.

Briggs's Philadelphia Telegraph

Alvin F. Harlow refers in his *Old Wires and New Waves* to a visual telegraph operated for a few years between New York and Philadelphia, mainly for the use of brokers. The proprietor was William C. Briggs, a Philadelphia broker, who established it about 1840 to convey stock prices and the drawn numbers in lotteries by means of code messages. Morse's telegraph put it out of business about the end of 1845.

The Philadelphia *Public Ledger* of 7 January 1846 announced the sale of the equipment. It reported that messages took about ten minutes to pass the length of the line. They were sent by 'elevating boards on a pole in a particular way...At night lamps of different colors are used. The whole concern has been sold for about $3000.' The lamps used at night were housed in a box with a door which was raised or lowered. Before the service ceased, it is said that daytime messages were being heliographed, though this seems unlikely if not impossible except in high summer. There is no indication of the route, save for Harlow's reference that it ran via Mount Holly, New Jersey.

Swan also refers briefly to the line but says it ran to Washington. He adds, inaccurately, that it proved

> so efficient that it was kept in operation for several years after a certain Bostonian with an iron horseshoe and a few miles of wire grounded every other visual telegraph tower and mast in the world.

Abortive Projects

In 1837 a memorial was presented to Congress praying for a telegraph to link New York and New Orleans. A Captain Hunter of Baltimore suggested that the Popham system would be the most suitable. It was said that a good telegraph could relay news between Washington and New Orleans in an hour. If such a line were extended to Boston, the 56 stations required between Washington and Boston would cost, say, $18,000 a year to operate.

As early as 1807 William Duane, editor of the *Philadelphia Aurora,* had tried in vain to interest President Jefferson in a telegraph between the seat of government and the main ports and in 1812 Colles had proposed a coast telegraph from Passamaquoddy to New Orleans.

Ennemond Gonon arrived in the United States in 1837 and gained the promise of Congress aid to erect visual telegraphs. His action seems to have spurred on Samuel Morse, who first demonstrated his electric telegraph successfully in January 1838. The Secretary of the Treasury was impressed by the possibilities of Morse's invention, whose triumph killed any further schemes for long-distance visual telegraphs in the United States.

The San Francisco Telegraph

In 1849, two years after its name had been changed from Yerba Buena, the growing town of San Francisco acquired a semaphore telegraph. *San Francisco: The Bay and its Cities* says:

> The visitor who boards a street car for North Beach will no more find an ocean beach at the end of the line than will the pedestrian who toils up Telegraph Hill find a telegraph station at the end of his climb...long since vanished is that telegraph station on the summit of the hill which was a city landmark for decades after it was connected by wire in 1853 with a look-out station at Point Lobos. The station replaced the still older semaphore of which Bret Harte wrote in *The Man at the Semaphore,* '...on the extremest part of the sandy peninsula, where the bay of San Francisco debouches into the Pacific, there stood a semaphore telegraph...it signified to another semaphore farther inland the "rigs" of incoming vessels, by certain uncouth signs, which were passed on to Telegraph Hill, San Francisco, where they reappeared on a third semaphore...and on certain days of the month every eye was turned to welcome those gaunt arms widely extended at right angles which meant "Side-Wheel Steamer" (the only steamer which carried the Mails) and "Letters from Home"' [Other combinations of the arms denoted screw steamer, ship, brig or schooner].

The service was begun by George F. Sweeney and Theodore F. Baugh, who ran a merchants' exchange in Sacramento Street which, like rival establishments, advised clients of the arrival of ships off the Golden Gate. The outermost station was on the south head of *Point Lobos,* directly communicating with another at the *Presidio House. Telegraph Hill,* the third and last station before the town, was an abrupt, rocky hill which was subsequently a favoured place of residence because of the fine views over the city which it commanded.

In *The Hills of San Francisco* Margaret Perkins Deering recounts an amusing incident to show the great impression which the semaphore made on the San Franciscans. At a theatre one evening a stupid actor rushed on to the stage with his arms stretched out awkwardly, asking 'What means this, my Lord?' The actor who was to respond hesitated, in ignorance of his part, but a newsboy in the third tier shouted: 'Side-wheel steamer.' The answer was so appropriate that there was long and loud applause.

The service was supported by public subscription and seems to have satisfied the community. But the occasional fogs of the Pacific Coast were a drawback which impelled the partners to replace it in 1853 by an electric telegraph over the same course, a considerable enterprise for the San Francisco of those days.

Bibliography

Admiralty Charts, especially Sheet II of the River Thames from Ramsgate to the Nore, 1844-56.
Admiralty Minutes (Public Record Office, London).
Albion, Robert Greenhalgh: *The Rise of New York Port, 1815-1860* (New York, 1939).
Allen, Thomas: *History of the Counties of Surrey and Sussex* (London, 1829).
Appleyard, Rollo: *Pioneers of Electrical Communication* (London, 1930).
Atlas des Lignes Télégraphiques Aériennes construites en France de 1793 à 1852 (produced under the direction of Jacquez–Kermabon) (Paris, 1892).

Bagshaw, Samuel: *History, Gazetteer and Directory of the County of Kent* (Sheffield, 1847).
Baines, Thomas: *History of the Commerce and Town of Liverpool* (London, 1852).
Baines, Thomas: *Liverpool in 1859* (London, 1859).
Balleine, G. R.: *A Biographical Dictionary of Jersey* (London, 1948).
Barham Papers, Vol III (Public Record Office, London).
Barratt, T. J: *The Annals of Hampstead* (London, 1912).
Bayne, A. D: *A Comprehensive History of Norwich* (Norwich, 1869).
Belloc, Alexis: *La Télégraphie Historique depuis les temps les plus reculés jusqu'à nos jours* (Paris, 1888).
Berry, William: *The History of the Island of Guernsey* (London, 1815).
Beutlich, F. C. G: *Norges Sjøvæbning 1750-1809* (Oslo, 1935).
Bowers, Robert W: *Sketches of Southwark Old and New* (London, 1905).
Brayley, E. W., with Britton, John and Brayley, E. W., junior: *A Topographical History of Surrey* (London, 1841-48).

Brett-James, Antony: *The Hundred Days: Napoleon's Last Campaign from Eye-Witness Accounts* (London, 1964).
Brink, ten, Dr. E.A.B.J. and Schell, C.W.L. *Geschiedenis van de Rijkstelegraaf, 1852-1952* (The Hague, 1954).
Bryant, Sir Arthur: *The Years of Endurance (The Napoleonic Wars, 3 vols.)* (London, 1942-50).

Cachemaille, J.L.V: *The Island of Sark* (London, 1928).
Carlyle, Thomas: *The French Revolution* (London, 1837).
Chappe, Ignace (Chappe l'Aîné): *Histoire de la Télégraphie* (Paris, 1824).
Clarke, Rev. C.C. (pseudonym of Sir Richard Phillips, q.v): *The Hundred Wonders of the Modern World* (London, 1837).
Clarke, Desmond J: *The Ingenious Mr. Edgeworth* (London, 1965).
Clarke, Marcus: *For the Term of His Natural Life* (London, 1885).
Clowes, Sir W. L: *The Royal Navy: A History* (London, 1897-1903).
Cobbett, William: *Rural Rides* (London, 1830).
Colles, Christopher: *The Numerical Telegraph* (New York, 1813).
Conolly, J: *Philanthropic Vocabulary and Code of Signals* (1821).
Cornish, J. S: *Stranger's Guide through Liverpool* (Liverpool, 1847).
D'Arblay, Madame (Fanny Burney): *The Diaries of Madame d'Arblay* (edited by her niece) (London, 1854).
Daru, Emile: *Le Télégraphe Aérien Chappe de Paris à Bayonne par Dax* (Auch, 1961).
Dean, C. G. T: *The Royal Hospital, Chelsea* (London, 1950).
De Camp, L. Sprague: *Lest Darkness Fall* (London, 1955).
Deeping, Warwick: *The Woman at the Door* (London, 1937).
Deering, Margaret P: *The Hills of San Francisco* (San Francisco, 1936).
Denham, H. M. (Commander, R.N.): *Mersey and Dee Navigation* (Liverpool, 1840).
Dictionary of American Biography
Dictionary of National Biography
Dumas, Alexandre (the elder): *The Count of Monte Cristo* (Paris, 1846).

East India Company: *Bengal and Bombay Despatches* (1816-29).
Edelcrantz, A. N: *Afhandling om Telegrapher och försök til en ny inrättning däraf* (Stockholm, 1796).

Edgeworth, Richard Lovell: *On the Telegraph* (1795). *A Letter to the Rt. Hon. the Earl of Charlemont on the Tellograph and on the Defence of Ireland* (1797). *On the Telegraphic Communication* (1801). *An Essay on the Art of Conveying Secret and Swift Intelligence* (1799). (and Edgeworth Maria): *Memoirs of Richard Lovell Edgeworth* (London 1820).
Encyclopaedia Britannica, 1797, 1824 (suppt. to 6th edition), 1929.
Engineer's and Mechanic's Encyclopaedia, 1836.
Evans, John: *An Excursion to Windsor in July 1810* (London, 1817).

Farington, Joseph: *Farington Diary [Diary of Joseph Farington]* (London, 1922-28).
Farwell, George: *The Outside Track* (Melbourne, 1951).
Feldhaus, F. M: *Die Technik der Vorzeit, der geschichtlichen Zeit und der Naturvölker* (Leipzig, 1914).
Feltham, J: *The Picture of London for 1808* (London, 1808).
Finch, W. C: *The Medway River and Valley* (London, 1929).
Fisher, T: *The Kentish Traveller's Companion* (Canterbury, 1799).
Freeling, A: *The South Western Railway Companion* (London, 1840).
Fürst, A: *Das Weltreich der Technik.* (Berlin, 1923).

Gachot, Prof. Henri: *Le Télégraphe Optique de Claude Chappe: Strasbourg-Metz-Paris et ses Embranchements* (Saverne, 1967).
Gamble, Rev. John: *Essay on the Different Modes of Communication by Signals* (1797).
Gates, W. G: *Illustrated History of Portsmouth* (Portsmouth, 1900). *History of Portsmouth: A Naval Chronology* (Portsmouth, 1931).
Gilder, Rodman: *The Battery* (Boston, 1936).
Goodsall, R. H: *Whitstable, Seasalter and Swalecliffe* (Canterbury, 1938).
Gosse, P: *St. Helena, 1502-1938* (London, 1938).
Gregory, G: *Dictionary of Arts and Sciences* (London, 1806).
Gunther, R. W. T: *The Life and Work of Robert Hooke* (Oxford, from 1920).

Handbook of Communication by Telegraph (London, 1842).
Hardy, Thomas: *The Dynasts* (London, 1904-8).
Hare, A. J. C: *The Life and Letters of Maria Edgeworth* (London, 1894).
Harlow, Alvin F: *Old Wires and New Waves* (New York, 1936).
Harper, Charles: *The Portsmouth Road* (London, 1895).
Harte, Francis Bret: *Complete Works (including 'The Man at the Semaphore')* (London, 1873).
Hayward, R: *Where the River Shannon Flows* (London, 1940).
Hennig, Dr. Richard: *Die Älteste Entwicklung der Telegraphie und Telephonie* (Leipzig, 1908).
Herdman, William: *Views in Modern Liverpool* (Liverpool, 1864).
Herdman, William G: *Pictorial Relics of Ancient Liverpool* (London, 1843).
Hicklin, John: *The Handbook to Llandudno and its Vicinity* (London, 1858).
Hodgkinson, H: *The Adriatic Sea* (London, 1955).
Hull Chamber of Commerce Reports 1838-39.
Hunt, Leigh: *The Town* (London, 1848).

Inglis-Jones, Elizabeth: *The Great Maria* (London, 1959).
Jackson, Thomas: *The Visitor's Handbook for Holyhead* (London, 1853).
Jacob, John: *Annals of Some of the British Norman Isles Constituting the Bailiwick of Guernsey* (Paris, 1830).
Joyce, Michael: *Ordeal at Lucknow* (London, 1938).
Karrass, T: *Geschichte der Telegraphie* (Brunswick, 1909).
Kavanagh, M. B: *Signal Stations in Jersey* (extract from The Bulletin of the Société Jersiaise) (St Helier, 1970).
Kaye, T.: *The Stranger in Liverpool* (Liverpool, 1831).
Knight, Charles: *London (Vol. V)* (London, 1843).
Lamb, H. A: *Theodora and the Emperor* (London, 1953).
Levy, M. C. I: *Governor George Arthur* (Melbourne, 1953).
Lloyd, C. C: *Mr. Barrow of the Admiralty* (London, 1970).
Longford, Elizabeth: *Wellington – The Years of the Sword* (London, 1969).
Lord, Lt. William: *A Telegraphic Vocabulary adapted for the Line of Semaphoric Telegraphs from Liverpool to Holyhead* (London, 1845).
Macdonald, Lt. Col. John: *A Treatise on Telegraphic Communication* (London, 1808). *A Treatise Explanatory of a new system of Naval, Military and Political Telegraphic Communication* (London, 1817).
Mackie, Charles: *Norfolk Annals* (Norwich, 1901).
Marchand, O: *Histoire de Montlhéry et de son Château* (Montlhéry, 1956).
Marryat, Frederick: *Peter Simple* (London, 1834).
Marshall-Cornwall, General Sir James: *Marshal Massena* (London, 1965).

Martineau, Gilbert: *Napoleon's St. Helena* (London, 1968).
Massie, J. F: *Le Télégraphe Aérien Système Chappe de Bordeaux à Bayonne et à Béhobie* (Aire-sur-l'Adour, 1968).
Maude, F. C: *Memoirs of the Mutiny* (London, 1894).
Maurice, J: *Montbazon et Veigné aux temps jadis* (Paris, 1970).
Mead, Com. Hilary P: *The Story of the Semaphore (collected articles from The Mariner's Mirror)* (London, 1939).
Mee, Arthur: *Surrey* (London, 1938).
Merryweather, F. S: *Half a Century of Kingston History, 1837-87* (Kingston, 1887).
Mollet, Ralph: *A Chronology of Jersey* (St. Helier, 1954).
Moore-Robinson, J: *Historical Brevities of Tasmania* (Hobart, 1937).
Moss, Dr. William: *A General and Descriptive History of the Town of Liverpool (published anonymously)* (Liverpool, 1795).
Murray's *Handbook to Egypt* (1847).
Murray's *New English Dictionary* (1914).
Nadaud, Gustave: *Chansons (includes Le Vieux Télégraphe)* (Paris, 1857).
Navy List
New Handbook to the Downs Neighbourhood (Deal, 1868).
New South Wales Post Office Directory 1832.
Nicolas, Sir Nicholas Harris: *The Despatches and Letters of Vice-Admiral Lord Viscount Nelson* (London, 1844).
Nisbet, John: *The Story of the One Tree Hill Agitation* (Nunhead, 1905).

O'Byrne, W.R: *Naval Biographical Dictionary* (London, 1849).
Ogilvy, James S: *A Pilgrimage in Surrey* (London, 1914).
Oman, Sir Charles: *A History of the Peninsular War* (Oxford, 1902-30).
Paço, Afonso do: *As Comunicações Militares de Relação em Portugal* (Lisbon, 1938).
Page, William: *St. Albans (The Story of the English Towns)* (London, 1920).
Palmer, C. J: *The Perlustration of Great Yarmouth* (Great Yarmouth, 1872-75).
Parker, Eric: *Highways and Byways in Surrey* (London, 1908).
Parker, John R: *A Treatise upon the Semaphoric System of Telegraphs* (Boston, 1842).
Pasley, Lt. Col. C. W: *Description of the Universal Telegraph for Day and Night Signals* (London, 1823).
Penny Cyclopaedia (London, 1833).
Philip, A. J: *History of Gravesend and its Surroundings* (London, 1914).

Phillips, Sir Richard: *A Morning's Walk from London to Kew* (London, 1817).
Pickering, John: *A Lecture on Telegraphic Language* (Boston, 1833).
Popham, Admiral Sir Home: *Telegraphic Signals* (London, 1800); *Telegraphic Signals or Marine Vocabulary* (London, 1812).
Potter, G.W: *Random Recollections of Hampstead* (London, 1907).
Pritchard, S: *The History of Deal* (Deal, 1864).

Raddall, Thomas H: *Halifax: Warden of the North* (London, 1950).
Rafto, Thorolf: *Telegrafverkets Historie* (Oslo, 1954).
Rees, Abraham: *The Cyclopaedia* (London, 1819).
Reid, James D: *The Telegraph in America* (New York, 1879).
Report of the Select Committee on Railway Communications (London, 1840).
Report of the Proceedings of a Court of Inquiry into the existing state of the Corporation of Liverpool (Liverpool, 1834).
Report of the Metropolitan Railway Commissioners (London, 1846).
Risberg, Einar: *Suomen Lennätinlaitoksen Historia* (Helsinki, 1958).
Risberg, N. J. A: *Svenska Telegrafverket Historisk Framställning (Vol 3): Den Optiska Telegrafens Historia i Sverige 1794-1881* (Gothenburg, 1938).
Royal Kalendar, The: (London, 1798).
Rudd, Lewis C: *The Duke of York's Royal Military School 1801-1934* (Dover, 1935).
Russell, Earl (John Francis Stanley): *My Life and Adventures* (London, 1923).

Saavedra, A. Suarez: *Historia Universal de la Telegrafía* (Madrid, 1880); *Tratado de Telegrafía* (Barcelona, 1882).
Sandes, Col. E. W. C: *The Military Engineer in India, Vol. 1* (London, 1933).
San Francisco: The Bay and its Cities (New York, 1940).
Schaffner, Taliaferro P: *The Telegraph Manual* (New York, 1859).
Scharf, J. T., and Westcott, T: *The History of Philadelphia* (Philadelphia, 1884).
Sidebottom, J. K: *The Overland Mail* (London, 1948).

Signals for the Better Communication between the Islands and the Squadron (St. Helier, 1806).
Signals to be used at the Signal Posts round the Island of Jersey (St. Helier, undated).
Signaux de la Ligne Sémaphorique de Bayonne à Dunkerque (Paris, 1807).
Smartt, Charles L: *The Hoogly River Code* (1833).
Smith, Albert E: *The Struggles and Adventures of Christopher Tadpole* (London, 1848).
Southampton Dock Company Minute Books, 1841.
Stead, John: *A Picture of Jersey, or Stranger's Companion through that Island* (St. Helier, 1809).
Stokes, I. N. Phelps: *The Iconography of Manhattan Island* (New York, 1926).
Stonehouse, J: *The Stranger in Liverpool* (Liverpool, 1849).
Surrey Archaeological Collections, Vols. 21-52.
Surrey County Council: Antiquities of Surrey, 5th ed.
Swanton, E. W., and Woods, P: *Bygone Haslemere* (London, 1914).
Swinnerton, J: *Guide to the Beauties of Anglesey and Caernarvonshire* (Macclesfield, undated).

Ternant, E: *Les Télégraphes* (Paris, 1884).
Through to 1970 *(Royal Signals Institution)* (London, 1970).
Tovey, Joan, and Maclean, Adeline: *Notes on the History of Worplesdon* (1951).
Turmine, H. T. A: *Rambles in the Isle of Sheppey* (London, 1843).

Vincent, W. T: *The Records of the Woolwich District* (Woolwich, 1888-90).
Walford, Edward: *Old and New London* (London, 1873-78).
Ward, S. G. P: *Wellington's Headquarters, 1809-14* (London, 1957).
Watson, Barnard L: *The Telegraphic Vocabulary* (Liverpool, 1827). *Telegraphic Vocabulary adapted for Semaphoric Telegraphs (4th ed.)* (London, 1840); *Code of Signals* (1827-42).
Webb, E. A., Miller, G. W., and Beckwith, J: *The History of Chislehurst* (London, 1899).
West, G. F. M. Cornwallis: *Life of Admiral Cornwallis* (London, 1927).
White, William: *History, Gazetteer and Directory of Norfolk and the City and County of Norwich* (Sheffield, 1845).
Wilkins, John: *Mercury, or the Swift and Secret Messenger* (London, 1645).
Worcester, Marquis of: *A Century of Inventions* (London, 1662).

List of Periodicals Consulted

Africana Notes and News, 1960
Annales des Arts, Les, 1801
Annual Register, 1816
Archaeologia Cantiana, 1932
Archiv für Post und Telegraphie, 1883, 1888, 1895, 1901, 1907, 1908, 1927, 1935

Berliner Zeitung, 1934
Besley's Devonshire Chronicle and Exeter News, 1845
Billinge's Liverpool Advertiser, 1827
Bristol Mirror, 1839
Builder, The, 1845
Bulletin de la Société de Borda, 1961, 1968

Cape Town Monthly Magazine, The, 1859
Chambers Journal, 1889
Columbian Centinel, The, 1825
Country Life, 1938

Daily Telegraph, The, 1963
Deutsche Postgeschichte, 1937
Devon and Cornwall Notes and Queries, 1920-21, 1936-37
Devonshire Chronicle and Exeter News, 1845
Diligence d'Alsace, No. 1, 1969

Esher News and Advertiser, 1961, 1963, 1964

European Magazine, The, 1794
Evening Standard, 1939

Faversham Institute Journal, The, 1900
Fighting Forces, The, 1924

Gentleman's Magazine, The, 1795, 1796, 1806
Graham's Town Journal, The, 1844
Guernsey Weekly Press, The, 1959
Guildford and District Outlook, 1926-27

Hampshire News, 1833
Hampshire Telegraph, 1913
Hants and Dorset Magazine, The, 1954
Historia, 1960
Hull Advertiser, The, 1839

Illustrated London News, The, 1842
Ingenieur, De, 1890-91

Kent County Journal, 1950
Kentish Gazette, 1841, 1842

Leatherhead Advertiser, 1963
Leslie's Illustrated Weekly, 1856
London Gazette, 1821

Mariner's Mirror, The, 1933, 1934, 1935, 1938, 1939

Mechanic's Magazine, The, 1825, 1827, 1834, 1836, 1838
Mercury, The (Hobart), 1940
Mersey, 1929
Moniteur, Le, 1822
Morgenpost, 1958

Nautical Magazine, The, 1840, 1842
Naval Chronicle, The, 1806
New England Palladium, The, 1801
Nicholson's Journal, 1798, 1799
Norfolk Archaeology, 1955-57
Norfolk Chronicle, The, 1803, 1805, 1807
Notes and Queries, 1909

Observer, The, 1803
Ons Leger, 1951

Papers and Procedings of the Royal Society of Tasmania, 1925, 1937
Papers and Proceedings of the Tasmanian Historical Research Association, Vol 2 No 4
Philadelphia Public Ledger, The, 1846
Philosophic Magazine, The, 1802
Philosophical Transactions of the Royal Society, 1684
Pictorial Times, The, 1843
Port of London, 1971
Proceedings of the Bostonian Society, 1933
Proceedings of the Dorset Natural History and Antiquarian Field Club, 1890

Railway Magazine, The, 1836, 1837, 1838, 1839
Railway Times, The, 1842, 1856
Revista de Telecomunicación, La, 1955
Revue des Transmissions, La, 1952

Royal United Service Institution Journal, The, 1892

Salem Gazette, 1802
Satirist, The, 1813
Scotsman, The, 1855
Shipping and Mercantile Gazette, The, 1840-42
Signalman, The, 1962
Soldier, The, 1956
Southwark and Bermondsey Annual, 1903
Star, The, 1796
Surrey Comet, The, 1961
Surrey Times and Weekly Press, 1947

Times, The, 1816, 1822, 1839, 1842, 1847, 1849, 1850
Transactions of the British Association for the Advancement of Science
Transactions of the Devon Association, 1908
Transactions of the Guernsey Society of Natural Science and Local Research, 1903
Transactions of the Royal Irish Academy, 1795, 1797
Transactions of the Royal Society of Arts
Transactions of the Society of Arts, 1816-17
Transactions of the Society for the Encouragement of Arts, Manufactures and Commerce, 1821
Trident, The, 1940

Vie du Rail, La
Vossische Zeitung, 1915

Zeitschrift für das Post- und Fernmeldewesen, 1958, 1965, 1966, 1971

Index

Aalsmeer, 155
Aarhus, 174
Abensberg, Battle of, 134
Åbo (Turku), 166, 167
A Branca, 185
Abro, 188
Abu-el-Gheyt, 188
Admiralty (Building), Whitehall, 2, 15, 17, 26, 27, 34-36, 38, 40, 45, 47, 49
   shutter telegraph at, 15, 27
   semaphore telegraph at, 34-36, 40, 45, 47, 49
   superintendent of telegraphs at, 17, 34
   first and second Secretaries at, 26
Aegeus, 4
Aeschylus, 2
Agamemnon, 2
Akra, 199
Akrell, Carl Fredrik, 170, 171
Åland Islands, 167, 170, 171
Alderney — see Channel Islands
Alexandria, 188, 189
   Cairo line, 188, 189
Alexandre, 147
Algeria, 139, 140, 146, 147
   Kabylie expedition, 147
   mobile telegraph used in, 147
   modified Chappe apparatus used in, 147
   telegraph adopted in, 146
   telegraph extended in, 147
Algiers, 139, 146
Algiers, Dey of, 139
Altona — see also Hamburg, 159, 160
Alto de Bandeira, 185
American Indians
   smoke signalling by, 3
Amiens, Peace of, 26, 94, 108, 130, 208
Amontons, Guillaume, 5, 120
   experiments of, 5, 120
Amis de l'Histoire des PTT d'Alsace, Les, 150

Amsterdam, 130, 155
   Weesperpoort, 155
   Westerkerk, 130, 155
      Ossendrecht line, 155
      Paris — Antwerp line extended to, 130
Ancona, 152, 153
Angoulême, 137
Angoulême, Louis-Antoine de Bourbon, Duke of, 136
Anglesey — see Holyhead-Liverpool line
Annapolis Royal (Nova Scotia), 207, 208
Antwerp, 130, 139, 155-7
   Citadel, 157
      Amsterdam line, 155, 156
      Brussels line, of Ferrier, 139
      Paris line, 130, 155, 156
Apelação, 185
Aranjuez, 183
Arbogast, Louis-François-Antoine, 122
Arc, Valley of the, 134
Arcangues, 187
Arendal, 175
Arethusa, HMS, 207
Arholma Beacon, 170
Aristotle, 3
Army, British, use of visual telegraphy by
   in Lines of Torres Vedras, 186
   in SW France, 187
   in South Africa, 190-3
   during Siege of Lucknow, 199-201
Arnemuiden, 157
Arras, 123
Arthur, Lt. Governor Colonel George, 203
Ashey Down, 92
Askersund, 170
Atalaia, 185
Athlone, 109-11
Atkins, Lt. 36
Audierne, 131

Aumale (Algeria), 146
Aumale, Henri-Eugène-Philippe d'Orléans,
    Duke of, 183
Auxerre, 129
Avignon, 137, 138, 140
    Bordeaux line,
        proposed, 140
        progress of, 140
        completion of, 140
    Perpignan line
        proposed, 138
Avranches, 128, 140, 146
    Paris/Brest line, 128
    Cherbourg line, 140
    Nantes line, 140
    Eu-Boulogne-Lille-Metz line proposed, 140, 146
Azores, 185

Baaring, 174
Babraham, 29
Badajoz, 184, 185, 187
Badbury Rings, 44
Baegdrup, 174
Bagwell, Lt. Colonel, 110
Balchik, 147
Bain de Bretagne, 150
Baines, Thomas, 191, 192
Baldock, 29
Bannicle (Banacle) Hill, 39, 40, 53, 62
Baltimore, 212, 216
Bangalore, 195
Barcelona, 184
Bard, 142
Barère, de Vieuzac, Bertrand, 123
Barfleur, HMS, 186
Barham, Admiral Lord (Charles
    Middleton, 1st Baron Barham), 30
Barham Downs, 20, 31, 36, 37
Barnham, 29
Barrackpore, 195
Barrow Hill, 19, 36, 37, 44, 87, 89
Barrow, John (later Sir John), 8, 26, 34, 42
Basle, 129
Basses-Pyrénées, 138
Bass Strait, 206
Bathurst (Cape Province), 190
Bath (Netherlands), 157
Batna, 146
Baugh, Theodore F, 216
Bavel, 155
Bayer, General, 136

Bayonne, 131, 137, 146, 148, 187
    St Etienne, 138
        Béhobie line, 138, 146
        Paris line, 137
        Perpignan line proposed, 138
Beachy Head, 36, 37
Beacon Hill (Faversham), 19, 20, 36, 37
Beacon, The (Beacon Hill, South Downs),
    15, 19, 22, 24, 39, 40, 54-6, 62
Beacon Hill (Devon), 44
Beatson, Lt. 198
Beauclerk, Rear-Admiral Lord Amelius, 66
Beaufort, Captain Francis (later
    Rear-Admiral Sir Francis), 109, 110
Beauharnais, Eugène de (Duke of
    Leuchtenberg, Prince of Eichstedt), 133
Bechtamy, 188
Bederkesa, 160
Bedhampton — see Camp Down
Bedlam (NSW), 202
Beer Head, 44
Begent, W. A., 206
Béhobie, 138, 146, 183
Belchalwell, 25
Belcherstown, 210
Bell-Lloch, 184
Belloc, Alexis, 126, 135
Belloy, 123
Belsey, Lt. Henry, 31, 59
Belt, Great, 173, 174
Belt, Little, 174
Benares, 196, 197
Bengal, 193, 195, 196, 199
    Presidency, 195, 199
Bere Regis, 44
Berg Heber, 164
Bergen, 175
    Bergenhus, 175
Bergen-op-Zoom, 155, 157
Bergsträsser, Johann Andreas Benignus, 5
Berkeley, Admiral George Cranfield, 186
Berlenga, 185
Berlin, 161, 162
    Alte Sternwarte, 162
        Koblenz line, 161-5
            technical features of, 161, 162
            operation of, 162
            personnel of, 161
            equipment of, 161, 162
            preserved stations of, 165
Berry, Caroline Ferdinande Louise de
    Bourbon, Duchess of, 142

Berthier, Alexandre (Prince of
    Neuchâtel), 134, 135
Bétancourt — see Béthencourt y Molina
Besançon, 140, 141
    stations at, 141
    Dijon line, 140
Béthencourt y Molina, Agustín de, 149
Betsham, 35
Betteshanger, 20, 37
Beydah, 188
Biarritz, 138
Bidston Hill, 68, 70, 71
Bilbao, 184
Binstead — see River Hill
Bishopstone — see Herne Bay
Biskra, 146
Black Ball Line, 214
Blackdown, 22
Black Sea, 147, 182
Black, Wm. F, 86
Blandford, 25
Blankenese, 159
Blaye, 137, 142
Blidah, 146
Blockülb, 135
Blücher, Field Marshal Gebhard
    Leberecht von (Prince of Wahlstadt),
    135, 136
Bluebell Hill, 87, 88
Blurton, Lt. George, 61
Bôa Vista, 185
Boaz, James, 118
Böckmann, Professor, 159
Bodebakken, 173
Boialvo, 185
Bøjden, 174
Bonaparte, François Charles Joseph (King
    of Rome), 135
Bonaparte, Napoléon — see Napoleon
Bonaparte, Louis, King of Holland, 155
Bône (now Annaba), 146
Bommel, 155
Bombay, 195-7
    proposal for lines to Madras and
        Calcutta, 195
    consideration of Calcutta line
        suspended, 196
Bombay Presidency, 195
Booth, Captain Charles O'Hara, 203-5
    telegraph of, 203-5
Borda, Société de, 137
Bordeaux, 137, 138, 141, 142
    St. Michel, 137, 138, 141
        Paris line, 137
        Bayonne/Béhobie line, 137
        Toulouse/Narbonne line, 137
        Blaye line, 137, 142
Borodkin, Mihail, 178
Boston, Mass, 73, 210-3, 216
    Boston Light, 211
    Central Wharf, 211
    Dorchester (Mass.), 210
    Exchange, 212
    Exchange Coffee House, 211
    Fort Independence, 211
    George's Island, 212
    Harbour, 211
    Hull (Mass.), 211-3
    Long Island Head, 211
    Merchants' Hall, 211
    Orange Street (No. 112), 210
    Point Allerton (Nantasket), 211
    Rainsford Island, 212
    State Street (No. 75), 210
        Grout's line to Martha's Vine-
            yard, 210, 211
        harbour lines of Topliff and
            Parker, 73, 211-3
Botha's Post Tower, 191
Bougie (now Bejaia), 146
Boulogne-la-Grasse, 123
Boulogne-sur-Mer, 131, 132, 136, 137,
    140
    Lille line, 132, 137
    on Calais-Eu extension, 140
    Lille-Boulogne-Eu-Avranches line
        proposed, 140, 146
Bramshaw, 24, 31
Brandenburg, 163
Brake, 160
Bramont, 134
Brandy Pots, 208
Breda, 155-7
    line to The Hague and 's
        Hertogenbosch, 156
    Flushing line, 157
Breguet, Abraham Louis, 121, 149
Breguet, Louis, 147
Bremen, 160
    Bremerhaven line, 160
Bremerhaven, 160
    Bremen line, 160
Breskens, 130
Brest, 128, 130, 131, 141
Briggs, William C, 215
    telegraph of, 215
Bristol 85
    Durdham Downs, 85

Bristol Channel, 85
British Association for the Advancement of Science, 81
Brown, F. A. S., 213
Brownrigg, Colonel, 107
Brûlon, 121
Brumaire, Coup d'État of, 130
Brunswick, Grand Duchy of, 164, 165
Brussels, 128, 130, 139, 155, 156
    Antwerp/Amsterdam line, 155, 156
    Antwerp line, of Ferrier, 139
    Paris line, 128, 130, 155, 156
Buçaco, 185
Bugeaud de la Piconnerie, General Thomas Robert (later Marshal and Duke of Isly), 142
Bugio, 185
Bundlecund, 196
Burgberg, 164, 165
Burket-Gheytas, 188
Burney, Fanny (Madame d'Arblay), 21
Bussy-la-Pesle, 142
Büttler, 174
Byzantine Empire, heliography in, 3
Byban, 188

Cabbage Hill, 21
Cadiz, 184
Cairo, 188, 189
    Boulak Fort, 188
    Citadel, 188
        Alexandria line, 188, 189
        Suez line, 188, 189
Calabria, 153
Calais, 137, 141
    Boulogne — Eu line, 140
Calcutta, 195-9
    Barrackpore, 195
    Bellary, 199
    Budge Budge, 199
    Exchange, 199
    Fort William, 195, 196, 198, 199
    Fulta, 199
    Gungra, 199
    Howrah, 198
    Jiggercolly, 199
    Kedgeree, 199
    Mahiari, 198
    Moyapoor, 199
    Nibria, 198
    Roypoor, 199
    Sundea, 199
        experimental line to Nagpur proposed, 196
    line to Bombay proposed, 195
    consideration of Bombay line suspended, 196
    line to Chunar proposed, 195
    line to Chunar surveyed, 196, 197
    Chunar line described, 198
    construction of Chunar line, 197, 198
    Saugor Island line, 199
Callum (Calham) Hill, 18, 19, 36, 37
Calshot Castle, 92
Cambon, Pierre Joseph, 122
Cambridge, Adolphus Frederick, Duke of, 63
Camden, Lord (John Jeffreys Pratt, 2nd Earl and 1st Marquis of Camden), 105-7
Camp Down, 39, 57, 58
Campbell, General Sir Archibald, 195
Campbell, Sir Colin (Lord Clyde), 200, 201
Cancale, 95
Canelas, 185
Canning, George, 8, 21
Cap Fréhel, 131
Cap St. Ignace, 208
Cape Carvoeiro, 185
Cape Clear, 71
Cape Colony, 190
Carhampton, Lord (Henry Lawes Luttrell, 2nd Earl Carhampton), 105
Carleton Rode, 29
Carlyle, Thomas, 124
Carnot, Lazare, 123, 124
Carpenter, Lt. 42
Carrington, John, 29
Castlereagh, Lord (Robert Stewart, 2nd Marquis of Londonderry), 21
Catalonia, 184
Cedarville, 210
Cefndu, 75
Ceiras, 185
Chalbury, 25, 31, 32
Chaleur, La, 142
Châlon-sur-Saône, 146
Chalton, 24
Chambéry, 133
Channel Islands
    Jersey,
        first signal stations in, 94
        D'Auvergne, Rear-Admiral Philip, Duke of Bouillon, 94-6, 99
        Mulgrave, Peter Arthur (see also Tasmania), 95, 96, 99, 100

Mulgrave devises telegraph to
 supersede signal system in, 96
successful inaguration of
 telegraph in, 96
Don, Lt. Gen. Sir George, 94-6,
 99, 100
signal system reorganised in, 94
 St. Helier, 94, 95, 97, 100
 Mont Orgueil, 94, 97
 Grosnez Point, 94, 95, 98
 other stations in, 95, 97, 100
Guernsey
 Saumarez, Sir James, 95
 first signal stations in, 95
 St. Peter Port, 94, 98
 Castle Cornet, 95
 Jerbourg Point, 98, 100
 Doyle, Lt. Gen. Sir John, 95, 96
 Fort George, 95, 98, 99
 other stations in, 95, 98-100
Sark, 94, 95
 station in, 96, 100, 101
Alderney, 94
 station in, 99, 100
telegraph service suspended in, 99
partial readoption of telegraph in, 100
final abandonment of telegraph in,
 100
Chappe, Claude, 2, 5, 120-8, 132, 133,
 150
birth of, 121
religious training of, 121
first interest in telegraphy, 121
experiments with opto-acoustic
 system, 121
experiments at Etoile, 121
experiments at Ménilmontant, 121,
 122
perfects 'T' type telegraph, 11, 121
Convention refers his telegraph to
 committee, 122
Romme reports favourably on his
 telegraph, 122
Lakanal supports, 122
completes first line near Paris, 122
trial of first line of, 122
Lakanal eulogises, 122
styled *ingénieur-télégraphe*, 122
Committee of Public Safety gives
 powers to, 123
his telegraph described, 8, 124-7
experiences difficulties with Lille
 line, 123, 124
Paris addresses of, 123, 128

triumph of Lille line of, 123, 124
instructions of as to Landau line,
 127
named *ingénieur-en-chef*, 128
proposes pan-Europe commercial
 telegraph, 132
commits suicide, 133
tomb of, 133
memorials to, 133, 150
Chappe, René (Chappe-Chaumont), 121,
 128, 139, 140, 146
appointed director of Brussels, 128
named Second Administrator, 139
retires, 140
retains interest in telegraph, 146
Chappe, Abraham (Chappe des Arcis),
 121, 123, 132, 133, 135, 137, 139,
 150
ordered to devise night telegraph
 across Channel, 132
devises mobile telegraph for Army,
 132
attached to Army Staff, 133
rank and functions of, 133
and Mainz line, 135
as Inspector General of telegraphs,
 133, 137
receives Legion of Honour, 137
named Second Administrator, 139
retires, 140
Chappe, brothers
special position of, 136
difficulties of, 140
 Louis-Philippe and, 140
Chappe d'Auteroche, Jean, 121
Chappe, Ignace Urbain Jean, 37, 121,
 128, 133, 137
named technical assistant, 128
named joint Administrator, 133
receives Legion of Honour, 137
retires, 139
Histoire de la Télégraphie of, 37,
 139, 140
Chappe, Pierre-François, 121, 128, 133,
 137, 139
named technical assistant, 128
named joint Administrator, 133
receives Legion of Honour, 137
retires, 139
Charlemont, James Caulfeild, 1st Earl of,
 108
Charles II, King, 29
Charles X, King, of France, 139
Charleston (South Carolina), 212

Chatau, Pierre Jacques, 181
   system of, in Russia, 181-2
Château de Larralde, 187
Château d'Urdains, 187
Chatham, 18, 19, 36, 87, 89
   Dockyard, 18, 19, 36
   Lines, 36
      possible station on Watson's line, 87, 89
Chatley Heath, 37, 39, 42, 43, 45, 50-3, 63
Chauncey Capt, 213
Chestford (Cheesefoot Head), 43, 45
Cherbourg, 95, 136, 140
   Avranches line, 140
Chessington, 21
Chichester, 57
Chichester, Lord — see Pelham, Thomas, 2nd Earl of Chichester
Chilterns, The, 27, 28
Chiltonville, 210
Christian August, Prince of Slesvig-Holsten-Sønderborg-Augustenborg, 175
Chunar, 196-8
   Calcutta line, 196-8
Church Bay, 70, 75
Clarence, Prince William Henry, Duke of (later King William IV), 8, 64
Clarke, General Henri Jacques Guillaume (later Count of Hunebourg, Duke of Feltre, and Marshal), 129, 135
Clarke, Marcus, 204
Claygate Hill, see Coopers Hill
Claremont, 43
Cleoxenus, 3
Clermont, 123
Clyde, River, 85
Clytemnaestra, 2
Coal Mines (Tasmania), 204, 205
Coast Signals, British
   recommendation of board as to, 64
   first posts established, 64
   original equipment, 64
   total of stations, 1803/4, 64
   extension of system, 1804, 64
   Irish lines, 65
   Shoebury — Tower of London line proposed, 65
   and link with London-Deal semaphore, 36
   re-equipment of, 65
   Deal-Yarmouth line, 65, 66
   stations in 1814, 66
   Scilly Isles station, 66
   stations given up in 1814, 66
   North Foreland — Lands End posts reinstated, 67
   converted to French-type semaphore, 65
   converted to Popham semaphore, 67
   East Coast stations, 65, 66
   Scottish stations, 66
Coast Signals, French, 130-2
Cobbett, William, 40
Cobh (Cove), 110
Cobham — see Chatley Heath and Queen Anne's Hills
Cochrane, Admiral Thomas (10th Earl of Dundonald), 131
Cocks, Colonel James, 194
Cod, Cape, 213
Coimbra, 185
Cohasset, 210
Coldred, 36, 37
Coleman, W. B., 72
Colles, Christopher, 213-6
   Numerical Telegraph of, 213-5
   experiments of, 213, 214
Collon, 104
Cologne, 161-5
   restored station at Flittard, 165
      Berlin line, 161-5
      Koblenz line, 161-5
Commercial Telegraph Association — see Watson's General Telegraph Association
Compton, 39, 56, 57
Concoeur, 137, 150
Condé-sur-l'Escaut, 124
Condorcet, Marie-Jean-Antoine-Nicolas de Caritat, Marquis of, 120
Congress of the United States, 216
Conolly, Joseph, 117, 118
   telegraph of, adopted in India, 118, 199
Constantine (Algeria), 146
Consulate, The (French), 130
Convention, The (French), 122, 128, 159
   adopts telegraph as national utility, 122
Conyngham, Elizabeth, Dowager Marchioness of, 89
Cooke, Sir William Fothergill, 60, 147
Coombe (Dorset), 44, 45
Coombe (Surrey) — see Kingston Hill
Cooper's Hill, 39, 42, 50, 52
Copenhagen, 173, 174

Battle of, 173
Round Tower, 173
Vesteport, 173
Quintus Fortress, 173
lines from, 173, 174
Corfe, Thomas, 39, 40
Cork, 105, 109-11
Cornwallis, Admiral Sir William, 130
Cornwallis, Charles, 1st Marquess, 115
Cossack, HMS, 177
Coste, 188
Côte d'Or, 142
Cove — see Cobh
Coventry, M., 86
Coyton, 44
Crawford, Colonel C., 195
Cree, Lt. David, 119
Cretet, 133
Crevecoeur, 155
Crimean War, 147, 148, 177, 182
French telegraph in, 147, 182
Varna — Balchik line, 148
Croix des Bousquets, La (see Béhobie)
Croker, John Wilson, 26, 36, 42-4, 56
Cuenca, 184
Cuttack, 195
Cuxhaven, 159, 160
Hamburg line, 159, 160

Dahlem, 163
Dalarö, 170-2
Dalhousie, George Ramsay, 9th Earl of, 209
Dalwood Common, 25
Damanhur, 188
Dans Hoogte, 191
Danton, Georges Jacques, 122
Danzig, 134
D'Arblay, Frances — see Burney, Fanny
Darley, George, 59
Dartmouth, 85, 92, 93
Start Point line, 85, 92, 93
Darwin, Dr. Erasmus, 105
Daugavpils (Dvinsk), 182
Daunou, Pierre Claude François, 122
D'Auvergne, Philip, Rear-Admiral, Duke of Bouillon — see Channel Islands
Deal, 20, 36, 37, 67, 90
Admiralty shutter line, 13-17
Admiralty semaphore line proposed, 36
coast signal line, 36, 65-7
Debstedt, 160
Deccan, 197

Decrès, Vice-Admiral Denis, 131
Dedersdorf, 160
Deeping, Warwick, 52
Delaunay, Léon, 126
draws up first French telegraphic vocabulary, 126
Delaval, Sir Francis, 102
Delft, 156
Dellis, 146
Democlitus, 3
Denham, Lt. Harry Mangles, RN, 73
Den Helder, 156
Dennys, Lt. Lardner, 61
Depillon, 5, 130-2, 149
invents *sémaphore*, 5, 130
adaptation of semaphore of, 130-2
Derwent River (Tasmania), 203
Dibdin, Charles, 11
Dickens, Charles, 18
Dickens, John, 56
Dicksen, 173
Dieppe, 131
Dieren, 155
Dijon, 129, 133, 140-42
stations at, 133, 141
proposed line to Strasbourg, 140
Paris-Lyons line deviated to serve, 142
Besançon line, 140
Diptford South, 44
Directory, The (French), 128, 149
and Paris-Strasbourg line, 128
orders Paris-Brest line, 128
Dixon, Lt. W. H., 61
Dnieper, River, 182
Dolland, 34
Domburg, 157
Don, Lt. Gen. Sir George — see Channel Islands
Dongen, 155, 156
Dorchester (Mass.) — see Boston (Mass.)
Dover
Admiralty semaphore line proposed, 36
Downs, The, 86-91
London line, 86-91
Downton, 43
Doyle, Lt. Gen. Sir John — see Channel Islands
Draguignan, 139
Drakenorth, 44
Drammen, 176
Dresden, Battle of, 135
Dreux, 128
Drogge, Horst, 165

Drottningholm, 169
Drottningskär, 170
Duane, William, 216
Dublin, 65, 104, 105, 109-11
   Pigeon Loft, 65
   Hospital, 109-10
   Phoenix Park, 106
      Edgeworth's proposed lines from, 105, 109
      Galway line, 110, 111
Ducos, Pierre-Roger, 130
Duckham, Alfred B., 93
Duckworth, Sir John, 66
Dumas, Alexandre, 143, 148, 149
Dumonceau, Marshal Jean-Baptiste, 155
Dundas, Rear-Admiral Sir Richard Saunders, 177, 178
Dunkirk (Dunkerque), 128, 131
Dunstable Downs, 14, 27, 28
Dunstan's Hill, 44
Duppenweiler, 135
Dupuis, 120
D'Urban, Colonel Benjamin (later Lt. Gen. Sir Benjamin), 110
Duxbury, 210

Eaglehawk Neck, 204, 205
East Bay Neck, 206
East Harling, 29
East India Company, 90, 188, 193, 196
Ebberup, 174
Eckmühl, Battle of, 134
Edelcrantz, Abraham Niclas Clewberg, 108, 167, 169, 173-5, 177
   first telegraph of, 167
   second telegraph of, 169
   design of stations of, 169
   meets Edgeworth, 108
Edgeworth, Richard Lovell
   versatility of, 102
   first experiments with rapid communication, 102, 103
   management of Irish estates, 103
   resumes interest in rapid communication, 103
   perfects *tellograph*, 103, 104
   tries tellograph in Ireland and between Ireland and Scotland, 104, 105
   corresponds with Dr. Erasmus Darwin, 105
      memorialises Lords Fitzwilliam and Camden, 105
      corresponds with Lord Carhampton, 105
   directed by Lord Camden to prepare experiment, 105
   builds improved apparatus, 106, 108
   estimates cost of Irish coast tellograph, 106
   experiments in Phoenix Park, 106
   experiments at Collon, 106
   devises portable tellograph, 106, 107
   government in Dublin decides not to adopt tellograph of, 107
   Letter of, to Earl of Charlemont, 108
   Essay on the Art of Conveying Swift and Secret Intelligence, 108
   visits Paris with Maria, 108
   meets Edelcrantz, 108
   renews offer of tellograph, 109
   submits plan for all-Irish system, 109
   builds Dublin—Galway line, 109, 110
   proposal to use variant of his system in India, 195
Edgeworth, Lovell, 107
Edgeworth, Maria, 102, 108, 109
Edgeworthstown (Meathas Truim), 103, 104
Edward Augustus, Prince (Duke of Kent and Strathearn), 207, 208
   introduces telegraph in Canada, 207, 208
   flogs soldiers 'by telegraph', 208
Eemnes-Buiten, 155
Egegaarden, 174
Ehrenbreitstein, 164
Ekerö, 167, 169
Elbe, River, 159, 160
Electric Telegraph Company, 189
Elford, James Maud, 212
   Code of, 212
Ellore, 195
Elmlohe, 160
Elspeet, 155
Elsfleth, 161
Elsinore — see Helsingør
Ems, River, 160
Encyclopaedia Britannica, 6, 32, 34
Epsom Downs, 38, 39
Eschkopf, 135
Eu — see also Avranches, 140
   extension to, from Calais and Boulogne, 140

Everest, Captain George (later Lt. Col. Sir George), 196, 197

Fahlenberg, 160
Faithful, Major H., 196
Falmouth, 30, 32, 91
    proposed shutter telegraph extension from Plymouth to, 30
    proposed line to The Lizard, 93
Fano, 152
Faraounyeh, 188
Farley Chamberlayne, 43, 45
Farrington Common, 43, 45
Fawcett, General Sir William, 107
Fedje, 176
Ferdinand IV, King of Naples, 153
Ferguson, Lt. R. B., 196, 197
Ferrier (de Tourettes), Alexandre
    experiments with night telegraph, 139
    proposes commercial telegraph network, 139
    system of, 139
    tries telegraph between Paris and Le Havre, 139
    removes to Belgium, 139
Fijnaart, 157
Filby, 30
Finland, Gulf of, 177, 181, 182
Fisker, Lorens Henrich, 173, 175
    telegraph of, 173
Fitzwilliam, Lord (William Wentworth, 2nd Earl Fitzwilliam), 105
Fleury-sur-Ouche, 142
Flamborough Head, 85
Flocon, Ferdinand, 141, 146
Flushing, 131, 155
    Paris — Brussels line extended to, 130, 155
Folda, 176
Foley, Vice-Admiral Thomas, 119
Fontenelle, Bertrand le Bovier de, 120
Fortunen, 173
Fort Beaufort, 190
Fort Loevestein, 155
Fort Peddie, 190, 191
Fort Selwyn — see Grahamstown
Fort William — see Calcutta
Forestier's Peninsula, 205
Foryd, 70, 71
Foster, E. R., 66
Foster, John, 104, 105
Four Marks — see Farrington Common
Foy, Alphonse, 140, 146
    -Breguet system, 147

Frankfurt-am-Main, 11, 159
Franklin, Benjamin, 120
Fraser, Lt. George, 61
Fraser's Camp, 191
Frauenhöfer, 161
Fredericia, 173
Fredericton, 207, 208
Frederikshald (Halden), 175
Frederiksvern (Stavern), 175
Freund, 161
Friesland, 160
Frondenberg, 165
Fundy, Bay of, 207
Funen, 173, 174
Furze Hill, 19

Gads Hill, 18, 35, 36, 44
Galgebakken, 174
Galleni, Adjutant-General, 153
Galsworthy, John, 50
Galway, 109-11
    Dublin line, 110, 111
Gamble, Reverend John, 11, 13, 39, 107, 114, 115, 187
    reports on telegraphy, 11
    tries out shutter telegraph at Portsmouth, 13
    chagrin of, 13
    Radiated Telegraph of, 39, 115
    essay by, in 1797, 114
    experiments of, in Woolwich, 115
    proposes London — East Coast telegraph, 155
    temporary telegraph of, between London and Army camp near Windsor, 115
Ganges, River, 196
Gareglwyd — see Church Bay, 70
Garderen, 155
Garonne, River, 138
Garrett, Lt. Henry, 61
Gatchina, 181
Gävle, 167, 170
Geisslingen, 134
Gelder, 155
Genoa, 152
George III, King, 26, 207
George Town (Tasmania), 206
George's Island — see Boston (Mass.)
Gibbon, Lieut., 31
Gilbert, Samuel, 211
Gilbert, Sir William Schwenk, 7
Gilbert-Cooper, Captain H. E., 197
Gizayr-Yssa, 188

INDEX 237

Gladman, Samuel, 36
Goddard, Thomas, 34, 37-9, 52
   surveys London – Portsmouth semaphore line, 38, 39
   surveys semaphore line to Plymouth, 42-45
   surveys London –Dover/Deal semaphore line, 36, 37
   and conversion of coast signals to semaphore, 67
Gogmagog Hills, 27, 29
Golden Grove – see Voel Nant
Golegã, 185
Gomshall, 22
Gonon, Ennemond, 150, 216
Goslar, 163
Gosport, 60, 61
   Clarence Victualling Yard, 60, 61
   electric telegraph to London, 60, 61
Gothenburg (Göteborg), 166, 169, 170, 172
Göttingen, 165
Governor's Kop, 190-2
Grand Junction Railway
   proposed telegraph along, 81
Granville, 95, 131
Gradignan, 129, 150
Grahamstown, 190-92
Graskop, 191
Great Belt – see Belt, Great
Great Fish River, 190
Great Haldon, 25, 26
Great Orme's Head, 69, 70, 78, 80
Great Western Railway, 147
Greenaway, Thomas, 52, 53
Greenhalgh, Robert, 214
Green Hill – see Sutton
Gregory, Charles Hutton, 7
Grenville, Lord (William Wyndham, Baron Grenville), 107
Griffith, Morris, 78
Grimsby, 83, 84
   Dock Company, 84
Grisslehamn, 167, 169, 171
Grosnez Point – see Channel Islands
Grodno, 182
Grout, Jonathan, Jnr, 210, 211
   Martha's Vineyard – Boston line of, 210, 211
Gubbins, Martin, 199-201
Guelma, 146
Guernsey – see Channel Islands
Guéthary, 187
Guildford – see also Pewley Hill, 38, 53

Guilleminot, General Armand Charles, Count, 134
Guillotin, Dr. Joseph-Ignace, 120
Guillotine, 120
Gungra, 199
Günther, Senator, 159
Gutzkow, Carl Ferdinand, 162
Guyot, Dr. Jules, 141
Gustav IV, King of Sweden, 169

Hagestein, 155
Hague, The, 155-57
   Residency, 156, 157
     proposed Brussels line, 155
     proposed Tournai line, 156
     line to Breda and 's Hertogenbosch, 156, 157
Halden – see Frederikshald
Halifax (Nova Scotia), 207, 208
   Bedford Basin, 207
   Chebucto Head, 207
   Citadel Hill, 207
   Dockyard, 207
   Harbour, 208
   York Redoubt, 207
     Prince Edward's telegraph at, 207, 208
Hall, Henry, 190, 191
Hälsingborg, 166, 170
Halstow, Lower, 18
Hambledon (Surrey), 140
Hamburg, 159, 160, 174
   Cuxhaven line, 159, 160
'Hampelmänner', 162
Hangö (Hanko), 177
Hanover, Kingdom of, 160, 164, 165
Hardwicke, Lord (Philip Yorke, 3rd Earl of Hardwicke), 109, 110
Hardy, Thomas, 25, 44
Hare Hatch, 102, 103
Harries, Lt. Edward, 42, 52, 53
Harris, Lord (George Harris, 1st Lord Harris of Seringapatam and Mysore), 37
Harrison, Lt. John, 35, 56, 57
Harte, Francis Bret, 216
Hartshill, 36
Harty, 89
Harz, The, 163
Hascombe, 22, 38, 39
Haslemere – see Haste Hill
Hassell, Edward, 54
Hassell, James, 54
Haste Hill, 38-40, 42, 53, 54

238   THE OLD TELEGRAPHS

Hatchford — see Chatley Heath
Haut-Barr
   station restored, 150, 151
Havelock, Major-General Sir Henry, 200
Hawkesbury, Lord (Robert Banks
   Jenkinson, Baron Hawkesbury, later
   2nd Earl of Liverpool), 110
Hazaribagh, 196
Heinkenszand, 157
Helgenaas, 174
Heliography, 3, 184, 185
Helsingfors (Helsinki), 177, 179
   and Gulf of Finland line in Crimean
      War, 177
   electric telegraph to Åbo and St.
      Petersburg opened, 179
Helsingør (Elsinore), 174
Helvoirt, 156
Herbert, Sidney, 90
Herne Bay, 87, 89, 90
Hestensbakke, 173
Het Loo, 155
Heukelom, 155
Heusden, 155
Highland Light, 213
High Stoy, 25, 26, 31
Hiitinen, 177
Hilbre Island, 69-71, 75, 76, 80
Hills, Lt. Thomas, 61
Hinchley Wood — see Cooper's Hill
Hindhead, 40
Hitterøy (Flekkefjord), 175
'HMS Pinafore', 7
Hobart, 202-6
   Mulgrave Battery, 200
   restored station, Battery Park, 206
   Mount Royal telegraph, 203
      line to Tasman's Peninsula, 203-6
      possible line to Launceston, 206
Hobbes Down — see Compton
Hofer, Andreas, 134
Hohenlinden, Battle of, 129
Hohenwedel vor Stade, 159
Høje Taastrup, 174
Holcombe, 44
Holder Hill - see Older Hill
Holl (of Navy Board), 52
Holstein, 160
Holt (Norfolk), 30
Holyhead, 69, 70, 80
   Holyhead Mountain, 69, 70, 80
Holyhead-Liverpool Line
   Act to build, 69
   first apparatus of, 69, 70, 212
   list of stations, 70
   1833 Report on, 71-3
   improvements to, 75
   surviving stations of, 75
   layout of stations of, 76
   operation of, 77-9
   second apparatus of, 76, 77
   codes, 70, 72, 75, 77, 78
   contemporary descriptions of, 78, 79
   end of service of, 80
Holzminden, 164
Honingham, 30
Hooghly, River, 199
Hoogstraaten, 155
Hooke, Dr. Robert, 4, 103, 120
Horten, 176
Howrah — see Calcutta
Høyerup, 174
Hoylake, 76
Hüggeberger, 174
Hugo, Victor, 138, 140
Hudson, Robert E. 213
Hull, 83-5, 170
   Incorporated Chamber of Commerce
      and Shipping engages Watson to
      survey line to Spurn Head, 83
   Spurn Head Line
      stations of, 83
      opening of, 83
      value of, 83, 84
      ceases operation, 85
Hull and East Coast Marine Telegraph
   Association, 84
Hull (Mass.) — see Boston (Mass.)
Humber, River, 83
Humes, 135
Humphrey, Alderman J. 91
Hundred Days, The, 34, 136
Huningen (Huningue), 129
Hunt, Leigh, 47
Hunter, Captain, 216
Hurtado, Don Francisco, 183
Hüttner, 174
Hyderabad, 195

Icklingham, 29
Ierseke, 155
Ile de Bas, 131
Ile de Bréhat, 131
Ile d'Orléans, 208, 209
Ile de Rhé, 131
Imperieuse, HMS, 131
Impression Bay, 205

India, trigonometrical survey of, 196, 198
Indian Mail, 146, 188
Inglis, Brigadier, 199
Inonmeni, 177
Irún, 183
    Madrid line, 183
Isabel II, Queen of Spain, 183
Iserlohn, 164-5
Islandshöi, 173
Isle Vert (Ile Verte), 208
Ivybridge, 25

Jackson, Captain, J. N., 198
Jamestown (St. Helena), 194
Jarsö, 177
Jay, Commander Charles Hawse, 17, 47, 61
Jefferson, President Thomas, 216
Jerbourg Point — see Channel Islands
Jersey — see Channel Islands
'Jesuit Telegraph', 154
Jooss, G., 199
Junquera, La, 184
Jutland, 173, 174

Kaffir Wars, 190, 191
Kalmar, 170
Kalundborg, 173
Kamouraska, 208
Karaouy, 188
Karl Friedrich, Margrave of Baden, 159
Karlsborg, 170
Karlskrona, 166, 170, 172
Karlsruhe, 159
Karlsten Castle, 169
Kat River, 190
Kaub, 135
Keith, Lt. John, 186
Kellerman, Marshal François-Christophe (Duke of Valmy), 135
Kelling, 30
Kent, Duke of — see Edward Augustus, Prince
Kerespertz, Count
    appointed administrator of French telegraphs, 139
    resigns, 140
Kieler, 174
Killingholme, 83
King, Peter, 7th Lord King (Baron King of Ockham), 38
Kingsdown — see Downs, The
Kingston Hill, 38, 39, 42, 43, 50, 52, 63
Kirby, Professor Percival R., 190-2

Kitching, G. C., 194
Klaaswaal, 155
Knockholt Beeches, 87, 88
Koblenz, 161, 164, 165
    Berlin line, 161, 164, 165
Korsö, 170
Korsør, 173
Korppoo, 177
Koudekerke, 157
Krauseneck, General-Leutnant Wilhelm Johann von, 161
Kreutznach, 135
Kristiania (now Oslo), 175, 176
    Akershus, 175
        lines from, round coast, 175, 176
        1848 proposal for line from, 176
        Drammen electric telegraph, 176
Kristianiafjord, 175
Kristiansand, 175
Kristianstad, 170
Kristiansund, 175
Kronborg Castle, 173
Kronstadt, 177, 181
Kruiningen, 157
Kudelstaart, 155
Kullen, 166
Kyrkslatt, 177

La Bédoyère, General Charles Angélique François Huchet, Count of, 136
Labouheyre, 138
La Croix des Bouquets — see Béhobie
Ladoga, Lake, 181
La Hogue, 95
Lair, César, 147
La Junquera, 184
Lakanal, Joseph, 122, 135
Lamberts Castle, 25
Lambton, Lt. Col. W., 196, 197
Lamstedt, 160
Landes, The, 138
Landau, 123, 127, 128
    see also Paris-Landau line
Landrecies, 123
Lands End, 66
Landshut, 135
Landsort, 167, 170-72
Lange, Dr. Augusta, 152
Langeland, 174
Langeraar, 155
Langesund, 175
Lanslebourg, 134
La Lanterna, 152

Lapenotiere, Lt. John, 32
Larena, Juan José, 183
Larvik, 175
La Tête de Buch, 131
Launceston (Tasmania), 206
   line to Georgetown, 206
   possible line to Hobart, 206
La Valette, Antoine Marie Chamans, Count of, 136
Law, Alexandre, 118, 119
Leaver, Lt. Charles, 54
Ledreborg, 174
Le Croisic, 131
Leerdam, 155
Le Havre, 95, 131, 139, 140
   and Ferrier's telegraph, 139
   Paris line proposed, 140, 146
Leitersweiler, 135
Le Quesnoy, 124, 159
Lérida, 184
Leryoun, 188
Les Sables d'Olonne, 131
Lewis, Lt. Col. Griffith George, 190
L'Hôpital (Mont Cenis), 134
Lidköping, 170
Liguria, 152
Lihons, 123
Lille, 123, 124, 128, 134, 136, 137, 141
   St Catherines, 123
      Paris line — see Paris-Lille line
      Extension to Brussels of Paris-Lille line, 128, 130
      Boulogne extension, 132, 137, 141
      line to Calais and to Eu 140
      Metz-Lille-Calais-Boulogne-Eu-Avranches line proposed, 140, 146
Lilley Hoo, 28, 29
Lillo, 157
Limerick, 109-11
L'Islette, 208
Lindesnes, 175
Lindsay, John, Jr., 86
Linguet, Simon Nicholas Henri, 120
Lipkens, A., 156, 157
Lisbon, 185, 186
   Penha de França, 185
      Oporto line, 185
      possible communication with Paris, 185
Little Belt — see Belt, Little
Liverpool
   Chapel Street, 68, 70, 78
   St. Nicholas's Church, 68
   Tower Buildings, 68, 70, 71
   Dock Trustees, 69
   Corporation, 71
Liverpool — Holyhead line — see Holyhead-Liverpool line
Lizard, The, 93
Llandudno — see Great Orme's Head
Llaneilian, 69, 70, 75, 78, 80
Llanrhyddlad — see Cefndu
Lloyds, 90, 93
Llysfaen, 70, 75, 76
Locker-Lampson, Frederick, 47
Lombardy, 133, 134, 152
London
   Admiralty — see Admiralty (Building) Whitehall
   Blackheath, 18
   Brixton, 38, 39
   Chelsea
      Duke of York's School, 15, 39, 49
      Hospital, 21, 27, 39
   Cornhill, 87
   Hampstead, 27, 28, 88
   Honor Oak Hill — see One Tree Hill
   Horse Guards, 17
   Imperial War Museum, 17
   Lambeth Church, 38, 39
   Lower Grosvenor Place, 20
   New Cross — see Nunhead
   Nunhead, 7, 16, 18, 34, 36, 37, 38
   One Tree Hill, 7, 87, 88, 91
   Plow Garlic Hill — see Nunhead
   Public Record Office, 2
   Putney Heath, 21, 38, 39, 49, 62
   Red Hill (Chislehurst), 35, 36
   St. Michaels, Chester Square, 49
   Shooters Hill, 18, 34, 36
   Southwark — see Toppings Wharf
   Toppings Wharf (Shot Tower), 87 91
   Tower of London, 65
   West Square (Lambeth), 15, 17, 34, 36, 37, 61
   Westminster Abbey, 6
   Whitehall — see Admiralty (Building) Whitehall
   Wimbledon Common, 13, 21, 38
   Woolwich, 115
London Conference of 1839, 157
London — The Downs line — see also Watson, Barnard Lindsay
   planned, 86
   progress and description of, 86-91
   station sites of, 87
   type of apparatus of, 87

INDEX 241

operation of, 87, 90
end of service of, 91
London & Birmingham Railway —
   proposed telegraph along, 81
London & Croydon Railway, 7, 86
London & Southampton Railway, 50, 91
London & South Western Railway, 60, 91
London to Paris telegraph proposal, 81
London Gazette, The, 39
Longford, Lord (Edward Michael
   Pakenham, 2nd Baron Longford), 104
Longwood (St. Helena), 194
Loop Head, 110
Lord, Lt. William, 75
   telegraphic vocabulary of, 8, 77
Lorient, 131
Louis XVIII, King, 136, 149
Louis XIV, King, 120
Louis Napoléon, Prince (Later Emperor
   Napoléon III), 142
Louis-Philippe, King, 138, 140, 142, 144,
   183
Low Head (Tasmania), 206
Lowe, Sir Hudson, 194
Lower Halstow, 18
Lowestoft, 85
Lucerne, 154
Lucknow
   Siege of, 199-201
   Residency, 199-201
   Alambagh, 199-201
      Residency — Alambagh line,
         199-201
   Martinière College, 199, 200
   Bibiapore, 201
   Dilkusha Palace, 210
   Chinhut, 201
   Gumti River, 201
Lumps Fort — see Portsmouth
Lunéville, 129
Luton, 28
Lyngør, 175
Lyons, 129, 133, 134, 141, 152
   stations at, 141
   Paris line, 129, 133, 141 — see also
      Paris-Lyon (Toulon) line
   Marseilles/Toulon line — see Paris —
      Lyons (Toulon) line
   Milan/Venice line, 133, 134, 152

Macdonald, Colonel John, 1, 7, 34, 73,
   112-4, 118
   Army Service of, 112
   importuning of Admiralty and War
      Office by, 34, 112
   obsessions of, 34, 112
   inventions of, 73, 113, 118
   publications of, 113, 114
Macdonald, Flora, 112
McIntyre, Major, 200
Mackenzie, Lt. Colonel C., 196
Macquarie, Governor Lachlan, 202, 203
Madeira, 185
Madras, 195
   Presidency, 195
Madrid, 183, 184
   lines from, 183, 184
   Paris despatches, 183
Magdeburg, 134
Maglebylille, 173
Magol, 142
Mahmoodeeh Canal, 188
Mainz, 135
   stations at, 135
      Strasbourg line proposed, 135
      Metz line, 135
Malin Head, 65
Malines (Mechelen), 155
Malmains, 36, 37
Malmö, 170
Mandal, 175
Mangalore, 195
Manston, 87, 90
Mantua, 133, 134
Marcel, 120
March, Lord (William Douglas, 3rd Earl
   of March, later 4th Duke of
   Queensberry), 162
Marchal, 140
María-Luisa, Doña, 183
Marie-Louise, Empress (Léopoldine
   Françoise Thérèse Joséphine Lucie),
   135
Mariestad, 170
Marley, 25
Marryat, Capt. Frederick, 47
   Code of, 47
Marseilles, 137, 141
   stations at, 141
   Paris line, 137
   Toulon line, 137
Marshfield (Mass.), 210
Marstrand, 171
Martha's Vineyard, 210, 211
   Boston line, 210, 211
Mary, 146
   Châlon-sur-Saône line, 146
Mascara, 146

Massachusetts, 210
Masters W. E., 206
Mathé, José Maria, 146, 183
    system of, 146, 183
Maudslay, Henry, 34, 39
Mead, Commander Hilary Poland, 2, 8, 18, 66
Médéa, 146
Melbourne, 202, 206
    proposed line to Sydney, 202, 206
Melville, Lord (Robert Saunders Dundas, 2nd Viscount Melville), 34, 113, 116
Menouf, 188
Merchants Exchange Company (of New York), 215
Merifield (Merryfield), 43, 45
Merrick, Mrs., 38
Mersey Docks and Harbour Company (later Board), 75, 76
Messangis, 146
    Tonnèrre line, 146
Metcalfe, Sir Charles, 198
Metz, 127-9, 135, 140, 141, 143, 146
    stations at, 129, 135, 141
        Lille/Boulogne/Avranches line proposed, 140, 146
        Paris line — see Paris-Strasbourg and Paris-Landau lines
        Strasbourg line —see Paris-Strasbourg and Paris Landau lines
        Mainz line, 135
Meyer, Domherr, Dr., 159
Milan, 133, 134, 152
    lines to Paris and Venice, 133, 134, 152
Miliana, 146
Millbrook (Mass.), 210
Millery, 137, 142
Miot, Andre François, Count of Melito, 5, 121
Mitcham, 38
Modane, 134
Mohammed Ali (Mehemet Ali), 188
Moncabrier, 149
Montureux, Count of, 141
Mont-Affrique, 151
Mont-Cenis, 133, 134
Montdidier, 123
Mont-Dol, 128, 150
Montedo, 185
Montlhéry, 148
Mont Orgueil — see Channel Islands
Montpellier, 140
Montpensier, Antoine Marie Philippe, Louis d'Orléans, Duke of, 183
Mont-St.-Michel, 128
Mont-Tonnèrre, 135
Monteleone, 153
Monte Agraça, 185
Monte Cerres, 185
Monte de Vez, 185
Monte Gordo, 185
Monte Santa Lucia — see Porto San Benedetto del Tronto
Mount Arthur, 206
Mount Augustus, 206
Mount Communication, 204, 206
Mount Direction, 206
Mount Donkin, 191
Mount Fortescue, 206
Mount George, 206
Mount Holly (New Jersey), 215
Mount Nelson, 203, 205, 206
Mount Royal, 203
Mount Wise — see Plymouth
Moorson, Lt. Henry, 199, 200
Moreau, Jean Victor, 129
Moreton Bay (Brisbane), 202
Morice, 141
Morrison, Lt. Richard James, 72, 73
Morse, Samuel, 215, 216
Mosebacke, 171
Mostaganem, 146
Muiden, 155
Munken, 175
Murat, Joachim, King of Naples, 152, 153
Murray, Lord George, 13, 17, 20
    proposes shutter telegraph, 13
    Admiralty adopts telegraph of, 13
Mulgrave, Peter Archer — see Channel Islands and Tasmania
Murado, 185
Muston, John, 56

Naarden, 155
Najaden, 176
Nagpur, 196-8
Nadaud, Gustave, 148
Nadir, 188
Nakkehoved, 173
Nakskov, 174
Nantes, 137, 140, 141
    stations at, 141
        Avranches line, 140
Nantucket Island, 210
Naples, 152, 153
Naples, Kingdom of, 152, 153

Napoleon (Napoléon Bonaparte, General, then First Consul, then Emperor of the French), 129, 130, 132-37, 152, 194
Narbonne, 138, 140, 141, 147
    station at, 141
        Perpignan line, 138, 140
        Flocon's experiments on, 147
    speed of transmission on, 147
Nautical Magazine, The, 47
Navesink Highlands, 215
Navy Board (British), 14, 38, 40, 42, 43, 54, 56, 99
Needles, The, 92
Negrelli, Chevalier de, 189
Netley Heath, 22
Nettlebed, 103
Nettlecombe Tout, 25, 26
Nevlungshavn, 176
New Brunswick, 207, 208
New Jersey, 215
New Orleans, proposed line to New York, 216
New Sand Light — see Spurn Head
New South Wales
    visual telegraph introduced in, 202
    electric telegraph introduced in, 202
New York City, 213-5
    Battery, The, 214
    Coney Island, 215
    Custom House, 213
    Governor's Island, 213
    Holt's Hotel, 215
    Navy Yard (Brooklyn), 213
    Sandy Hook, 215
    Staten Island, 213, 215
    Tontine Coffee House, 214
    Wall Street, 215
        first telegraph in, 213
        Thompson's telegraph in, 214
        telegraph sponsored by Merchants Exchange Company in, 215
        proposed line to New Orleans, 216
        line to Philadelphia, 215
Newburyport, 210
Newcastle (New South Wales), 202
Newmarket, 29, 102
    Edgeworth's proposed 'telegraph' to London, 102
    shutter telegraph at, 29, 102
Ney, Marshal Michel, Duke of Elchingen and Prince of the Moskowa, 136
Nibria — see Calcutta
Nicholas I, Tsar, 177, 181

Nicolas, Sir Nicholas Harris, 7, 116, 118
    telegraph of, 7, 118
Niersen, 155
Nieuwekerk, 155
Nielsen, Lt. Carsten Tank, 176
Niersen, 155
Nile, River, 186
Nimes, 138
    station at, 141
Nops, Lt., 37
Nord-Libre — see Condé-sur-l'Escaut
Nore, The, 19, 65, 66, 89, 90
Norfolk Bay, 206
North Downs, 51
North Foreland, 67, 85, 87, 90
Northumberland, HMS, 194
Norwich, 29, 30, 86
    Thorpe, 29
        proposed lines to Yarmouth, 30, 86
        shutter line to London and Yarmouth, 29
Nossa Senhora de Socorra, 186
Nova Scotia, 207, 208
Nunhead — see London
Nunziate, General Vito, 153
Nya Varvet, 169, 171
Nyborg, 173
Nystad (Uusikaupunki), 177

O'Etzel, Major August (later General August von Etzel) 142-4, 161
Oldenburg, Grand Duke of, 160
Oldenburg, Grand Duchy of, 160
Ohlsen, Captain Ole, 175, 176
    system of, 175, 176
Older Hill (Holder Hill), 38, 54, 62
Old Wives Lees, 20, 236
One Tree Hill (London) — see London
One Tree Hill (New South Wales), 202
One Tree Hill (Tasmania), 204
Oporto (Porto), 185
    Lisbon line, 185
Oran, 146
Orange 137
Orange, Prince William of — see William I, King of Holland
Öreborg. 170
Orford Ness, 85
Orleans, 137, 141, 149
Orleans, Duke of — see Louis-Philippe, King of the French
Orléansville (now El Asnam), 146
Oscar I, King of Sweden and Norway, 176

Oslebshausen, 160
Oslo — see Kristiania
Ospringe, 19
Ossendrecht, 155
Ostend, 128
Østrupgaard, 174
Otschakov, 182
Otterndorf, 160
Ouderkerk, 155
Oudenarde, 128
Oudinot, Marshal Charles Nicolas, Duke of Reggio, 134
Outram, Lt., General Sir James, 199-201
Outrup, 174
Overland Mail and Route (Egypt), 188-9
Overschie, 156

Pace, Lt. George, 34, 37
Pacific Ocean, 216, 217
Paderborn, 164
Palatinate, Rhenish, 123
Pallas, HMS, 131
Pamplona, 183
Papal States, 152
Parcé, 121
Paris, 11, 120-4, 128, 129, 132, 139-42, 146, 147
    Athis-Mons, 141
    Bac, Rue du, 150
    Bagneux, 120
    Bailly, 150
    Belleville, 11, 120, 123, 141
    Boulevard St. Germain, 150
    Ecouen, 122, 123, 141
    Ercuis, 141
    Etoile, 121
    Fontenay-aux-Roses, 141
    Gagny, 141
    Grenelle, Rue de, 133, 140, 150
    Hôtel des Télégraphes, 133, 140, 150
    Louvre, 123, 132
    Luxembourg Gardens, 120
    Ménilmontant, 121
    Montmartre, 123, 141, 143, 150
    Mont Valérien, 141, 149, 150
    Palais-National, 128
    Passy, 141
    Père-Lachaise, 133
    Place de la Concorde, 109, 128
    St. Eustache, 141
    St. Martin du Tertre, 122, 123, 141
    St. Sulpice, 141
    Trou d'Enfer, 150

Tuileries, 128
Vaugirard Cemetery, 133
Villejuif, 141
Bordeaux/Bayonne (Béhobie) line
    first considered, 137
    begun, 137
    completed, 137, 146
    Blaye branch of, 137, 142
    speed of messages on, 138, 142
    difficulties of operation in the Landes, 138
    Madrid despatches, 183
    possible communication with Lisbon, 185
    last message relayed by, 147
Le Havre line proposed, 140, 146
Brest line
    ordered by Directory, 128
    route of, 128
    opening of, 129
Lyons (Toulon) line
    ordered, 129
    route of, 129
    work on suspended, 133
    work on resumed, 133
    deviated west of Dijon, 133
    completion of, 133
    extension Lyons-Milan-Venice, 133, 134
    Lyons-Toulon extension first proposed, 136
    relays news of Napoleon's escape from Elba, 136
    attacks on stations of, 137
    relays news of capture of Algiers, 139
    relays news of elevation of Pius VII, 140
    extended to Marseilles and Toulon, 137
    deviated to serve Dijon, 142
    as link in Indian Mail, 146
    Tonnèrre branch of, 146
    Châlon-sur-Saône branch of, 146
Strasbourg line — see also Paris-Landau line)
    Directory orders completion of, 128
    opening of, 128
        extension to Huningen, 129
        Luneville branch of, 129
        importance of in 1809, 134
        Mainz branch of, 135
        relays news of Louis Napoleon's

attempted coup, 142
Lille line
  decided on, 123
  budget for, 123
  initial constructional difficulties of, 123
  early completion of ordered, 123
  route of 123
  opening of, 123, 124
  relays news of capture of Le Quesnoy and Condé, 124
  extended to Brussels, etc., 128, 130
  extended to Boulogne, 132
  messages over, in 1806-7 134
  relays news of Waterloo, 136
  Calais branch of, 137
  Calais branch extended to Boulogne and Eu, 140
  relays news of Napoleon's death, 137
Landau line (see also Paris-Strasbourg line)
  proposed, 123
  decided on, 127
  Chappe's instructions for, 127
  special apparatus proposed for, 127
  work suspended, 128
  resumed as Paris-Strasbourg line, 128
Paris-Rouen Railway, 147
Paris-Versailles Railway, 147
Paris, Peace of, 31
Parker, John Rowe, 73, 211-3
  takes over and extends Boston Harbour line, 211-2
  apparatus of 73
  advocates United States national telegraph, 73
  later years of Boston system of, 73
  possibly inspires B. L. Watson's system, 212
Parker, Richard, 19
Parramatta, 202
Pasley, Lt. Col. Sir Charles William 7, 8, 115-7
  service of, 115
    publications on telegraphy, 116, 117
  first telegraph of, 116
  impressed by French semaphore, 116
  second telegraph of, 113
  final telegraph of, 117

adoption of telegraph of, by Royal Navy, 117
adoption of telegraph of, in South Africa, 191
Universal Telegraph of, tested in Sweden, 170
possible use of in Australia, 202, 203
Patton, Col Robert, 193
  establishes 'telegraph' in St. Helena, 193
Pau, 138
Paull, 83
Pelham, Thomas (later Lord Chichester), 105-8
Penaud, Rear-Admiral Charles, 177
Peniche, 185
Peninsular War, 184, 186, 187
Penmark, 131
Pentridge, 25
Perdriau, Lt. Stephen, 17, 37
Perpignan, 138, 140, 141
  station at, 141
    Narbonne line, 140, 141
Peterhead, 92
Pewley Hill, 39, 53
Pforzheim, 134
Philadelphia, 215, 216
  line to New York City, 215
Philippines, The, 184
Phillips, Sir Richard, 21
Pickle HMS, 32
Piddletown (Puddletown) Heath, 44, 45
Piedmont, 133, 134, 152
Pier Appel's Tower, 191
Pistle Hill, 25
Pitkapaasi, 177
Pistor, Geheime Postrat Carl, 114, 161-163
  telegraph of, applied between Berlin and Koblenz, 161, 162
  code of, 162
Pisy, 142
Pius VII, Pope, 140
Pizzo 153
Plantation House (St. Helena), 194
Playfair, Major W. D., 196
Plymouth, 24-6, 30, 31, 44, 45
  Mount Wise, 25, 26, 44, 45
    Admiralty shutter line to London, 24-6
    Falmouth line proposed, 30, 31
Poad, Lt., James, 53, 61, 62
Pocasset, 210

Point Lynas — see Llaneilian
Pointers — see Chatley Heath
Pointe d'Ailly, 131
Poitiers, 137, 141
    station at, 141
Pollard, Lt., Charles, 49
Polybius, 3
Poona, 195
Pope, Joseph, 211, 213
Popham, Rear-Admiral Sir Home Riggs,
    6, 7, 34, 35, 41, 113, 186, 194, 199,
    202, 216
    early life of, 34, 35
    sea semaphore of, 6, 7
    land semaphore of, 6, 7, 34, 41, 113
    trial of semaphore of, between
        London and Chatham, 34
    use of semaphore of, in Lines of Torres
        Vedras, 186
    proposed use of semaphore of, in
        USA, 216
    possible use of semaphore of, in New
        South Wales, 202
    semaphore of, submitted to Society of
        Arts, 34
    'Telegraphic Signals' of, 34, 186
    and St. Helena, 194
    use of semaphore of, in Siege of
        Lucknow, 199
Porkkala, 177
Port Arthur (Tasmania), 203, 204
Portland (Maine), 210, 212
Porto San Benedetto del Tronto, 152
Portsdown Hill, 13, 22, 57
Portsmouth
    Clarence Pier, 23
    Dockyard, 13
    Fortifications of, 22
    Lumps Fort, 39, 58
    Platform, the, 58, 59
    Port Admiral's office, 22, 59
    Sail Loft and Rigging House, 59
    Shutter station on Southsea
        Common, 22, 23
    Spithead, 23
    Spur Ravelin, 22, 23
    Square Tower, 58, 59
        shutter line to London — see
            Telegraph, Shutter, Admiralty
    semaphore line to London — see
        Telegraph, Semaphore, Admiralty
    electric telegraph to London, 60-2
Portsmouth (New Hampshire), 210
Port Macquarie, 202

Port Vendres, 131
Portugalete, 184
Post Down — see Camp Down
Potsdam, 163
Poulton T. A., 88
Poyle Hill, 43, 45
Prestatyn — see Voel Nant
Pskov, 182
Public Building and Works, Ministry of
    (now Department of the
    Environment), 51
Public Instruction, Committee of
    (French), 122
Public Safety, Committee of (French),
    122, 123, 127, 128
Puffin Island, 70, 75, 78
Punta Ravenna, 152
Putten, 155
Puttershoek, 156

Quarterly Review, 26
Quebec City, 208, 209
    Cape Diamond station, 208
Quebec Province, 207-9
    St. Lawrence River telegraph, 208,
        209
Queen Anne's Hills (Cobham, Surrey), 43
Queenborough-in-Sheppey, 19
Quemigny, 142

Rabaut (Pomier) Jacques Antoine, 128
Railways — see individual systems
Rainsford Island — see Boston (Mass.)
Ramstedt, Captain, 177
    description of system of, 177
Ranganathpur, 196
Ranmore, 38, 39
Rastatt, Congress of, 128, 129
Raven and Hill, Messrs, 188
Ravnebjerg, 174
Reculver — see Herne Bay
Red Sea, 188
Redlynch, 43
Redoubt 30, 186
Reid, Samuel C., 214
Remilly, 142
Remmington, Samuel, 86
Rennes, 140, 141
    stations at, 141
        Avranches line, 140
        Nantes line, 140
Restorations, First and Second (France),
    136, 137
Revolution, French, 120

(see also under Chappe, Claude, and Paris)
Revolution of 1830 (French), 140
Revolution of 1848 (French), 147
Rhine, River, 164
Rhyl — see Foryd
Rimini, 152
Ringland, 30
River Hill, 43, 45
Rivière du Loup, 208
Rivière d'Ouelle, 208
Rivoli, 134
Robertson, Lt. J. B., 38, 42
Robinson, Lt. Edward, 61
Rochefort, 131, 137
Rockbeare Hill, 25
Roebuck, George, 13, 24, 26, 30-2
   appointment of, 13
   surveys and completes Deal/Sheerness and Portsmouth lines, 13
   surveys and completes Plymouth line, 24
   surveys and completes Yarmouth line, 26, 30
   and discontinuance of shutter telegraph, 31
   'private' code of, 32
Roman Empire, signalling in, 3
Rome, King of — see Bonaparte, François Charles Joseph
Romme, Charles Gilbert, 122
Ronalds, Sir Francis, 34
Ross, James, 206
Rotterdam, 155, 156
Roubaix, 128
Rouen, 139
   Paris line (Ferrier's telegraph), 139
Rowe Hill, 35, 36
Rowley, Rear-Admiral Sir Charles, 34
Royal Engineers, 190, 191, 201
Royal Hospital, Chelsea — see London
Royal Mail Steam Packet Company, 91
Royal Signals, 187
Royal Society, 4
Royan, 131
Royston, 29
Rushmore, 44
Russell, Earl (John Francis Stanley Russell, 2nd Earl Russell), 55
Russell, Bertrand, 55
Russell, Percy, 92
Russo-Swedish War of, 1808-9, 169

St. Albans, 15, 28

St. Brieuc, 128
St. Catherines (Isle of Wight), 92
St. Cyrus Hill, 25
St. Denis (Quebec), 208
St. Fargeau, Lepelletier de, 121
   park of, 121, 122
St. Georges, 150
St. Haouen, Rear-Admiral Yves Lecoat de, 149
   trials of telegraph of, 149
St. Helena, 137, 193, 194
St. Helier — see Channel Islands
St. Jean-de-Luz, 187
   telegraph at, in 1813-14, 187
St. Jean-de-Maurienne, 134
St. John (New Brunswick), 207
St. Laurent, Julie de (Alphonsine Thérèse Bernadine Julie de Montgenet, Baronne de Fortisson), 207
St. Lawrence River, 208, 209
   telegraph alongside, 208, 209
St. Malo, 95, 128, 136, 141
St. Petersburg, 179, 181, 182
   Winter Palace, 182
      first lines from, 181
      Warsaw line opened, 181
      Helsingfors and Åbo electric telegraph opened, 179
St. Michel (Quebec), 208, 209
St. Nicholas (Belgium), 130
St. Omer, 128, 132, 137
St. Pierre d'Oléron, 131
St. Peter Port — see Channel Islands
St. Roch, 208, 209
Sally Phil's Poem, 194
St. Vallier, 208, 209
St. Thomas (Quebec), 208
Saltram, 25, 26, 31, 32
Salt Water River, 206
San Ildefonso, 183
San Benigno, 152
San Francisco, 216, 217
   Golden Gate, 216
   Port Lobos, 216
   Presidio House, 216
   North Beach, 216
   Telegraph Hill, 216
      telegraph in, 216, 217
São Juliao da Barra, 185
Sandhamn, 170-2
Sandwich, John William Montagu, 7th Earl of, 44
Santarem, 185-6
Sark — see Channel Islands

Sas, 130, 155
Saugor Island, 196, 199
Saverne, 150
Savona, 152
Schäferberg, 163
Scherholm, 135
Schleuninger, Johann Nepomuk, 154
Schlüsselburg (Petrokrepost), 181
Schmidt, Johann Ludwig 159, 160
Schwyz, 154
Scilly Isles, 66
Scituate, 210
Scott, Sir Walter, 131
Scrabster, 92
Sealand, 173, 174
Sebastopol, Capture of, 147
Selling, 20
Selwyn's Battery — see Grahamstown, 192
Semaphore (see also Telegraph, Semaphore)
    invention of, 5, 130
    adoption of for French coast service, 130-132
    distinction between *sémaphore* and *télégraphe*, 5, 6, 130
    postes electro-sémaphoriques, 132
Semur, 142
Senegallia, 152
Seringapatam, 195
Serou, 188
Sétif, 146
's Gravenwezel, 155
's Hertogenbosch, 156
Shannon, River, 110, 111
Shepherd, George, 28
Sheerness, 19, 34, 36, 89
Sheppey, 19, 36, 89
Sherbrooke, Sir John Coape, 208
Sherfield English, 43, 45
Sherghati, 196, 197
Shipping Telegraphic Association — see Watson's General Telegraph Association
Shoeburyness, 65
Shottenden Hill, 20, 36
Shutter Telegraph — see Telegraph, Shutter
Sibley, Captain, 200
Sidi-bel-Abbès, 146
Sidmouth — see South Down
Siegburg, 164
Siemens, Werner von, 165
Siersberg, 135

Sieyès (Abbé), Emmanuel Joseph, Count, 130
Signalsskär, 167, 169
Signal Hill (Tasmania), 204-6
Simmonds, T. C., 86
Sittingbourne, 19
Skagerak, 175
Skaw The, 173
Skirsa Head, 92
Skibnes, 175
Skibbereen, 71
Skreene, 104
Slesvig (Schleswig), 173, 174
Smartt, Charles L, 199
Smith, Lt. James, 53
Sligo, George Ulick, 6th Marquess of, 54
Slopen Island, 206
Smith, John T., 213
Slots Bjergby, 174
Snow, Lt. William John, 62
Sobral, 186
Society of Arts (later Royal Society of Arts), 34, 116, 167
Söderarms Beacon, 170
Söderköping, 170
Soeven, 165
Solthoeren-bei-Wreden, 160
Sønderborg, 174
Sontag, General, 186
Sønder Højerup, 174
Sori, 152
Soult, Marshal, Nicolas Jean-de-Dieu, Duke of Dalmatia, 136, 142, 184
South Down, 44
South Downs, 22, 40, 54
Southampton, 24, 91-3
    Dock Company, 91, 92
    Isle of Wight line, 91, 92
South Foreland — see Downs, The
South Head (NSW), 202
South Knighton, 25
South Plymouth (Mass.), 210
Southsea — see Portsmouth
Southwark — see London
Souto Redondo, 185
Spencer, Lord (George John, 2nd Earl Spencer), 38, 107
Spike Island, 111
Spurn Head, 83-5
Spithead — see Portsmouth
Sprang, 155
Sprogø, 173
Stadlandet, 176
Stavern — see Frederiksvern

Stavanger, 175
Standaarbuiten, 155
Steenbergen, 155, 157
Stiftrup, 174
Stierberger, 174
Start Point, 85
Steepleton, 44, 45
Stettin, 134
Stewart, Lt. P., 201
Stockholm 166, 167, 169, 170, 172
   Castle, 169
   Drottningholm, 169
   St. Catherines Church, 169
   Tullports Street, 169
   Stora Ässingen, 169
     lines from, 167, 170-2
Strasbourg, 123, 127-9, 134, 135, 141, 144
   station on cathedral, 129, 141, 144
     proposed line to Landau, 123, 128
     proposed line to Mainz, 135
     Vienna flag telegraph, 134
     Paris line, 128, 129
     Huningen line, 129
Strumpshaw, 29, 30
Suez, 188, 189
   Cairo line, 188, 189
Sunninghill, 34
Surrey County Council, 51
Susa, 134, 152
Sutton, 38, 39
Sveaborg, 177
Swaddling Down — see Swarling
Swanscombe, 18, 34, 35, 36, 56
Swarling, 36
Sweeney, George F., 216
Swingate, 43, 45
Sydney, 202, 206
   stations at, 202
     projected line to Melbourne, 206

Telegraph (French) — see also Chappe
   description of, 124-7
   operation of, 126, 127, 141
   'Landau' type, 127, 128
   used in service of lottery, 132
   cost of, 132, 146
   vote for, 132
   trials of night operation of, 132, 141, 150
   mobile apparatus, 132
   position of through political changes, 136
   organisation of, 127, 142-4
   officials of, 127, 144, 145
   operators of, 143

   speed of transmission of, 122, 124, 138, 139, 140, 143, 147
   proposed use of by public, 141
   comparison with Prussian, 143, 144
   codes of, 126, 127
   divisional system of, 143
   greatest extent of, 146
   in literature, 148, 149
   later improvements to, 146
   as national consolidator, 148
   O'Etzel's report on, 142-144
   survivals of, 150, 157
   transition to electric telegraphy, 147
   last message relayed by, 147
   end of, 147
Tachygraphe, 5, 121
Tamar River (Tasmania), 206
Tankerton — see Whitstable
Tarancon, 184
Tarbes, 138
Tarka Post, 190
Tasmania, 202-6
Tasman's Peninsula, 203-6
Tatham, W. G., 194
Telbany, 188
Telecommunication, microwave, 2, 151
   between Paris and Lyons, 2, 151
Telegraph
   derivation of name, 5, 121
   distinction between semaphore and, 5, 6, 130
   coding versus spelling messages, 8
   Act for establishing signal and telegraph stations in the United Kingdom (1815), 34
Telegraph, electric
   alongside English railways, 62, 147
   alongside French railways, 146, 147
   international function of, 146
   in Crimean War, 147
Telegraph, Semaphore — see also Semaphore
   derivation of word semaphore, 5
   invented by Depillon, 5, 130
   distinction between telegraph and semaphore, 5, 6, 13
   on board ship, 7
   and influence on railway signalling, 7
Telegraph, Semaphore, Popham's
   tried between London and

Chatham, 34
  tested against shutter system, 34
  description of, 41
  ordered to be installed between London and Sheerness, 34
  London-Chatham trial line installed, 34
  permanent lines decided on, 35, 36
  operating and maintenance regulations of, 35
  London-Chatham line re-surveyed, 36
  sites for Deal/Dover extension chosen, 36
  tested against French-type telegraph, 37
  London-Chatham line given up, 38
  London-Portsmouth line surveyed, 38, 39
  cost of sites, 37, 39
  types of station houses, 40, 51
  Portsmouth line described, 40, 41 45-59
  operation of, 40, 41, 59
  Plymouth line surveyed, 42-4
  difficulties of routing through Dorset, 44, 45
  Plymouth line abandoned on Hants/Wilts border, 45
  functioning of London-Portsmouth line, 59
  list of last officers of London-Portsmouth line, 60
  end of London-Portsmouth line, 60 61
Telegraph, Shutter
  popularity of, 11
  Macdonald's, 112, 113
  Gamble's, 13
  Murray's, 13
    adoption of by Admiralty, 13
    appointment of Roebuck, 13
    building of Deal/Sheerness and Portsmouth lines, 14
    description of Deal/Sheerness and Portsmouth lines, 17-23
    construction of, 14
    station complement, 15
    fan showing, 15
    operation of, 15, 16, 21
    description of Plymouth line, 24-6
    first proposal for Yarmouth line, 26
    final proposal for Yarmouth line, 26
    description of Yarmouth line, 26-9
    Plymouth-Falmouth extension proposed, 30, 31
    discontinuance of, in 1814, 31
    cost of operation of, in Kent, 31
    rental of land for sites for, 31
    fate of buildings of, 31, 32
Tellograph, Edgeworth's — see Edgeworth, Richard Lovell
Tenes, 146
Texel, 66
Terheijden, 156
Thélus, 123
Theseus, 4
Thompson, Jeremiah, 214, 215
  Battery — Staten Island telegraph of, 214, 215
Thurso, 92
Tilsit, 134
Tisted, 43
Tlemcen, 146
Tolker, 174
Toller Down, 25, 31
Tonge, 19
Tonnerre, 146
Toot Hill, 24
Topliff, Samuel, 211
Torbay, 45
Torres Vedras, Lines of — see also Army, British, and Portugal, 186, 187
Tossal de Mollerusa, 184
Toulon 136, 137, 139, 141
  stations at, 141
    Paris/Marseilles line, 137
Toulouse, 140
  Bordeaux-Avignon line, 140
Tournai, 156
Tours, 137, 141
Tower of London — see London
Town Hill, 24
Trafalgar, Battle of, 1, 32, 61, 62
Trebjerg, 174
Trieste, 134, 152
  Venice line, 134, 152
Trinity House, 85, 90
Trinity House, Hull, Corporation of, 83
Tromborn, 135
Trondheim, 175
  Kristiansten Fortress, 175
Troy, 2
Tsarskoe Selo (now Pushkin), 181

INDEX 251

Tuck, Instructor Captain Oswald T., RN, 2
Turin
    Paris line, 133, 134
    Venice line, 133, 134
Turku — see Åbo
Tuskar Rock, 71

Utrecht, 155

Valby, 174
Valence, 137
Valencia, 183
Valenciennes, 140
Valloire, 134
Vänersborg, 170
Vass, Nicholas, 64, 65
Van Diemen's Land — see Tasmania
Varna, 147
Västervik, 170
Vaxholm, 170
Veen, 155
Vegesack, 160
Vegetius, 3
Venetia, 133, 152
Venice, 133, 134, 152
    Milan/Paris line, 133, 152
    Trieste line, 134, 152
    Porto San Benedetto del Tronto line, 134, 152
Versailles, 150
Vianen, 155
Viborg (Viipuri), 177, 179
Vienna, 134
    Strasbourg flag line, 134
Vic, 129
Vieux Mongerre, 107
Victoria, Queen, 90, 207
Victoria (State of), 206
Vigigraph,
    proposed between Paris and Le Havre, 149
Vila Nova dos Fusos, 185
Villaret de Joyeuse, Vice-Admiral Louis Thomas, 130
Vilnyus (Vilna), 182
Vitrolles, Eugène François d'Arnauld, Baron de, 136
Voel Nant, 70, 71, 73, 75, 76, 80
Volta do Monte, 185
Voltaire (François-Marie Arouet), 169
Vorbruch, 160
Vougy, Viscount de, 147
Vreeland, 155

Waarde, 157
Waddinxveen, 155
Waghorn, Lt. Thomas, 188
Walcheren, 116, 130, 157
War of 1812, British-American, 213, 214
Ward, Henry, 26
Warsaw, 181, 182
    Royal Palace, 182
        St. Petersbourg line, 181, 182
Warteberg, 174
Washington, DC, 215, 216
Waterloo, Battle of, 1, 32
Watson, Barnard Lindsay, 7, 69, 70-73, 75, 81, 84, 86-92, 114, 213
    possibly inspired by John Rowe Parker, 212
    code signals of, 70
    proposes Liverpool-Manchester line, 71, 81
    gives evidence before Government commissioners, 72, 73
    dispute of, with Lt. Morrison, 72, 73
    dispute of, with Liverpool Dock Committee, 75
    surveys Hull-Spurn Head Line, 81
    residences of, in Liverpool and London, 86
    establishes Watson's General Telegraph Association, 84, 86
    and London-The Downs line, 86-91
    and Southampton-Isle of Wight line, 91, 92
    and proposed London-Paris line, 81
    and the Admiralty, 90
    death of, 92
    (see also Holyhead-Liverpool line)
Watson's Code of Signals, 6th edition, 1842, 92
Watson's General Telegraph Association, 84, 86
Watt, Superintendent, 208, 209
Weager, Sergeant, 193
Wedge Bay, 205
Wellington, Arthur Wellesley, 1st Duke of, 26, 136, 184, 186, 187
    telegraph sections formed by, 187
Weser, River, 160
West Down, 44
West, George, 53
Westbroek, 155
West Barendrecht, 156
West Falmouth (Mass.), 210
Weston, Captain C. T. G., 197, 198

Weybourne, 30
Weymouth (Dorset), 26
Weymouth (Mass.), 210
Wexford, 71, 105, 106, 109
Wheatstone, Sir Charles, 60, 147
White, Andrew, 56
Whitelands, 44
Whitstable, 87, 89
Wick (Caithness), 92
Wick (Surrey), 43, 45
Wickham, 24
Wight, Isle of, 91, 92
    line to Southampton, 91, 92
Wildey, Lt. John, 61
Wilkins, John, 4, 103
Willemsdorp, 156
Willemstad, 155
William, Prince of Orange — see William I, King of Holland
William Henry, Prince — see Clarence, Duke of,
William I, King of Holland, 156
Williams, Lt. Joseph, 61, 62
Williamson, Lt., 57, 58
Winchester, 43
Windham, William, 115
Windmill Hill (Tasmania), 206
Windsor (Berks), 115
Windsor (Nova Scotia), 207
Winterfold, 38, 39

Wissembourg, 135
Witley, 40, 53
Womenswold, 20
Woodcock Hill, 28
Woodfalls — see Woodfield Green
Woods Hole, 210
Woody Island, 204, 206
Woest Hill — see Grahamstown
Worcester, Edward Somerset, 2nd Marquis of, 4
Worplesdon, 37, 39, 43-5, 59
Woubrugge, 155
Wrench, Henry, 86
Wreningham, 29
Wrotham, 87, 88
Wyke Regis, 26

Yarmouth, Great, 15, 26, 29
    South Gate, 15, 29
        proposed lines to Norwich, 30, 86
        London shutter line, 15, 26-30
Yerba Buena — see San Francisco
York, Prince Frederick Augustus, Duke of, 11, 107, 115

Zaiyat, 188
Zaragoza (Saragossa), 183
Zevenbergen, 157
Ziegelsdorf, 163
Ziffet-Chalahan, 188

R1